GEOLOGY EMERGING

Frontispiece of Josias SIMLER, Vallesiae et Alpium descriptio, 1633

GEOLOGY EMERGING

A Catalog Illustrating the History of Geology (1500-1850) from a Collection in the Library of the University of Illinois at Urbana-Champaign

By

Dederick C. Ward, ***Geology Librarian, University of Illinois at Urbana-Champaign***

Albert V. Carozzi, ***Professor of Geology, University of Illinois at Urbana-Champaign***

Robert B. Downs Publication Fund No. 8
University of Illinois Library
and
The Graduate School of Library and Information Science

Distributed by the Publication Office,
Graduate School of Library and Information Science
University of Illinois

1984

PRINTED IN THE UNITED STATES OF AMERICA

LIBRARY OF CONGRESS CATALOGING IN PUBLICATION DATA

University of Illinois at Urbana-Champaign. Library.
Geology emerging.
(Robert B. Downs Publication Fund; no. 8)
Includes bibliographical references and index.
1. Geology—History—Bibliography—Catalogs.
I. Ward, Dederick C. II. Carozzi, Albert V. III. Title.
IV. Series: Robert B. Downs Publication Fund (Series); no. 8.
Z6035.U55 1984 [QE11] 016.55'09 84-8556
ISBN 0-87845-071-8

THIS PUBLICATION HAS BEEN MADE POSSIBLE
THROUGH THE GENEROSITY OF

GEORGE W. and **MILDRED K. WHITE**

UNDER THE AUSPICES OF
THE UNIVERSITY OF ILLINOIS LIBRARY FRIENDS
AT URBANA-CHAMPAIGN

Contents

List of Illustrations

Acknowledgments

The idea of this catalog originated with Harriet E. Wallace, Professor Emerita of Library Administration and Geology Librarian from 1962 to 1979. She identified and selected a major portion of the titles included in this catalog. We are deeply indebted to her infinite patience in the tedious task of bibliographic verification of authors, titles, and editions.

Marguerite Carozzi devoted countless hours to this project by finding many titles to which there was no lead, by examining book by book all the uncataloged entries, and by drawing attention to the geological content of many works, particularly in foreign languages. The final editing and typing of the introduction, indexing, and proofreading are among the other tasks she efficiently undertook on our behalf. She deserves our deepest gratitude.

N. Frederick Nash, Rare Book Librarian, and the other members of his staff: Mary S. Ceibert, Gene K. Rinkel, and Louise M. Fitton kindly helped on numerous occasions to trace difficult bibliographic entries, and guided us through the various collections. Their help has been invaluable.

Rebecca Finney did a superb performance at standardizing and typing most of the entries.

We would like to thank Norman B. Brown, Assistant Director for Bibliographic Services, who has actively supported for numerous years the growth of the history of geology collection and authorized the purchase of many titles.

Joan M. Hood, Library Director of Development and Public Affairs handled in the most gracious way the arrangements with the University of Illinois Library Friends and the Donors which made the publication of this book possible.

Special thanks are due to Diana L. Walter who supervised the transfer of hundreds of rare books from the Geology Library to the Rare Book Room which took place during the preparation of the catalog. Lois Johannsen and Lore Raether from the Library Friends prepared the books for transfer.

The help of the Research and Publication Committee of the Library in the form of grants for typing and photocopying is gratefully acknowledged.

May 1984

D. C. Ward
A. V. Carozzi

Introduction to the Catalog

Geology Emerging is the first catalog of the Library's rare book collection in geology. Its publication concludes a three years' bibliographical inquiry following three decades of book collecting. Our aim is to call attention to the outstanding history of geology collection at the University of Illinois at Urbana-Champaign.

Very few catalogs on collections of books in early geoscience exist. Some announcements of special collections in the history of geology began as annotated catalogs to exhibits. For instance, in conjunction with the 1971 Annual Meeting of the Geological Society of America in Washington, D.C., the United States Geological Survey Library prepared an exhibit on the landmark "Historic Books in Geology." In 1962, a catalog entitled "Early Geology in the Mississippi Valley" described an exhibit of works selected by George W. White and Barbara Slanker at the University of Illinois. These two catalogs, however, only hint at the richness of historical material in the collections of both libraries.

Within the last decade, catalogs of two special collections of interest to historians of geology were published. At Oklahoma, the DeGolyer-inspired history of science collection, strong in geology, was the subject of *The Catalog of the History of Science Collections of the University of Oklahoma Libraries* (Mantell, 1976). In 1980, the Claremont Colleges' Libraries published the *Biblioteca De Re Metallica, the Herbert Clark Hoover Collection of Mining and Metallurgy.* This superb, annotated work owes its existence to the efforts of a host of bibliographers, librarians, and historians of science over a period of many years. The Hoover work appears to be the first modern catalog of a book collection restricted to the history of geoscience. Thus *Geology Emerging* continues a very young tradition.

From the vast array of scientific, natural history, theological, agricultural, and travel literature in which early geology appears, we chose to include those works which identify with the emergence of geology as a separate discipline. From the earliest books we selected works with important geological descriptions and authors who applied scientific reasoning to problems of earth history and to geological phenomena.

From naturalists, in particular of the eighteenth century, and some later scientists who published on a wide range of subjects, we selected only their works of geological significance. We included geological travels, but other travel narratives only if they were sustained by a reasonable amount of geological observation. We chose, for example, to include Fremont's 1842 expedition to the Rocky Mountains, but not the expeditions of Pike or Lewis and Clark.

Western geology began in Europe and from there spread to America. Geology became a definable science in Europe at the end of the eighteenth century, whereas in colonial America, geological observation was confined mainly to travel narratives. Therefore, for early American geology we were more liberal in our criteria for selection. The authors were fortunate to have consulted the bibliographical notes of Dr. George W. White, Emeritus Professor of Geology at the University of Illinois. His notes pointed to important geological observations in works of travel and natural history that otherwise might have escaped our attention. Similarly, Carozzi's research leading to his translations into English of early landmark books in geology helped to identify European contributions.

Every book in the catalog was personally inspected by the authors or our associates. Moreover, we checked the Library's holdings against published bibliographies, reference lists, and works cited in the following books on the history of geology: Karl Alfred von Zittel, *Geschichte der Geologie und Paläontologie, bis Ende des 19. Jahrhunderts* (R. Oldenbourg, 1899); Archibald Geikie, *The Founders of Geology* (Macmillan, 2nd ed., 1905); Charles Davidson, *The Founders of Seismology* (Cambridge Univ. Press, 1927); Frank Dawson Adams, *The Birth and Development of the Geological Sciences* (Dover, 1938); K. F. Mather and S. L. Mason, *A Source Book in Geology* (McGraw-Hill, 1939); Asit K. Biswas, *History of Hydrology* (Elsevier, 1970); and H. Faul and C. Faul, *It Began with a Stone — A History of Geology from the Stone Age to the Age of Plate Tectonics* (Wiley, Interscience, 1983).

We chose 1850 as our limiting year because it represents the advent of specialized fields in geology as well as the writing of textbooks which in North America had just started while in Europe, Lyell's *Principles* was in its eighth edition. We have included a geologist's post-1850 works if his productive period preceded or averaged out to 1850. Similarly, all editions of a work are shown, if the Library owns them; in the case of popular texts, publication dates may reach well into the second half of the nineteenth century. A few geologists whose post-1850 works are not included appear as cross-references because they have appended notes to works in the catalog.

The catalog contains chiefly monographic works, defined as books, separately-bound reports, and a small collection of individual geologic maps. Articles disbound from journals, or reprinted as separates, sometimes with changes from the original or signed as presentation copies, are included if the Library has cataloged them as separates. Their historical value varies greatly, in particular if they belonged to the first half of the nineteenth century.

We chose, however, to include a small, select group of journal articles, in particular from the Académie Royale des Sciences at Paris and the Royal Society of London, because these articles mark the first introduction of fundamental concepts which may or may not have appeared later in book form. The catalog, augmented in this way, thus becomes more indicative of the true strength of our holdings in the landmark works of early geology.

With the exception of landmark papers, we made no attempt to include early serial literature. Bibliographies have been published which classify and index this literature. Some modern examples are John Challinor's *The History of British Geology* (David and Charles, 1971), Robert and Margaret Hazen's *American Geological Literature, 1669-1850* (Dowden, Hutchinson & Ross, 1980), and Harold Pestana's *Bibliography of Congressional Geology* (Hafner, 1972). The University of Illinois Library's holdings in early scientific serials is substantial, dating from v.1 in most cases; this catalog, therefore, can be considered complementary to and not duplicating existing bibliographies in the history of geology.

The illustrations and the index mark the two special features in the catalog. The illustrations are reproductions of title pages, thus departing from illustrative matter in existing histories of geology which generally consists of the reproduction of frontispieces or plates within a work, much of it repetitive from book to book. The authors felt that readers might like to see the title pages of some of the early books in geology as expressions of graphic art, printers' emblems, other artistic concerns, and as indicative of the intellectual climate of the times. We designed an index in depth so that statistics could be developed for any one aspect of the collection. The index is subject-geographical on the first and second levels; the third level is reserved for the names of authors.

Introduction to the Collection

The collection originates from geological publications which were at first part of the Library of the land-grant Illinois Industrial Institute, predecessor of the University of Illinois at Urbana-Champaign, geology being considered a science basic to the state's agricultural and mineral industry. The collection grew over the years and in the early 1950's, the Library, under the encouragement of Dr. George W. White, then Head of the Department of Geology, gave special attention to the purchase of books important to the history of geology. White's numerous visits to book dealers in North America and Europe for almost 30 years, and his extensive correspondence with them were instrumental in the acquisition of many of the titles.

The evaluation of this collection can be done in at least two ways. One is to leaf through the index which combines specific topics with individual authors and allows one to appreciate, at a glance, the size of the coverage as well as the quality of the collection. A second more historical way is addressed to those readers who are interested in reading a synopsis of the history of the major fields in geology between 1500 and 1850. This history is based on the holdings of the collection we felt were of great importance; the enumeration of titles, we hope, will provide a fascinating flavor. Since this kind of approach might be linked to a so-called "tunnel vision" where the interpretations of naturalists of the past are studied and sometimes compared with modern concepts, we would like to explain why we have chosen this approach.

We believe that the history of geology is nothing more than a succession of different interpretations of the same phenomena and facts. In that respect it differs from the history of biology, for instance, where living things cannot be studied unchanged over hundreds of years. Since rocks undergo only minor modifications due to weathering, we can, therefore, imagine the Earth as a set of unchanging outcrops which are visited in turn by naturalists from various countries over many centuries, say from Avicenna through de Maillet, Werner, Hutton, Lyell, Suess, and finally modern geologists.

What has taken place during this succession of studies has usually occurred as follows. A given naturalist made a set of observations, reached some conclusions, criticized or refuted some of his predecessors, and generalized his deductions into a synthesis that could be, for instance, a theory of the earth. Our first task as historian of geology is to understand the man and his work in the context of his time, namely his training, his regionalism, his potential of observation, his preconceived ideas, his social and religious constraints. As an example of preconceived ideas we might mention de Saussure who in 1779, after a careful description of erratic boulders on glaciers high up in Alpine valleys, and those found in front of the Alps, on the Swiss plateau, and the slopes of the Jura Mountains, could not realize that the presence of all these boulders pointed to a former greater extent of Alpine glaciers. In the context of his time, we must accept the fact that he was probably under the influence of the Wernerian concept of a universal ocean when he stated that huge floods, during their final drainage into cracks of the earth's crust, would be the best mechanism to explain the presence of huge boulders in the plains.

If we now examine a second naturalist, who several decades later observed the same set of data with improved techniques of observation and increased general knowledge, we will find his conclusions probably more encompassing and his theory accounting for more facts than before. It is quite possible that at this stage he has "solved" certain geological problems in the sense that until today all subsequently collected evidence has confirmed his theory. This happened, for instance, for the hydrological cycle demonstrated by Perrault in 1674, or the volcanic origin of basalt proposed by Desmarest in 1771; no scientist would deny their interpretations today. Although our approach could be interpreted as a history of winners, it is undeniable that scientific knowledge has progressed in an irreversible manner by the application of sets of more sophisticated techniques, leading to new theories, to scientific revolutions if you wish, or toward an ultimate truth never to be reached.

Returning to our naturalists, let us assume that we have analyzed each individual contribution in its context and understand the major factors which led to their concepts or theories. We then realize that these contributions have all been influenced by previous thoughts, as in the case of de Saussure, and have led, by positive or negative reactions, to new interpretations. Hence we are inevitably led into a comparative history of geology in time, that is, into an analysis of how concepts have evolved, have dominated geology for a while, and were modified or rejected by subsequent ideas. This evolutionary process, irreversible and discontinuous, brings us inevitably to the present which is just another step. It associates physics and chemistry with geology, adds subsurface data not available to our predecessors, and is dominated by the theory of plate tectonics.

We feel that it is only in the light of modern geology that earlier interpretations can be adequately understood today. Of course it would be inappropriate to emit on the basis of modern science a judgment on past observations or theories as being "right" or "wrong." It would be anachronistic and of no historical value to state that Werner was "wrong" when he thought that basalt was an aqueous precipitate in the hot universal ocean. His interpretation was not "wrong" by any means; it was the best he could reach within his intellectual context. Two examples will show what we mean. Basalt may appear interlayered as submarine lava flows between limestone, sandstone, and shale beds; it may show some kind of internal layering, even channeling related to flow processes; it may even display in some cases phenocrysts of plagioclases with a size sorting due to gravity settling in the melt. All these characteristics of liquid lava simulate the action of water. No wonder Werner believed basalt to be an aqueous precipitate. Similarly with porphyritic granites altering to soil profiles, the so-called "regenerated granites" sometimes show a crude stratification because the more resistant crystals which are scattered among finer weathering products look like angular pebbles dispersed among sand. No wonder that Maillet, for instance, interpreted granite as water-laid gravel and sand. Now we understand the reasons for his interpretation. Such deceptive aspects of geology are of utmost interest for finding the dilemma which hampered or the reasons which dictated some early naturalists in their conclusions. In other words, our approach tries to assess within the science what naturalists saw and interpreted in the past.

Lapidaries and Museum Catalogs

We shall leave aside Aristotle, Lucretius, Seneca, and Pliny, that is, what our classical heritage said about stones, with the exception of THEOPHRASTUS, who wrote about 314 B.C. one of the earliest known treatises on minerals and their uses. In our collection we have *De lapidibus liber . . .* with Latin and Greek texts, compiled by Daniele Furlano in 1605 and 1613, and *De gemmis et lapidibus libri duo . . .* annotated by Joannes de LAET, 1647, translated into English in 1774 and into French in 1754. Many names given by Theophrastus to rocks and minerals have survived. Furthermore, unlike the medieval lapidaries, Theophrastus did not necessarily believe in the superstitions related to stones or their supernatural powers although he recorded them. Medieval lapidaries, on the whole, only recorded these superstitions describing each mineral, stone, and gem in regard to its medicinal, magical, or mythical character. Such treatises were written by ALBERTUS MAGNUS and copied many years later as *De mineralibus libri quinque* . . . in 1519, and by MARBODE, while Bishop of Rennes in the eleventh century, later printed as *De lapidibus pretiosis Encheridion* . . . 1531. A later lapidary was written by LEONARDUS, *Speculum lapidum clarissimi artium et medicine* . . . in 1516, and was translated into English as *The mirror of stones* . . . in 1750. STELLA in *Libonothani interpraetamenti gemmarum* . . . 1517, treated gems more scientifically and showed little belief in their magical property.

The middle of the seventeenth century saw the publication of the voluminous works by CESI, *Mineralogia* . . . 1636, and by ALDROVANDI, *Musaeum metallicum in libros IIII, distributum* . . . 1648, which attempted to present the entire body of knowledge of the time on the mineral kingdom (metals, earths, rocks, concretions, fossils, artifacts, etc.). Although these works retain a certain concern about magic which characterized the preceding lapidaries, their encyclopedic character is well shown by the remarkable organization of their presentation including definitions, origin, properties, location, uses, and historical background.

The Renaissance witnessed a renewed interest in nature, and collections of natural objects such as minerals, fossils, shells, feathers, dried plants, stuffed animals, which had been gathered from many parts of the world, were displayed in the houses of rich collectors. These "cabinets" were shown to important visitors and were forerunners of our modern museums. Some collectors wrote detailed illustrated accounts of the holdings or catalogs while others hired curators to do so. These catalogs were distributed to the curious or scientifically inclined persons and are fundamental today for the understanding of "geological" endeavors during that particular period, endeavors which continued to expand well into the eighteenth century. The earliest catalogs were written by KENTMANN, *Catalogus rerum fossilium* in Dresden and by GESNER, *De rerum fossilium* . . . and both were published together at Zürich in 1565. These were followed by the CALZOLARI Collection, *Musaeum Franc. Calceolari* . . . described by Benedetto Ceruti and Andrea Chiocco at Verona in 1622, by *Historia naturale di Ferrante IMPERATO Napolitano* . . . 1672, and *De testaceis fossilibus Musaei Septalliani et Iacobi Grandi* . . . published at Venice in 1676 by QUIRINI. The catalog of the Royal Museum of Natural History of Copenhagen was written by OLIGER, *Muséum Regium seu catalogus* . . . in 1696, while the BOCCONE Collection, *Sylvii Bocconis Curiöse Anmerckungen* . . . was printed in Frankfurt in 1697.

In the eighteenth century, SCHEUCHZER wrote catalogs of his famous collections, *Herbarium diluvianum collectum* . . . 1709, 1723, and *Museum diluvianum* . . . 1716, with the intention to prove that the biblical deluge had transported fossil shells and other relics. The Papal collection was described by MERCATI in *Metallotheca* . . . 1719, while the famous cabinet of Amboine was presented by RUMPF in *D'Amboinsche rariteitkamer* . . . 1741. Also published in Amsterdam was the enormous collection of KNORR, first described by himself and then by WALCH in five spectacular volumes, *Sammlung von Merckwürdigkeiten der Natur* . . . 1755 to 1773 with French and Dutch translations in 1768-1778 and 1773 respectively. Earlier SPADA had written a catalog of his collection in Verona, *Corporum lapidefactorum agri Veronensis catalogus* . . . 1744.

A more systematic approach to cataloging had already started in the description of WOODWARD's great collection of fossils, *An attempt towards a natural history of the fossils of England; in a catalogue of the English fossils in the collection of J. Woodward, M.D.* . . . 1729, a trend which was continued by BUFFON in *Histoire naturelle, générale et particulière, avec la description du Cabinet du Roy* . . . 1749, v.3., and LAMARCK's *Système des animaux sans vertèbres . . . suivant l'arrangement établi dans les galeries du Muséum d'Histoire Naturelle* . . . 1801. Clearly "cabinets" had already become museums.

Mining and Metallurgy

While collectors provided entertainment for the curious, princes were searching for gold, silver, and minerals. Mines were so abundant in the Erz Mountains that the mining business was soon booming and mining communities sprang up in many places. St. Joachimsthal (today in Czechoslovakia) hired AGRICOLA as town physician. He wrote the first technical books on mining: *Bermannus; sive de re metallica* . . . 1530, followed by *De ortu et causis subterraneorum, Lib. V* . . . 1546, 1558 (translated into Italian in 1550), and *De re metallica Libri XII* . . . 1556, 1557, 1621. BIRINGUCCI studied in Germany, and back in Italy, he wrote an influential book based on a pragmatic view of the entire field of metallurgy according to his observations and experiments in foundries and smelters: *De la pirotechnia libri X* . . . 1540, 1550, 1556, 1558, 1678. The superintendent of mines of Rudolf II of Austria, Lazarus ERCKER tried to put together all the different methods in a manual: *Beschreibung aller fürnemisten mineralischen Ertzt und Bergkwercksarten* . . . 1580, which was translated into English by John PETTUS in his *Fleta minor. The laws of art and nature in knowing, judging, assaying, fining, refining . . . metals* . . . in 1683, 1686, while he was serving a term for debts in Fleet prison. Meanwhile the Church sent Alvaro Alonso BARBA to Peru to teach religion and study mineral deposits, mining, and in particular amalgamation techniques pertaining to silver ores. His book published after his death (Madrid, 1640) became famous and was translated into English as the *First and second book of the arts of mettals* . . . in 1670, 1674, 1739, 1740, as well as into French in 1730. Now the field was open for later works on metallurgy by WEBSTER, *Metallographia: or, an history of metals* . . . 1671; LEHMANN, *Abhandlung von den Metall-Müttern und der Erzeugung der Metalle* . . . 1753; HENCKEL, *Pyritologia* . . . 1757 (translated into French in 1760); DELIUS, *Abhandlung von dem Ursprunge der Gebürge und der darinne befindlichen Erzadern, oder der sogenannten Gänge und Klüfte* . . . 1770; FERBER, *Physicalisch-Metallurgische Abhandlung über die Gebirge und Bergwerke in Ungarn* . . . 1780, and BORN, *New process of amalgamation of gold and silver ores* . . . 1791 to which RASPE contributed. Finally, WERNER in *Neue Theorie von der Entstehung der Gänge* . . . 1791 (translated into French in 1802, and into English in 1809), although strongly influenced by his ideas on the neptunian origin of the earth, prepared the background for modern mining geology. It is evident from the above enumeration of titles that early metallurgy was strongly affected by the need for gold and silver and that naturalists involved with mining operations were soon to explain what they thought about the interior of the earth and the origin of mountains.

Mineralogy and Petrology

The study of minerals *per se* was greatly enhanced by the desire for the value and the magic of gems but hampered by two problems, one, the controversy on the origin of "fossils," a term applied both to minerals and fossilized organic remains, and two, difficulties in understanding the basic concepts of crystallography. In *De natura fossilium* . . . 1546, AGRICOLA, under the heading of *De ortu et causis subterraneorum,* classified minerals by their physical properties. A comparable approach was proposed by ENTZELT (ENCELIUS) in his *De re metallica* . . . 1551. Entzelt's misleading title suggests a possible competition with Agricola or some mercantile reason while he was, in fact, discussing physical properties of minerals and not ores. GESNER, in his *De rerum fossilium* . . . 1565, went a little further, comparing some organic fossils to plants and animals. However, he did not conceive that they were actually the remains of once living animals or plants; magic and heavenly influences provided a perfectly acceptable alternative explanation. Other systematic descriptions were given by CESALPINO in *De metallici libri tres* . . . 1602; by Francisco IMPERATO in *De fossilibus opusculum* . . . 1610, and by BOODT, *Gemmarum et lapidum historia* . . . 1609, 1636, 1647, with a translation into French in 1644 which reads: *Le parfait ioaillier, ou histoire des pierreries: où sont amplement descrites leurs naissance, juste prix, moyen de les cognoistre, & se garder des contrefaites, facultez medicinales, & proprietez curieuses.* . . . Along similar lines is DOLCE, *Tratatto delle gemme che produce la natura; nel quale si discorre della qualità, grandezza, bellezza et virtù loro* . . . 1617.

Crystallography in its infancy, preceded by the earliest microscopic observations of HOOKE in *Micrographia* . . . 1665, 1667, can be related to STENO, *De solido intra solidum naturaliter contento dissertationis prodromus* . . . 1669, who pointed out the consistency of interfacial angles in crystals (the work was translated into English in 1671), and BARTHOLINUS, who in *Experimenta crystalli Islandici* . . . 1670, presented the concept of double refraction in calcite. BOYLE in turn undertook some experimental and chemical work on minerals and gems and presented *An essay about the origin & virtues of gems* . . . 1672 (with a Latin edition of 1673).

Early attempts at classification of minerals led to a long-lasting controversy which pitted the use of external characters versus the chemical composition of minerals as a basis for classification. The major naturalists involved were: WALLERIUS in *Mineralogia* . . . 1747 in Swedish (with a translation into French in 1759 and into German in 1750 and 1763), and *Systema mineralogicum* . . . 1772-1775, 1778, and CRONSTEDT whose work *Försök til mineralogie* . . . appeared in 1758 (with translations into German in 1760 and 1780, into English in 1770, 1772, 1788, and into French in 1771). In France, VALMONT DE BOMARE wrote *Minéralogie* . . . 1762, 1774, and in Prague, SCOPOLI published *Principia mineralogiae systematicae et practicae* . . . 1772 (translated into German in 1775). In spite of compromises proposed by Wallerius, the classification of minerals by external characters won a temporary victory with WERNER in *Von den äusserlichen Kennzeichen der Fossilien* . . . 1774. The library owns translations in Hungarian (1784), in French (1790), and English (1805). Werner's work created such an impact that countless translations and adaptations, known as "mineralogies after Werner" appeared as, for instance, by VANBERCHEM-BERTHOUT and STRUVE, *Principes de minéralogie; ou exposition succinte des caractères extérieurs des fossiles, d'après les leçons du professeur Werner* . . . 1795, and WAD, *Tabulae synopticae terminorum systematis oryctognostici Werneriani, latine, danice, et germanice* . . . 1798, and finally by Werner's disciple JAMESON in *System of mineralogy* . . . 1804-1808, 1816, 1820, and *A treatise on the external characters of minerals* . . . 1805, 1816, 1817. Werner's approach was rather weak in chemistry and the new advances would be precisely combining chemical composition with external characters. This trend is shown by KIRWAN, *Elements of mineralogy* . . . 1784, 1794-1796, 1810, with a classification based on qualitative chemical tests (with translations into

French in 1785, Spanish in 1789, and German in 1796-1799). An even better combination of chemical composition and external characters was proposed by the Swedish Torbern BERGMAN in *Sciagraphia regni mineralis* . . . 1782, of which we own a French translation (1784), and in *Meditationes de systemate fossilium naturali* . . . 1784, of which the collection has an English translation in *Essays, physical, and chemical* . . . under the heading "Thoughts on a Natural System of Fossils" (1791).

Increased sophistication in blending external character of minerals with chemical composition could not lead to modern mineralogy without an understanding of the internal structure of minerals, namely crystallography. This theoretical background was opened by ROMÉ DE L'ISLE in *Essai de cristallographie* . . . 1772, and *Cristallographie, ou description des formes propres à tous les corps du règne minéral* . . . 1783, a work followed by HAÜY, *Essai d'une théorie sur la structure des crystaux* . . . 1784, and *Traité de minéralogie* . . . 1801. Subsequent progress in instrumentation paved the way for the many textbooks in the nineteenth and twentieth century starting with DANA, *A system of mineralogy; including an extended treatise on crystallography; with an appendix containing the application of mathematics to crystallographic investigation* . . . 1837.

Petrology as an outgrowth of mineralogy had and still has its problems of classification and nomenclature. For instance, the word "marble" has been used and is still being used in a popular sense for any rock which may be polished as an ornamental stone. This usage goes back at least to the eighteenth century and has caused much confusion since the designation may include white and red granites and diorites of all kinds (porphyries), black gabbros and green serpentines, alabaster gypsum, fossiliferous to massive limestones of all colors, sedimentary and tectonic breccias, and eventually real marbles (e.g., Carrara marble) which are limestones recrystallized by metamorphism. A historian of geology will have to find out exactly what rock an author was describing as "marble." A similar point can be made for "variegated breccias and marbles" which are striking-looking rocks consisting of angular fragments set in a matrix of different color and which have attracted the attention of naturalists for many centuries. Numerous interpretations were proposed to explain the brecciated aspect: submarine reworking by powerful currents, ancient deposits of talus or screes, collapsing or crushing along faults and overthrusts, products of weathering processes, etc. At present we can distinguish types of brecciation which resulted from sedimentary or tectonic mechanisms, and this distinction is fundamental to understand what led early observers to their individual conclusions.

Paleontology

The presence of fossil fish and fossil shells far away from the sea puzzled many naturalists. They wondered whether the sea had once covered these lands and mountains, whether these fossils were merely figured stones, that is stony material which resembled some animal or plant, whether some seeds had been washed up through fissures in the mountains, or whether it was perhaps the Deluge as mentioned in the Bible which had carried them to their present location. Those who were in favor of a marine origin were generally living close to the Mediterranean Sea and had a good knowledge of some marine organisms. These were CARDANO, in *De subtilitate libri XXI* . . . 1550 (translated into French in 1584) and STENO whose work we already mentioned above, *De solido intra solidum*. . . . In his earlier work, *Elementorum myologiae specimen* . . . 1667, 1669, he described his dissection of the head of a shark washed up on the beach and showed that Glossopetra were nothing else but fossil shark teeth. The "colorless" SCILLA in *La vana speculazione disingannata dal senso* . . . 1670 (with a Latin translation in 1752), could not be any clearer in his comparison between the abundant living shells on the beach and those in the Pliocene coastal coquinas of Sicily. Another believer in the marine origin of fossils was VALLISNIERI in *De' corpi marini, che su' monti si trovano* . . . 1721, 1728. He described them as remains of organisms who had lived in a former widespread ocean. It should be stressed that all these Italian naturalists were discussing Cenozoic fossils around the Mediterranean which occur in uplifted and mostly flat-lying strata adjacent to the sea.

In France, PALISSY in *Discours admirables* . . . 1580 (included also in his *Le moyen de devenir riche* . . . 1636) stated clearly that fossils found inland must have lived and died there. He was in favor of some salty lakes to explain the presence of fossils found far away from the sea. Nobody read him until the beginning of the eighteenth century. The theory of a marine origin of fossils was proposed by JUSSIEU in a memoir to the Academy of Sciences at Paris in 1718: *Examen des causes des impressions des plantes marquées sur certaines pierres des environs de Saint-Chaumont dans le Lionnois*. He had examined exotic-looking fossil ferns in Carboniferous coal beds and, based on the evidence that they were surrounded by marine shells, and because similar plants existed in India, or in other warm countries, he deduced that currents from an ocean extending from India or thereabouts must have floated these plants to France. This discovery created such an upheaval in France that the Royal Academy of Sciences at Paris, which obviously set the tone, accepted the marine origin of fossils openly, and in fact much earlier than British geologists.

There is, however, another reason for the reluctance of the British to accept the marine origin of fossils. Whereas in Italy Cenozoic fossils can be easily compared to their abundant living counterparts in a warm sea, in Great Britain those fossils belonging to extinct forms are most often found in folded Mesozoic and Paleozoic rocks adjacent to a cold sea with a poor fauna and make any direct comparison difficult if not impossible. PLOT, in *The natural history of Oxford-shire* . . . 1677, 1705, and *The natural history of Stafford-shire* . . . 1686, studied "formed stones" with great care and concluded that they were so different from living shell fish that he considered them to be mineral crystallizations which simulated organisms. He agreed with LISTER who in *Historia animalium Angliae* . . . 1678 denied the organic origin of fossil shells despite their slight resemblance to shells of living molluscs. Lister was aware that this similarity was only superficial and that no such species were still alive. RAY in *Miscellaneous discourses* . . . 1692 and *Three physico-theological discourses* . . . 1693, reasoned along similar lines for the many different species of ammonites while he accepted the organic origin for less puzzling fossils. LHWYD, in *Lithophylacii Britannici Ichnographia* . . . 1699, proposed that most fossils had grown *in situ* from the same seed as the living organisms which they resembled; these seeds had been swept up into the rocks through fissures. HOOKE was an exception among his contemporaries. In his posthumous works containing *Lectures and discourses of earthquakes* . . . 1705, he accepted the organic origin of fossils in the belief that nature "does nothing in vain." WOODWARD, whom we shall meet in our discussion of theories of the earth, wrote *Fossils of all kinds* . . . 1728, and *An attempt towards a natural history of the fossils of England* . . . 1729. He had decided earlier that the Deluge was responsible for the occurrence of fossils in rocks.

Two naturalists in Switzerland were greatly influenced by British thought. LANG in Lucerne accepted Lhwyd's concept and wrote in *Historia lapidum figuratorum Helvetiae curiosae* . . . 1708 and in *Tractatus de origine lapidum figuratorum* . . . 1709, that all fossils found in rocks derived from seeds which had been transported by subterraneous currents and had been lifted through porous rocks to their present location where they germinated under the snow. SCHEUCHZER followed Woodward's diluvial theory. While he had remained undecided in *Specimen lithographiae Helveticae* . . . 1702, he changed his mind after corresponding with Woodward and accepted the diluvial theory in *Herbarium diluvianum* . . . 1709, in *Helvetiae Historie Naturalis* . . . 1716-1718, and the monumental *Physica sacra* . . . 1731-1735 and *Kupfer-Bibel* . . . 1731-1735. Obsessed by his diluvial theory, he finally explained in 1726 in *Homo diluvii testis* . . . that the remains of what we now know to be a large fossil salamander were the skeleton of one of the infamous men who had brought about the calamity of the Flood.

The fossil controversy spurred many descriptions and illustrations of fossils, many of them beautiful works of art. BOURGUET wrote *Traité des pétrifications* . . . 1742, 1778, while BERTRAND published an extremely valuable compendium in Latin, French, and German of fossils entitled *Dictionnaire universel des fossiles propres, et des fossiles accidentels* . . . 1763, followed by *Elémens d'oryctologie ou distribution méthodique des fossiles* . . . 1773. The most spectacular illustration of fossils is by DEZALLIER D'ARGENVILLE in *L'histoire naturelle éclaircie dans une de ses parties principales, l'oryctologie* . . . 1755. Earlier, the Dean of the medical school of the University of Würzburg, BERINGER, fell victim to his infatuation with fossils. Some of his colleagues fabricated exotic-looking petrifications and arranged for him to find them in the field. Accepting them as real fossils, he wrote *Lithographiae wirceburgensis* . . . 1726, 1767, now considered a classic among hoaxes in science. Eventually, the debate on the origin of fossils relented. With the exception of BUCKLAND, in *Reliquiae diluvianae* . . . 1823, 1824, who believed in the Deluge, the concept of extinction became accepted in the nineteenth century as illustrated by CUVIER in *Recherches sur les ossemens fossiles de quadrupèdes* . . . 1812, whose contribution to geology shall be mentioned in the next section. In D'ORBIGNY's *Cours élémentaire de paléontologie et de géologie stratigraphique* . . . 1849-1852, two disciplines, paleontology and stratigraphy, had emerged.

Theories of the Earth

Theories of the earth were proposed by numerous naturalists and philosophers over many centuries to explain the origin of the earth in relation to mankind, the presence of fossils in mountains, the vicissitudes of the sea, and the ever baffling problem of the origin of mountains. The following questions were asked: has everything always remained the same, have great revolutions taken place, or have slow changes occurred over a long period of time? DESCARTES was particularly interested in changes which had occurred on the surface of the earth and he suggested in *Principia philosophiae* . . . 1644, a buckling and collapse of the earth's crust together with the existence of an internal layer of water to account for present mountains and seas. His concept of a broken earth crust was incorporated in the theory by BURNET, who in his *Telluris theoria sacra* . . . 1681, 1689 (translated into English in 1684, 1690, and 1699) suggested that the earth before the Deluge had a smooth surface which broke and opened the way for the water underneath to gush out and, together with the cataracts from the skies, to cover the earth completely during the time of the Deluge. This theory was refuted by WOODWARD in *An essay towards a natural history of the earth* . . . 1695, 1702, 1723 (translated into French in 1735). Woodward argued that mountains had existed before the Deluge; in order to explain fossils inside rocks he advocated a complete dissolution of the first earth during the Deluge followed by a settling of all the materials by gravity. Another English diluvialist, WHISTON, explained in his *A new theory of the Earth* . . . 1696, 1722, 1737, that the waters of the Deluge originated from the tail of a comet. Diluvial theories were repeated on the continent by Scheuchzer as mentioned earlier, by BOURGUET in *Lettres philosophiques . . . avec un mémoire sur la théorie de la terre* in 1729, by BERTRAND in *Recueil de divers traités* . . . 1766, and later by DE LUC in *Lettres physiques et morales sur l'histoire de la terre* . . . 1779-1780, and in *Lettres sur l'histoire physique de la terre* . . . 1798. Each author had a different interpretation; the only thing they had in common was an attempt to reconcile their theories with the orthodox chronology in the Bible and, in particular, the catastrophic event of the Deluge.

STENO in *De solido intra solidum* . . . mentioned above, wrote in 1669 on the geology of Tuscany which led him to formulate the law of superposition and to explain tilted beds by collapse processes. He managed thus to reconstruct six stages in the history of the earth, episodes which may or may not have coincided with the Bible. KIRCHER, on the other hand, advocated in *Mundus subterraneus* . . . 1665, 1678, that although the earth had visibly undergone great changes since its creation, mountain chains circling the globe from pole to pole and from East to West must have necessarily existed since the beginning. He saw a central fire and fire- and water-filled caverns leading respectively to active volcanoes and thermal springs. MORO was profoundly influenced by volcanic processes observed at Vesuvius and believed that mountains were the result of volcanic eruptions in *De' crostacei e degli altri corpi marini che si truovano su'monti* . . . 1740 (translated into German in 1751). LEIBNIZ combined fire and water and wrote a synthesis in 1693, the *Protogaea* which was published in 1749. He adopted Descartes' broken earth crust by internal forces from a molten center, followed by huge inundations that deposited sediments. He did not mention any Deluge and was even in favor of freshwater fossils. On the whole, he was little cited by his contemporaries. It remains to be studied which one of the above mentioned naturalists had the greatest impact on his contemporaries.

HOOKE, we know, remained ignored for quite some time. His *Lectures and discourses of earthquakes* . . . 1705, where he proposed vertical movements which were related to earthquakes and volcanoes, were rediscovered by RASPE in the eighteenth century who glorified him in *Specimen historiae naturalis globi terraquei* . . . 1763. He expanded Hooke's ideas of vertical movements and applied them to the origin of continents and islands.

Great fame was achieved by DE MAILLET in *Telliamed* . . . 1748, 1749, 1755 (translated into English and published in London in 1750 and in Baltimore in 1797). He introduced the concept of slow changes over a great period of time and that of a diminishing universal ocean as well as the evolution of marine life into terrestrial life. Everybody read *Telliamed* which had circulated before in manuscript form among the forbidden literature in France. The all-encompassing role of the sea was taken over by BUFFON in *Histoire naturelle* . . . which included his theory of the earth in the first volume in 1749. The library owns translations of Buffon's work into English in 1775-1776, 1785, 1797, and into German in 1771-1774. Buffon's later additions and the famous *Époques de la nature* were added in 1778 where he proposed a division of the history of the earth into six epochs. In his first theory of 1749 Buffon had proposed that all mountains were formed by the sea; in the *Époques* and his additions he changed his mind and accepted Leibniz's concept of both fire and water.

The concept of a universal ocean developed in various ways and led, for instance, to the *Théorie de la terre* . . . 1795, 1797 by LA MÉTHERIE. He thought that mountain peaks of granite as in the Alps were huge crystals grown from the waters of the universal ocean. WERNER's geognosy attributed an even greater role to the ocean. Most of his ideas became known through his numerous students and disciples; they are summarized in *Kurze Klassifikation und Beschreibung der verschiedenen Gebürgsarten* . . . 1786. According to Werner, any geological formation had a worldwide extent so that his system applied to the entire earth. The concept of rock formations called "mountains," Gebürge or Gebirge in German, and "monti" in Italian, and their subdivision into distinct groups had been initiated by LEHMANN in *Versuch einer Geschichte von Flötz-Gebürgen* . . . 1756, by ARDUINO in *Raccolta de memorie chimico-mineralogiche* e *orittographiche* . . . 1775, and by BERGMAN in *Physikalische Beschreibung der Erdkugel* . . . 1780, originally written in 1766. PALLAS in *Betrachtungen über die Beschaffenheit der Gebürge und Veränderung der Erdkugel* . . . 1778 (translated into French in 1782), based his theory on observations made during many travels in the Urals and other parts of Russia. He found that primitive rocks were always granites with an adjacent band of schists, secondary rocks consisted of fossiliferous limestones, and tertiary rocks were gravels, sand, and marl. The presence of frozen elephants in Siberia puzzled him and he proposed that the waters of the Indian Ocean during volcanic activity were forced in huge floods toward the

North where they swept elephants, and remains of other animals and plants.

The concept of some universal ocean was still in the mind of DE SAUSSURE when he formulated in *Voyages dans les Alpes* . . . 1779-1796, a brief theory of the earth. According to him the Alps had been uplifted vertically, pushed by lateral forces related to the explosion of elastic fluids. As a consequence, the universal ocean disappeared into huge cracks in the earth's crust. The more he observed the complexity of the Alps, the less he felt capable of solving the many problems. He decided to leave them to his successors.

It is not surprising that the concepts of primary, secondary, and tertiary rocks given by Arduino, Lehmann, Pallas, and de Saussure are often misinterpreted because these naturalists were not talking about the same rocks. They all agreed essentially that each mountain chain consisted of several distinct portions, groups of rock formations, or "mountains." A central axis with the highest peaks displayed granites, schists, and slates often mineralized: these were the primary or primitive mountains. On this central part they found resting, first at high angle, and then flattening away, sequences of fossiliferous limestones and shales, forming spectacular cliffs such as the Calcareous Alps: these were the secondary mountains. Overlying the latter and further away were lower hills formed of a variety of weakly consolidated rocks which seemed to have originated from the destruction of earlier mountains and consisted of sandstones, marls, clays, often with plant debris and remains of terrestrial animals: these were the tertiary mountains. In some places, a fourth division could be recognized which was made of volcanic products associated with alluvials. Due to regionalism, Arduino and de Saussure were looking at the Alps, Lehmann at the Harz Mountains and the Erz Mountains from Thuringia, and Pallas at the Urals, hence at mountain ranges formed at different times. Consequently, their subdivisions into primary, secondary, and tertiary mountains do not coincide. The most striking discrepancy lies in the fossiliferous limestones of the secondary mountains which attracted the greatest attention. There Arduino was looking at Jurassic and Lower Cretaceous limestones (our present Secondary or Mesozoic) while Pallas and Lehmann were observing Permian limestones (our present Primary or Paleozoic). Naturally, further discrepancies ensued in the overlying tertiary mountains. For historians of geology it does not really matter that Arduino, for example, arrived at a set of divisions which more or less correspond with our present stratigraphic system. What matters to them is the fact that primary, secondary, and tertiary mountains described by various authors correspond to rocks of different ages, hence with different fossils. Therefore, an adequate appreciation of early descriptions and interpretations can be reached only if one is aware of the real nature of the rocks and fossils these authors found in their respective regions.

De Saussure's observations provided a wealth of data for HUTTON who wrote *Abstract of a dissertation . . . concerning the system of the earth* . . . 1785 and *Theory of the earth* . . . 1788, 1795-1799, a work of fundamental importance. Hutton provided the concept of time, the principle of uniformity in nature, the geological cycle, an early concept of metamorphism related to burial of sediments, and, in *Observations on granite* . . . 1794, the crystallization of granite from melt. These and many other basic concepts in geomorphology were restated and amplified by PLAYFAIR in *Illustrations of the Huttonian theory of the earth* . . . 1802. One would assume today that Hutton had many followers. We are told, on the contrary, that it took two generations for the Huttonian theory to override the Wernerian school in England.

Meanwhile in France a slightly similar competition went on. LAMARCK wrote *Hydrogéologie* . . . 1802, based on the idea that during the long history of the earth, ocean basins were displaced around the globe. This process combined with an equatorial bulge generated high continental plains from which mountains would be carved. Lamarck's ideas on geology remained in the shadow of CUVIER's theory of the earth written in the first volume of *Recherches sur les ossemens fossiles de quadrupèdes* . . . 1812, 1821-1824, 1834-1836, a theory which he updated in *Discours sur les révolutions du globe* . . . 1828, 1830, 1850, and 1875 (the theory was translated into English in 1813, 1815, 1817, 1818, 1827, and 1831). Cuvier's concept of catastrophism which implied extinction and creation of species by divine intervention had a tremendous appeal. Studies have shown, however, that his concept was not as extreme as generally considered since he accepted survivors and migrations.

Theories of the earth had run their course in the early part of the nineteenth century. Any further progress in the understanding of mountain building required a more sophisticated stratigraphic knowledge, systematic mapping, and specialized studies dealing with deformation processes of the earth's crust, namely structural geology and geotectonics. For the moment, let us return to some spectacular geological events witnessed by naturalists in the past.

Seismology

Earthquakes and their frightening destruction had been included in many theories of the earth, but often they were mentioned alone. It was asked whether they were caused by the wrath of God, or by secondary causes, or both. The early works include BEROALDO, *De terremotu & pestilentia* . . . 1510; TELESIO, . . . *De terraemotibus* . . . 1570;

ROMEI, *Dialogo delle cause universali del terremoto* . . . 1587; KECKERMANN, *Meditatio de insolito et stupendo ille terrae-motu* . . . 1602; DU THOUM, *Le tremble-terre ou sont contenus ses causes, signes, effets, & remedes* . . . 1616; SCHWELING, *Memorabilium ab orbe condito in nostra tempora usq: terrae-motum historias physicè explicatas* . . . 1684. Each period of violent earthquakes led to publications such as WETTERSTEINT, *A full account of the great and terrible earthquake in Germany, Hungary and Turky. Which was one of the greatest and most dreadful earthquakes that ever was in the world* . . . 1673, and SHOWER, *Practical reflections on the late earthquakes in Jamaica, England, Sicily, Malta, etc. anno 1692* . . . 1693, with a second edition in 1750, during another period of strong tremors. Until that time the physical causes assumed did not differ very much from those proposed by the ancients: subterranean winds, collapse of underground caves, explosion of combustible matters, and volcanic origin.

In 1750, numerous earthquakes shook Western Europe and culminated on November 1, 1755 in the great Lisbon earthquake which killed ca. 60,000 persons. Thereafter many naturalists began to discuss these events in earnest. HALES wrote *Some considerations on the causes of earthquakes* . . . 1750; STUKELEY, *The philosophy of earthquakes, natural and religious* . . . 1750, 1756 (in regard to natural causes he was in favor of electric shocks); FLAMSTEED, *A letter concerning earthquakes written in the year 1693* . . . 1750. A few days after the Lisbon earthquake, New England was shaken in turn and WINTHROP wrote *A lecture on earthquakes, read in the Chapel of Harvard-College in Cambridge, N.E. November 26th 1755. On occasion of the great earthquake which shook New-England the week before* . . . 1755; this sermon was followed by a purely scientific account to the Royal Society of London: *An account of the earthquake felt in New England, and the neighbouring parts of America* . . . 1757. The reports from Portugal and Spain were numerous. For instance, FEJÓO Y MONTENEGRO wrote *Nuevo systhema, sobre la causa physica de los terremotos, explicado por los phenomenos electricos* . . . 1756. GENDRON published *Tratado da conservaçam da saude dos povos . . . consideraçoens sobre os terremotos* . . . 1757, and MOREIRA DE MENDONÇA, *Historia universal dos terremotos* . . . 1758.

A major contribution to the understanding of earthquakes came from MICHELL in *Conjectures concerning the cause and observations upon the phenomena of earthquakes* . . . 1761, where he assumed that steam generated from water, when entering subterranean fires, produced shocks propagated by waves along bedding planes. Michell guessed that the epicenter of the Lisbon earthquake was located under the ocean.

Interesting catalogs of earthquakes were published by KRÜGER,

Geschichte der Erde in alleraltesten Zeiten . . . 1746, translated into French in 1752 as *Histoire des anciennes revolutions du globe terrestre. Avec une Relation chronologique et historique des tremblemens de terre, arrivés sur notre globe depuis le commencement de l'ère chrétienne jusqu'à present,* as well as BERTRAND in *Recueil: Mémoires historiques et physiques sur les tremblemens de terre* . . . 1766, covering these events from 563 A.D. to 1754. Meanwhile, some optimistic naturalist was hoping for the end of earthquakes: CASTILLON, *Les dernières révolutions du globe, ou, conjectures physiques sur les causes de la dégradation actuelle des tremblemens de terre, & sur la vraisemblance de leur cessation prochaine* . . . 1771.

As with other beginning disciplines in geology, earthquakes had to be understood in their global context, as stressed by LYELL, *Principles of geology* . . . 1830-1835, in numerous editions, and by HUMBOLDT, *Kosmos* . . . 1845-1862, with many translations. The understanding of the propagation of shock waves in solids, the development of accurate instrumentation, and the general use of arbitrary scales of intensity as proposed by EGEN in *Ueber das Erdbeben in den Rhein und Niederlanden* . . . 1828, had to occur before the study of earthquakes could become the science of seismology.

Volcanology and the Origin of Basalt

The earliest descriptions of Vesuvius or Etna erupting came from Italians. SANTORELLI wrote *Discorsi della natura, accidenti, e pronostici dell'incendio del monte de Somma dell'anno 1631* . . . 1632, and RECUPITO, *Avviso dell'incendio del Vesuvio* . . . 1635, and *De Vesuviano incendio nuntius* . . . 1639. By PARAGALLO we have *Istoria naturale del monte Vesuvio* . . . 1705; by SORRENTINO, *Istoria del Monte Vesuvio* . . . 1734; by TORRE, *Storia e fenomeni del Vesuvio* . . . 1755 (translated into French in 1760, 1776); by SERAO, *Istoria dell'incendio del Vesuvio accaduto nel mese di maggio del'anno 1737* . . . 1738, 1740; by HAMILTON, British Ambassador at Naples, *Observations on Mount Vesuvius, Mount Etna, and other volcanos* . . . 1772, 1773, 1774; and by GIOENI, *Saggio di litologia vesuviana dedicato a S.M. la regina delle Due Sicilie* . . . 1790 (with a German translation in 1793).

Among the contributions to the study of Etna are BORELLI, *Historia, et meteorologia incendii Aetnaei anni 1669* . . . 1670; WINCHILSEA, *A true and exact relation of the late prodigious earthquake & eruption of Mount Aetna or, Monte Gibello* . . . 1669, 1775, and DOLOMIEU, *Mémoire sur les Isles Ponces, et catalogue raisonné des produits de l'Etna* . . . 1788.

It took a while until naturalists recognized extinct volcanoes and, regardless of their ideas about the causes — combustion of coal, pyrite, bituminous matters, etc. — were able to compare them with active ones. It needed the long and assiduous work of GUETTARD, *Mémoire sur quelques montagnes de la France qui ont été des volcans* . . . 1752, to make the important discovery of ancient volcanism in France. The origin of basalt, however, remained a puzzle although it was observed all over Europe. COLLINI in *Journal d'un voyage qui contient différentes observations minéralogiques; particulièrement sur les agates, et le basalte* . . . 1776, assumed that basalt columns were crystals; GESNER in 1565 had held a similar view. The influential Wernerian school believed basalt to be a chemical precipitate in the hot universal ocean. After having visited Vesuvius and Auvergne, DESMAREST, in *Mémoire sur l'origine & la nature du basalte à grandes colonnes polygones* . . . 1771 (accompanied by a map), realized that prismatic basalt was indeed part of lava flows. He later recognized a succession of three volcanic periods in Auvergne when he wrote: *Mémoire sur la détermination de trois époques de la nature par les produits des volcans* . . . 1806. Independently, RASPE studied basalt in *An account of some German volcanos and their productions* . . . 1776 (originally written in 1771), and FAUJAS DE SAINT-FOND in *Recherches sur les volcans éteints du Vivaray et du Velay* . . . 1778. DOLOMIEU, in *Mémoires sur les volcans éteints du Val di Noto en Sicile* . . . 1784, described sheets of basalt interbedded between limestones and interpreted them as submarine lava flows. Other contributions came from BREISLAK, *Essais minéralogiques sur la Solfatare de Pouzzole* . . . 1792, and *Voyages physiques et lithologiques dans la Campanie* . . . 1801, and FORTIS, *Mémoires pour servir à l'histoire naturelle et principalement à l'oryctographie de l'Italie* . . . 1802.

Werner's ideas on basalt were eventually challenged by two of his best students. AUBUISSON DE VOISINS, at first in favor of Werner's concept in *Mémoire sur les basaltes de Saxe* . . . 1803, later adopted the volcanic origin in his *Traité de géognosie* . . . 1819. VON BUCH, in *Geognostische Beobachtungen auf Reisen durch Deutschland und Italien* . . . 1802-1809, also recognized the volcanic origin of basalt. From then on volcanism was considered as a fundamental process to be accounted for in all theories of the earth as shown by SCROPE, *Considerations on volcanos; the probable causes of their phenomena, the laws which determine their march, the disposition of their products and their connexion with the present state and past history of the globe; leading to the establishment of a new theory of the earth* . . . 1825. This is also the time when the importance of volcanic rocks intercalated among many ancient formations was recognized in the works of DE LA BECHE, *Researches in theoretical geology* . . . 1831, and SEDGWICK, *On the classification of the fossiliferous slates of North Wales* . . . 1847.

Geological Mapping and Stratigraphy

The industrial revolution in Western Europe was certainly one of the major causes for the advancement of many geological fields. In the search for ores, fuels, and other necessities for men, canals were dug and highways were opened and spilled out hidden treasures. Geological maps pioneered by GUETTARD, *Mémoire et carte minéralogique sur la nature et la situation des terreins qui traversent la France et l'Angleterre* . . . 1746, became a fundamental aspect of investigation. SMITH wrote *A memoir to the map and delineation of the strata of England and Wales* . . . 1815, and *Strata identified by organized fossils* . . . 1816, followed by *Stratigraphic system of organized fossils* . . . 1817. These efforts led to stratigraphic works by CONYBEARE and PHILLIPS, *Outlines of the geology of England and Wales* . . . 1822; MURCHISON, *The Silurian system* . . . 1839; SEDGWICK and MURCHISON, *On the physical structures of Devonshire and on the subdivisions and geological relations of its older stratified deposits* . . . 1840, and *On the distribution and classification of the older or Palaeozoic deposits of the North of Germany and Belgium, and their comparison with formations of the same age in the British Isles* . . . 1842. This era introduced a flood of geological textbooks in many countries: AUBUISSON DE VOISINS, *Traité de géognosie, ou, exposé des connaissances actuelles sur la constitution physique et minérale du globe terrestre* . . . 1819; DE LA BECHE, *A geological manual* . . . 1832, 1833, and *Researches in theoretical geology* . . . 1834, 1837 (with French translation in 1838); OMALIUS D'HALLOY, *Éléments de géologie* . . . 1838, later changed to *Précis élémentaire de géologie* . . . 1843, then to *Abrégé de géologie* . . . 1862, and finally to *Géologie* . . . 1863; NAUMANN, *Lehrbuch der Geognosie* . . . 1858. The most influential textbooks by far were those by LYELL, *Principles of geology, being an attempt to explain the former changes of the earth's surface, by reference to causes now in operation* . . . 1830-1835, with too many editions to mention here, and also *Elements of geology* . . . 1838, 1841, subsequently called *A manual of elementary geology* in 1852.

Structural Geology

Structural geology as a separate discipline emerged at the level of experimental geology such as in HALL's *On the vertical position and convolution of certain strata and their relation with granite* . . . 1815, and at the level of orogenic theories as in VON BUCH's *Geognostische Beobachtungen auf Reisen durch Deutschland und Italien* . . . 1802-1809, where the theory of "craters of elevation" is presented, and in ÉLIE de BEAUMONT's *Recherches sur quelques-unes des révolutions de la surface du globe* . . . 1830, of which we own the English translation, *Researches on some of the revolutions which have taken place* . . . 1831.

Elie de Beaumont considered all mountains of same age arranged parallel to one another. At first he recognized twelve systems, then twenty-one, which intersected along a "pentagonal network." Later works showed a better understanding of folded belts; ROGERS and ROGERS, *On the physical structure of the Appalachian chain, as exemplifying the laws which have regulated the elevation of mountain chains generally* . . . 1842, and the more experimental work on the processes of deformation by DAUBRÉE, *Rapport sur les progrès de la géologie expérimentale* . . . 1867, and *Études synthétiques de géologie expérimentale* . . . 1879, translated into German in 1880.

Let us pause for a while and try to understand the many interpretations given in the past for folded and tilted beds. Steno interpreted them as the results of crustal movements due to the sudden rise of subterranean exhalation or the falling in of the roofs of cavernous spaces inside the earth. De Maillet assumed that folded beds resulted from the depositional action of submarine currents which underwent complex movements between rocky obstacles on the sea floor. He believed that tilted beds were due to the collapse of earlier cemented layers by the undermining action of marine currents. According to Raspe and his interpretation of Hooke, folded and tilted beds were the result of vertical uplift of the basement by earthquakes. De Saussure looked at these structures in the Alps and proposed that folded and tilted beds were caused by horizontal compressional efforts due to the explosion of elastic fluids during mountain-building. Today, enough observational data have been gathered to distinguish very early deformation of semi-consolidated beds of aqueous nature, as de Maillet probably observed, from beds which underwent later horizontal forces of mountain-building, as de Saussure probably suspected, and even from beds which are related to dissolution of soluble rocks by weathering processes leading to the distortion and collapse of underlying ones as mentioned by Steno. In other words, we have tools today which allow us to understand the reasons for the many different interpretations of rock deformation in the past.

Hydrology

Although the discussion of the numerous physical processes which are active at the earth's surface was an important part of many theories of the earth previously described, we shall now step back again a few centuries to find out what naturalists said about the origin of springs. They wondered whether there existed a perpetual cycle of rainwater supplying rivers which in turn flowed to the sea and evaporated, or whether sea water infiltrated the land through subterranean caverns, became distilled, and reappeared at the surface through complicated passages, and what happened to cause the saltiness of sea water. CARDANO, in *De subtilitate libri XXI* . . . 1550, and *De rerum varietate libri XVII* . . . 1557 (with a German translation of 1559), believed in a perpetual hydrological cycle, whereas AGRICOLA, in *De ortu et causis subterraneorum, Lib. V* . . . 1546, 1558, assumed that groundwater came from two sources: infiltration of surface water — rain, rivers, sea — and condensation of subterranean steam. PALISSY, in *Discours admirables* . . . 1580, and earlier in his *Recepte véritable* . . . 1563, of which apparently only one copy exists at Paris, clearly stated the theory of a perpetual cycle of water. He was followed by BESSON in *L'art et science de trouver les eaux et fontaines cachées sous terre, autrement que par les moyens vulgaires des agriculteurs & architectes* . . . 1569, and PAPIN, *Raisonnemens philosophiques touchant la salûre, flux et reflux de la mer, & l'origine des sources, tant des fleuves que des fontaines* . . . 1647.

DESCARTES contradicted these beliefs in *Discours de la methode . . . les meteores* . . . 1658, 1688. He assumed the infiltration of sea water in large caves at the base of mountains, evaporation by the heat of the earth's interior, deposition of salt and condensation of freshwater at the top of the caves, followed by the emergence of streams. Many versions of this "alambic theory" were subsequently presented. According to KIRCHER, in *Mundus subterraneus* . . . mentioned earlier, underground systems of caverns conducted sea water uphill to caves in mountains from which rivers originated. SCHOTT in *Anatomia physico-hydrostatica fontium ac fluminum* . . . 1663, disagreed with Kircher and proposed that some rivers originated from condensation of subterranean vapor and others from infiltration of rain and snow-water. HERBINIUS, in *Dissertationes de admirandis mundi cataractis supra & subterraneis* . . . 1678, believed that water was driven from subterranean caves to the surface but that precipitations ought to be disregarded. The idea that rainfall was an insufficient source for springs lingered in PLOT, *De origine fontium, tentamen philosophicum* . . . 1685. He gave an interesting classification of streams but thought that sea water ascended to mountain tops through fine canals. So did WOODWARD, in *An essay towards a natural history of the earth, and terrestrial bodies* . . . 1695, 1702, 1723.

A return to the theory of a hydrological cycle is shown by BARTHOLIN, in *De fontium fluviorumque origine ex pluviis* . . . 1689, 1698. PERRAULT in *De l'origine des fontaines* . . . 1674, demonstrated that enough rain falls on the catchment basin of a given river to insure its flow. He did not believe in general infiltration of rain into the ground, and hence in the recharge of aquifers. Apparently, his work remained unknown to his contemporaries. MARIOTTE, in *Traité du mouvement des eaux et des autres corps fluides* . . . 1686, 1700, initiated the trend toward hydraulics and hydrostatics. His work was translated into English in 1718 and into Latin in 1728-1729. It was followed by PITOT, *Description d'une machine pour mesurer la vitesse des eaux courantes* . . . 1732, and by DU BUAT, *Principes d'hydraulique* . . . 1786.

With the important works by HALLEY, *An estimate of the quantity of vapour raised out of the sea by the warmth of the sun* . . . 1678, *An account of the circulation of watry vapours of the sea, and of the cause of springs* . . . 1691, and *An account of the evaporation of water* . . . 1694, the part played in the hydrological cycle by water evaporation from the sea to sustain sufficient rainfall and insure flow of rivers was proven. Thus Halley complemented Perrault's earlier demonstration. VALLISNIERI took the results of Perrault and others to the Alps of Italy and demonstrated that in complexly folded areas of mountains, water derived from rain and melting snow not only flowed into rivers but also ran underground between impervious layers to reappear unexpectedly as springs when the structures were breached by the complex topographic surface or could be tapped by drilling artesian wells. His work *Lezione accademica intorno all'origine delle fontane* . . . written in 1715 and of which we have an edition of 1726, was refuted by GUALTIERI in *Riflessioni sopra l'origine delle fontane* . . . 1725.

The many works on springs reflect not only the desire to understand their origin but also the growing importance of water supply for expanding cities and industries, as well as of the beauty of fountains. FRANÇOIS wrote *L'art des fontaines; c'est à dire, pour trouver, esprouver, assembler, mesurer, distribuer, & conduire les sources dãs les lieux publics, & particuliers* . . . 1665. In England, a similar work was written by SWITZER, *An introduction to a general system of hydrostaticks and hydraulicks, philosophical and practical. Wherein the most reasonable and advantageous methods of raising and conducting water, for the watering of noblemens and gentlemens seats, buildings, gardens, &c. are carefully . . . laid down* . . . 1729. When RAMAZZINI wrote a treatise on the artesian wells of Modena in *De fontium mutinensium admiranda scaturigine. Tractatus physico-hydrostaticus* . . . 1691, it was translated into English by an irate opponent, Robert St. Clair, M.D., as *The Abyssinian philosophy confuted; or, Telluris theoria neither sacred, nor agreable to reason. Being for the most part, a translation of Petrus Ramazzini, of the wonderful springs of Modena* . . . 1697. This shows that interests in philosophical questions were as important as those in public works.

We have in our collection a wealth of works on the alleged medicinal value of thermal springs. A few examples are: FALLOPPIO, *De medicatis aquis, atque de fossilibus* . . . 1564, 1569; JORDEN, *A discourse on naturall bathes, and mineral waters* . . . 1632, 1673; ROCHAS, *La vraye anatomie spagyrique des eaux mineralles* . . . 1636-1637; SIMPSON, *Hydrologia chymica; or, the chymical anatomy of the Scarbrough, and other spaws in York-shire* . . . 1669; RUTTY, *A methodical synopsis of mineral waters, comprehending the most celebrated medicinal waters, both cold and hot, of Great-Britain, Ireland, France, Germany, and Italy, and several other parts of the world* . . . 1757; BECHER, *Neue Abhandlung vom Karlsbade* . . . 1772. North-America had its spas too: MEADE, *An experimental enquiry into the chemical properties and medicinal qualities of the principal mineral waters of Ballston and Saratoga, in the state of New-York* . . . 1817.

In the search for veins of mineral deposits and for springs, the dowsing-rod was mentioned by AGRICOLA, in *De re metallica Libri XII* . . . mentioned above; by an anonymous author, in *La verge de Jacob, ou l'art de trouver les tresors, les sources, les limites, les metaux, les mines, les mineraux, & autres choses cachées, par l'usage du Bâton fourché* . . . 1693; and by VALLEMONT, *La physique occulte ou traité de la baguette divinatoire, et de son utilité pour la découverte des sources d'eau, des miniéres, des tresors cachez, des voleurs & des meurtriers fugitifs* . . . 1696.

Engineering Geology

Engineering geology emerged from technical preoccupations on water- and land-rights after floods in alluvial plains as described, for instance, by AIMO in *De alluvionibus* . . . 1599; FABRICIUS, *Hydrotheologie* . . . 1734; in works on drainage and stream control by DUGDALE, *The history of imbanking and drayning of divers fenns and marshes, both in foreign parts and in this kingdom* . . . 1662, 1772; in BONINI, *Il Tevere incatenato ovvero l'arte di frenar l'acque correnti,* 1663; in GUGLIELMINI, *Della natura de' fiumi* . . . 1739; in SILBERSCHLAG, *Théorie des fleuves, avec l'art de bâtir dans leurs eaux et de prévenir leurs ravages* . . . 1769; in FRISI, *Traité des rivières et des torrens* . . . 1774, and FABRE, *Essai sur la théorie des torrens et des rivières* . . . 1797. Observations of landslides produced works by DELANGES, *Esperienze ed osservazioni intorno alla pressione delle terre, ed alla resistenza de' muri, che le medesime terre sostengono* . . . 1779, and by DONDI DALL'OROLOGIO, *Dello sprofondamento di una costa di monti negli Euganei* . . . 1787.

Marine Geology

The study of the sea, in particular its salinity, appears in PAPIN, *Raisonnements philosophiques touchant la salûre, flux & reflux de la mer* . . . mentioned above; its behavior in PANAROLO, *Il mare essaminato* . . . 1656, and in VOSSIUS, *De moto marium et ventorum* . . . 1663. These studies led to the forerunner works on oceanography and marine geology by MARSIGLI, *Brieve ristretto del saggio fisico intorno alla storia del mare* . . . 1711, followed by *Histoire physique de la mer* . . . 1725, and DONATI's *Della storia naturale marina dell'Adriatico* . . . 1750, of which we own the French translation *Essai sur l'histoire naturelle de la mer Adriatique* . . . 1758. Donati's work contains research on present-day sediments of the Adriatic and their comparison with ancient sedimentary rocks in the Apennines. CHAMBERS described more modern aspects of marine geology in *Ancient sea-margins, as memorials of changes in the relative level of sea and land* . . . 1848.

Glaciology

Glaciology began with early observations of Swiss Alpine glaciers by SIMLER, *Vallesiae descriptio libri duo* . . . 1574, and *Vallesiae et Alpium descriptio* . . . 1633, followed by SCHEUCHZER, ΟΥΡΕΣ ΙΦΟΙΤΗΣ HELVETICUS, *sive itinera per Helvetiae alpines regiones* . . . 1723; ALTMANN, *Versuch einer historischen und physischen Beschreibung der helvetischen Eisberge* . . . 1753, and GRUNER, *Die Eisgebirge des Schweizerlandes* . . . 1760, of which we own the French translation, *Histoire naturelle des glaciers de Suisse* . . . 1770. The most complete account of glaciers at the time is in DE SAUSSURE, *Voyages dans les Alpes* . . . 1779-1796. BOURRIT, in *Nouvelle description des vallées de glace et des hautes montagnes qui forment la chaîne des Alpes* . . . 1783, and in *Nouvelle description des glacières et glaciers de Savoye* . . . 1785, actually walked in de Saussure's footsteps, taking the same walks and hiring the same guides, while a schoolteacher from Solothurn, HUGI, studied glaciers independently and wrote *Naturhistorische Alpenreise* . . . 1830. An early discussion of the physical behavior of ice was presented by the French MAIRAN in *Dissertation sur la glace* . . . 1749, translated into German in 1752.

The presence of erratic boulders in the plains led to many speculations on how they had been transported. Volcanic eruptions, huge floods, sometimes related to the biblical Deluge as in BUCKLAND's *Reliquiae diluvianae* . . . mentioned above, and drifting by icebergs according to LYELL's *Principles* . . . were assumed to have carried these huge boulders. HUTTON suggested the action of glaciers in *Theory of the earth* . . . but remained unnoticed by his contemporaries, although PLAYFAIR in *Illustrations of the Huttonian theory of the earth* . . . (both works already mentioned above) repeated Hutton's beliefs. Glacier transport was also mentioned by BERNHARDI, *Wie kamen die aus dem Norden stammenden Felsbruchstücke und Geschiebe welche man in Norddeutschland und den benachbarten Ländern findet an ihre gegenwärtigen Fundorte?* . . . 1832.

While mountaineers and hunters of chamois and crystals had already talked about a greater extent of glaciers in the past, the concept of the Ice Age was best described by AGASSIZ, *Études sur les gla-*

ciers . . . 1840, followed by GODEFFROY, *Notice sur les glaciers, les moraines et les blocs erratiques des Alpes* . . . 1840, and a few months later by CHARPENTIER, *Essai sur les glaciers et sur le terrain erratique du bassin du Rhône* . . . 1841, and finally again by AGASSIZ, GUYOT, and DESOR in *Système glaciaire* . . . 1847. It took some thirty years before the glacial theory was accepted in North America. It involved, in particular, HITCHCOCK, *Report on the geology, mineralogy, botany, and zoology of Massachusetts* . . . 1833, 1835, and his *Final report* . . . 1841; MATHER, *Geology of New York* . . . 1843, ST. JOHN, *Elements of geology* . . . 1851, and a rather limited contribution by AGASSIZ, *Lake Superior* . . . 1850.

Coal and petroleum exploration

Applied geology such as petroleum exploration had some interesting early mentions by GOEBEL in *De succino libri duo* . . . 1558, with a second part entitled *De generatione succini, petrolei, ac reliquis bituminosis succis.* Oil seeps in the foothills of the Pyrenees in France were mentioned by RIVIÈRE, *Mémoire sur quelques singularités du terroir de Gabian, et principalement sur la fontaine de l'huile de pétrole qui y coule* . . . 1717; similar conditions were described in Alsace by HOEFFEL, *Historia balsami mineralis Alsatici, seu petrolei Vallis Sancti Lampert* . . . 1734. Closer to our times, WILKINSON wrote *Enquiry into the natural history, chemical properties and medical virtues of the rock oil, or green mineral naphta, of Barbados* . . . 1830.

Coal and peat have played an important role in the daily life of man and are mentioned casually in many early works on ore deposits and mining enumerated above. However, the following works display a more advanced technical character: PATIN, *Traité des tourbes combustibles* . . . 1663; SINCLAIR, *The hydrostaticks . . . a short history of coal, and of all the common, and proper accidents thereof; a subject never treated before* . . . 1672, and *Natural philosophy improven by new experiments . . . a large discourse anent coal, coal-sinks, dipps, risings, and streeks of coal, levels, running of mines, gaes, dykes, damps, and wildfire* . . . 1683. The importance of peat and coal increased with the development of metallurgy and other aspects of industrialization as shown by DEGNER, in *Dissertatio physica de turfis* . . . 1729; DE TILLY, *Mémoire sur l'utilité, la nature et l'exploitation du charbon minéral* . . . 1758; ZANON, *Della formazione ed uso della torba e d'altri fossili combustibili* . . . 1767, and MORAND, *L'art d'exploiter les mines de charbon de terre* . . . 1768. The impact of industrialization certainly marked the following contributions: BERNARD, *Mémoire sur les avantages et les inconvéniens de l'emploi du charbon de pierre et du bois dans les fabriques* . . . 1780, and WILLIAMS, *The natural history of the mineral kingdom. Part I. Of the natural history of the strata of coal, and of the concomitant strata* . . . 1789, 1810, translated into German in 1798.

The New World

Up to this point, we have divided our historical essay by subject; now we must turn to a geographical orientation to discuss the origins of geology in the New World.

Of rather folkloric and "wondrous" nature was the early information on the New World by MONARDES, *Dos libros, el uno que trata de todas las cosas que traen de nuestras Indias Occidentales* . . . 1569, translated into English in 1577, 1596, as *Joyful newes out of the newe founde worlde, wherein is declared the rare and singuler vertues of diverse and sundrie hearbes, trees, oyles, plantes, and stones;* by BRY in his Latin translation of Hariot's *A brief and true report of the new found land of Virginia,* 1590; and OVIEDO Y VALDÉS in his . . . *Summario de la naturale et generale historia de l'Indie Occidentali* . . . , a compilation published in 1534.

With respect to North America, early geological information is scattered among a number of travel accounts. The missionary HENNEPIN wrote *Description de la Louisiane* . . . 1683 (translated into Dutch in 1688 and into German in 1689), followed by *Nouvelle decouverte d'un tres grand pays situé dans l'Amérique, entre le Nouveau Mexique et la Mer Glaciale* . . . 1697, with the first description of Niagara Falls (other editions under changed titles in 1698 and 1704, a translation into English in 1698, and into Dutch in 1699). KALM, a Finnish botanist and naturalist, wrote *En resa till Norra America* . . . 1753-1761, translated into English in 1770-1771 as *Travels in North America* . . . and into Dutch in 1772. He described erratics without theorizing on their origin and speculated on a land bridge between North America and Europe.

In 1755, Lewis EVANS, a Welshman who spent his professional career as a surveyor and cartographer in America, produced a map of the Middle British Colonies attached to essays entitled *Geographical, historical, political, philosophical and mechanical essays. The first, containing an analysis of a general map of the Middle British Colonies in America.* . . . This analysis contains all the data that he could not represent on his map including his geological notes. The notes show Evans' clear understanding of the physiographic and geologic provinces in the eastern United States. Had their substance, ". . . nature of the soil, substrata, and particular fossils . . . ," been transferred to his map of 1755, Evans would have compiled the first geological map of the country. The map was reprinted at least twenty-six times during a period of more than forty years. POWNALL reprinted the map in 1776, accompanied by *A topographical description of such parts of*

North America as are contained in the map of the Middle British Colonies . . . to which he annexed extracts from Evans' journal of travel. It is from this journal that we learn the extent of Evans' geological ideas: fossil shells are remains of marine organisms; the exhumation of mountains from a level surface by stream erosion; the linear character of the Appalachian chain; the former extent of the Great Lakes, and the crustal uplifting after their draining. Evans died tragically one year after publication of his map, but his geological thinking can be traced in the later works of John BARTRAM and KALM, in MITCHILL's *A sketch of the mineralogical history of the State of New York . . .* 1798-1801, and VOLNEY.

An exception to the travel accounts of this period was GUETTARD's *Mémoire dans lequel on compare le Canada à la Suisse, par rapport à ses minéraux . . .* 1752. Guettard never traveled to America, but through the acquisition of specimens and notes from Canada, and from the reading of Jesuit relations, he produced a proto-geological map of the continent. The map, developed from his 1752 memoir, appeared in France in 1756. It was actually a locality map for thirty-nine kinds of "earths," fossils, minerals, and rocks. In addition, Guettard delineated three lithologic units (bandes) along the continent in an attempt to correlate them with similar units he recognized in Europe.

Other early travelers and their works include: ROGERS, *A concise account of North America . . .* 1765; MÄRTER, *Letters . . .* 1785; CHASTELLUX, who unlike Kalm speculated on erratics in *Voyage de M. le marquis de Chastellux dans l'Amérique Septentrionale dans les années 1780, 1781 & 1782 . . .* 1786, 1788-1791, and his *Remarks on the travels . . . in North America . . .* 1787, and finally SCHÖPF, who wrote comprehensively on American geology in *Beyträge zur mineralogischen Kentniss des östlichen Theils von Nord-Amerika und seiner Gebürge . . .* 1787, with much of his geology repeated in *Reise durch einige der mittlern und südlichen Vereinigten nord-amerikanischen Staaten . . .* 1788. On his travels in the South after the Revolutionary War, this young Hessian surgeon, who had studied mineralogy at the University of Erlangen, examined the field relationships between rock types and landforms, and realized that a map showing these relationships in color could be produced. Unfortunately, time did not allow him to begin the task.

It was left to VOLNEY to produce, in France, the first colored geological map of the United States. A French traveler in the United States for three years, Constantin François Chasseboeuf, later Comte de Volney, returned to Paris where he wrote *Tableau du climat et du sol des États-Unis d'Amérique . . .* 1803 (translated into English in 1804) which is actually an organized description of the physical features of the country. To accompany his account, Volney produced a map of the country on which he delineated, by means of color, the major lithologic units of granite, sandstone, calcareous rocks, marine sands, and river deposits. Although it was not Volney's intent, the map was sold to the public uncolored with only sparse geological notation. The colored version was limited to a few large paper copies, one of which resides in our collection.

Another copy of Volney's colored map, as we know now, was seen by MACLURE and formed the basis for his famous geological map which accompanied *Observations on the geology of the United States* . . . 1809 (translated into French in 1811) and his book bearing the same title, published in 1817.

Geological investigation in the "western country" or the Mississippi Valley states started with pioneer observers, first French, then English. By the beginning of the nineteenth century, geology in scientific reports was becoming increasingly organized. Descriptions of natural features by CUTLER, DRAKE, BROWN, THOMAS, and BRADBURY were followed by the more explicit geological narratives of KEATING, SCHOOLCRAFT, FEATHERSTONHAUGH, and HALL. It is interesting to note that the western writers were less influenced by European geological dogmas than their eastern counterparts; they made their own speculations about the origin of geological phenomena and used their own rock terminology instead of borrowing that from Europe.

Geological mapping moved westward too. In 1832 a remarkable map appeared in HINTON's *The history and topography of the United States.* It extended geology from the Mississippi Valley to the Rocky Mountains, using observations probably borrowed from the work of Edwin JAMES, geologist for Long's 1819 expedition to the Rocky Mountains. His *Account of an expedition from Pittsburg to the Rocky Mountains, performed in the years 1819, 1820,* was published in London in 1823 and included a route map showing prominent geological boundaries. The legend on Hinton's anonymous map reflected James' geological nomenclature.

Thirty years before the Federal Government would establish a national geological survey, the states were supporting geological surveys of their own. The mid-1830's were marked by publications of state geologists: EMMONS, OWENS, HITCHCOCK, JACKSON, MATHER, William Barton ROGERS, Henry Darwin ROGERS, and HALL. This period was also characterized by the appearance of geological textbooks. Americans were reading those of EATON, *Geological textbook, prepared for popular lectures on North American geology; with applications to agriculture and the arts,* 1830 (with an expanded version in 1832), and editions of BAKEWELL's *An introduction to geology* . . . 1829, 1833, 1839, edited by Professor Benjamin SILLIMAN.

Charles LYELL came to evaluate American geology and wrote *Travels in North America, in the years 1841-42; with geological observations on the United States, Canada, and Nova Scotia . . .* 1845, and *A second visit to the United States of North America,* 1849, 1855-1868. He spent several days with James Hall around Niagara Falls and was impressed by Hall's mastery of stratigraphy. In fact, he placed the work of American geologists in a favorable light knowing that they learned their trade the hard way: on the job.

Epilogue

We could not finish the evaluation of this collection without referring to some unusual items, such as the ptolemaic spheres described by BONARDO, *La grandezza, larghezza, e distanzia di tutte le sfere, ridotte a nostre miglia, cominciando dall'inferno, fino alla sfera, dove stanno i beati . . .* 1584; the plea to the King of France to study and use the mines of his country to better profit by Martine de BERTEREAU, *La restitution de Pluton . . . Des mines & minières de France, cachees et detenües jusques à présent au ventre de la terre, par le moyen desquelles les Finances de sa Majesté seront beaucoup plus grandes que celles de tous les Princes Chrestiens, & ses sujets plus heureux de tous les Peuples . . .* 1640; how geology fared with religion according to COLE, *Popular geology subversive of divine relevation! A letter to the Rev. Adam Sedgwick . . . being a Scriptural refutation of the geological positions and doctrines promulgated in his lately published commencement sermon, preached in the University of Cambridge, 1832 . . .* 1834; and finally, a touch of poetry in SCAFE, *A geological primer in verse; with, a poetical geognosy; or feasting and fighting, and sundry right pleasant poems, with notes. To which is added, A critical dissertation on "King Coal's levee," addressed to the professors and students of the University of Oxford . . .* 1820.

Organization of the Catalog

Arrangement

The basic arrangement of this catalog is alphabetical by author. Where the author is corporate (e.g., Indiana. State Geologist.), cross-reference is made from the individual who prepared the report (e.g., Owen, David Dale, 1807-1860). The heading "Anonymous" is used for the works which lack clues to authorship, and a few works of multiple authorship are entered by title.

Cross-references are given for joint authors and other contributors to the intellectual content of a work, but not for patrons, illustrators, translators, or founders of schools of geology (e.g., Werner).

The arrangement of entries under author is primarily chronological, but collected editions of an author's complete works, or of some of them, are placed first. Successive editions of a work immediately succeed the first, and translations follow the original work.

Atlases to works are given separate entry if their publishing or printing history differs from that of the text; otherwise, they are incorporated into the entry for text.

Bibliographical descriptions

Each entry in the catalog has been brought into conformity with basic Library of Congress format. This format will be familiar to most library users. Following each citation, in smaller type, additional bibliographical information is provided, including specifics on the Library's holdings and the physical condition of the volumes. Non-bibliographical annotation is limited to clarification of a title which may be unclear as to its geological significance.

We have provided the Library's call (or location) number for every title in the catalog. The abbreviation "uncat." means that the book is part of the uncataloged collection in the Rare Book Room. Quarto and folio size books are identified by "q" or "f" preceding either the call number or the abbreviation "uncat."

Index

Subdivision of index entries is by subject or place, then by author. The index is keyed to the catalog by entry number.

Catalog

1 **Accum, Friedrich Christian,** 1769–1838.
Elements of crystallography, after the method of Haüy; with, or without, series of geometrical models, both solid and dissected; exhibiting the forms of crystals, their geometrical structure, dissections, and general laws . . . London, Longman, Jurst, Rees, Orme, and Brown, 1813.
396p., illus. 548/Ac21e

2 **Acosta, Joseph,** 1539–1600.
Historia natural y moral de Indias . . . Sevilla, Juan de Leon, Compañia de Jesus, 1590.
7 books in 1. x970/Ac7dS/1590

Lacks title page.

3 **Acosta.**
De natura novi orbis libri duo et de promulgatione evangelii apud barbaros, sive, de procuranda indorum salute, libri sex. Colinae Agrippinae, In officina Birckmannica, Sumptibus Arnoldi Mylij, 1596.
7 books in 1. x970/Ac7d/1596

4 **Acosta.**
Historia naturale, e morale delle Indie; nella quale si trattano le cose notabili del Cielo, & de gli Elementi, Metalli, Piante, & Animali di quelle: i suoi riti, & ceremonie: Leggi & governi, & guerre de gli Indiani. Tradotta da Gio. Paolo Galucci Salodiano. Venetia, Bernardo Basa, 1596.
7 books in 1. x970/Ac7d1g and uncat.

5 **Acosta.**
Histoire Naturelle et Moralle des Indes, tant Orientalles qu'Occidentalles. Où il est traicté des choses remarquables du Ciel, des Elémens, Métaux, Plantes & Animaux qui sont propres de ce païs. Ensemble des moeurs, cérémonies, loix, gouvernements & guerres des mesmes Indiens. Traduite par Robert Regnault Cauxois. Paris, Marc Orry, 1598.
7 books in 1. uncat.

6 **Acosta.**
The Naturall and Morall Historie of the East and West Indies.

Intreating of the remarkeable things of Heaven, of the Elements, Mettalls, Plants, and Beasts which are proper to that Country: Together with the Manners, Ceremonies, Lawes, Governments, and Warres of the Indians. Transl. by E. G. London, Val:Sims, 1604.
7 books in 1. x970/Ac7dEg

7 **Acosta.**
Histoire Naturelle et Moralle des Indes, tant Orientalles qu'Occidentalles. Où il est traicté des choses remarquables du Ciel, des Elémens, Métaux, Plantes & Animaux qui sont propres de ce païs. Ensemble des moeurs, cérémonies, loix, gouvernemens & guerres des mesmes Indiens. Traduite par Robert Regnault Cauxois. Paris, Marc Orry, 1606.
7 books in 1. uncat.

8 **Acosta.**
Historie naturael en morael van de Westersche Indien. Waer inne ghehandelt wort van de merckelijckste dinghen des Hemels/ Elementen/ Metalen/ Planten ende Gedierten van dien: Als ook de Manieren/ Ceremonien/ Wetten/ Kegeeringen/ ende Vorlogen der Indianen. Overgheset door Ian Huyghen van Linschotten. De tweede Editie. Amsterdam, Broer Iansz, 1624.
7 books in 1. uncat.

9 Admiranda narratio, fida tamen, de commodis et incolarum ritibus Virginiae . . . Anglico scripta sermone à Thoma Hariot, *See* BRY, THEODOR DE, ed., [America, pt. 1 Latin] Admiranda narratio . . . 1590.

10 **Agassiz, Louis Jean Rodolphe,** 1807-1873.
Recherches sur les poissons fossiles. Neuchâtel, Aux frais de l'auteur. Imprimerie de Petitpierre et H. Nicolet, 1833-43.
5v. in 3 (text), plus 5v. in 4 (atlases). q567/Ag1r
xf567/Ag1r/atlas

11 **Agassiz.**
Monographies d'échinodermes, vivans et fossiles. Neuchâtel, Aux frais de l'auteur, 1838-1842.
[490p.] 63 plates. q593.95/Ag1m

This volume consists of 5 separately-paged monographs: no. 1 . . . Des Salénies by L. Agassiz; no. 2 . . . Des Scutelles by L. Agassiz; no. 3 . . . Des Galérites by Ed. Desor; no. 4 . . . Des Dyaster by Ed. Desor; no. 5 . . . Anatomie du genre *Echinus* by G. Valentin.

12 **Agassiz.**
Études sur les glaciers. Ouvrage accompagné d'un atlas de

GEORGII AGRICOLAE

De ortu & causis subterraneorum Lib. v
De natura eorum quæ effluunt ex terra Lib. IIII
De natura fossilium Lib. x
De ueteribus & nouis metallis Lib. II
Bermannus, siue De re metallica Dialogus Lib. I
Interpretatio Germanica uocum rei metallicæ, addito duplici Indice, altero rerum, altero locorum.

Omnia ab ipso authore, cum haud pœnitenda accessione, recens recognita.

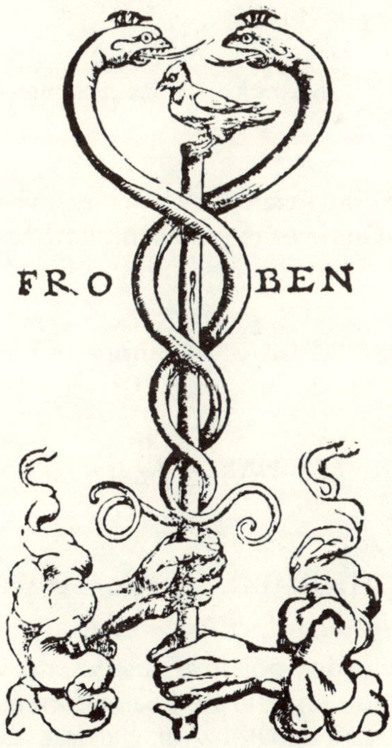

BASILEAE MD LVIII

Cum Imp. Maiestatis renouato priuilegio ad quinquennium.

Cat. No. 29

32 planches. Soleure. Aux frais de l'auteur. Jent & Gassmann, 1840.
346p. x551.31/Ag1e, 551.31/Ag1e and xf551.31/Ag1e/plates

13 **Agassiz.**
Untersuchungen über die Gletscher. Nebst einem Atlas von 32 Steindrucktafeln. Solothurn. Auf Kosten des Verfassers. Jent & Gassmann, 1841.
326p. x551.31/Ag1u and 551.31/Ag1u

14 **Agassiz.**
Untersuchungen über die Gletscher. Nach der Natur gezeichnet und lithographirt von Jph. Bettannier. Lithographie von H. Nicolet. Neuchâtel, 1840.
f551.31/Ag1u/atlas
Library owns one copy of the atlas.

15 **Agassiz.**
Études critiques sur les mollusques fossiles. Neuchâtel, Aux frais de l'auteur, 1840-45.
2v., 99 plates. q564/Agie
Contents: 1. Mémoire sur les trigonies, 1840. 2. Monographie des myes, 1842-45.

16 **Agassiz.**
Récit d'une course faite aux glaciers en hiver, par MM. Agassiz et E. Desor. Genève, Tiré de la Bibliothèque Universelle, 1842.
36p. uncat.

17 **Agassiz.**
See DESOR, ÉDOUARD, Compte rendu des recherches de M. Agassiz . . . 1843.

18 **Agassiz.**
See DESOR, ÉDOUARD, Agassiz geologische Alpenreisen . . . 1844.

19 **Agassiz.**
See DESOR, ÉDOUARD, Excursions et séjours . . . Alpes, 1844.

20 **Agassiz.**
Monographie des poissons fossiles du vieux grès rouge ou Système dévonien (Old Red Sandstone), des Îles Britanniques et de Russie. Neuchâtel. Aux frais de l'auteur. Imprimerie H. Wolfrath, 1844.
171p. q567/Ag1m

21 **Agassiz.**
Monographie des poissons fossiles du vieux grès rouge ou Système dévonien (Old Red Sandstone), des Îles Britanniques

et de Russie. Neuchâtel. Aux frais de l'auteur, A. Sorel, 1844-1845.
43 plates. q567/Ag1m/atlas

22 **Agassiz.**
*S*ee DESOR, ÉDOUARD, Nouvelles excursions . . . de M. Agassiz et de ses compagnons . . . 1845.

23 **Agassiz.**
*S*ee VOGT, KARL CHRISTOPH, Agassiz' und seiner Freunde geologische Alpenreisen . . . 1847.

24 **Agassiz.**
Système glaciaire ou Recherches sur les glaciers, leur mécanisme, leur ancienne extension et leur rôle qu'ils ont joué dans l'histoire de la terre, par MM. L. Agassiz, A. Guyot & E. Desor. — Première Partie. Nouvelles études et expériences sur les glaciers actuels, leur structure, leur progression et leur influence physique sur le sol . . . avec un atlas de 3 cartes et 9 planches. Paris, Victor Masson, Leipzig, Léopold Voss, 1847.
598p. and atlas with 12 plates. x551.31/Ag1n and xf551.31/Ag1n

Only the first part of a projected work with Guyot and Desor was published by Agassiz alone. The text is bound with manuscript page by Professor Silliman, dated Febr. 16, 1849 entitled "my opinion of this volume."

25 **Agassiz.**
Lake Superior: its physical character, vegetation, and animals, compared with those of other and similar regions. With a narrative of the tour, by J. Elliot Cabot. And contributions by other scientific gentlemen. Boston, Gould, Kendall & Lincoln, 1850.
[9]-428p., front., illus., plates, map. 574.929/Ag1ℓ

26 **Agassiz.**
*S*ee MILLER, HUGH, The foot-prints of the Creator . . . with a memoir on the author by L. Agassiz . . . 1850 and 1866.

27 **Agricola, Georg,** 1494-1555.
Bermannus; sive de re metallica. Basileae, in aedibus Frobenianis, 1530.
135p. x669/Ag8b

28 **Agricola.**
De ortu & causis subterraneorum, Lib. V. De natura eorum quae effluunt ex Terra, Lib. IIII. De natura fossilium, Lib. X. De veteribus & novis metallis, Lib. II. Bermannus, sive de re metallica, Dialogus, Interpretatio Germanica uocum rei me-

tallicae, addito Indice foecundissimo. Basilaea, per Hieronymum Frobenium et Nic. Episcopium, 1546.
538p. uncat.

29 **Agricola.**
De ortu & causis subterraneorum, Lib. V. De natura eorum quae effluunt ex terra, Lib. IIII. De natura fossilium, Lib. X. De veteribus & novis metallis, Lib. II. Bermannus, sive de re metallica Dialogus, Lib. I. Interpretatio Germanica uocum rei metallicae, addito duplici indice, altero rerum, altero locorum. Omnia ab ipso authore cum haud poenitenda accessione, recens recognita. Basileae, per H. Frobenium & N. Episcopium, 1558.
470p. x549/Ag8d

30 **Agricola.**
De La generatione de le cose, che sotto la terra sono, e de le cause de' loro effeti e nature, Lib. V. De La Natura di quelle cose, che la terra scorrono. Lib. IIII. De la Natura de le cose Fossili, e che sotto la terra si cavano. Lib. X. De le Minere antiche e moderne. Lib. II. Il Bermanno, ò de le cose Metallice, Dialogo, Recato tutto hora dal Latino in buona lingua volgare. Vinegia, Michele Tramezino, 1550.
467p. x551/Ag8dI/1550

31 **Agricola.**
De re metallica Libri XII. Quibus Officia, Instrumenta, Machinae, ac omnia denique ad Metallicam spectantia, non modo luculentissimè describuntur, sed & per effigies, suis locis insertas, adiunctis Latinis, Germanisque appellationibus ita ab oculos ponuntur, ut clarius tradi non possint. Eiusdem De animantibus subterraneis Liber, ab Autore recognitus: cum indicibus diversis, quicquid in opere tractatum est, pulchrè demonstrantibus. Basilea apud H. Frobenium et N. Episcopium, 1556.
502p., illus. xq669/Ag8d

32 **Agricola.**
De re metallica Libri XII. Quibus Officia, Instrumenta, Machinae, ac omnia denique ad Metallicam spectantia, non modo luculentissimè describuntur, sed & per effigies, suis locis insertas, adjunctis Latinis Germanisque appellationibus, ita ab oculos ponuntur, ut clarius tradi non possint. Quibus accesserunt hâc ultimâ editione, Tractatus ejusdem argumenti, ab eodem conscripti, sequentes. De Animalibus Subterraneis, Lib. I. De Ortu & Causis Subterraneorum. Lib. V. De Natura

eorum quae effluunt ex Terra, Lib. IV. De Natura Fossilium. Lib. X. De Veteribus & Novis Metallis, Lib. II. Bermannus sive de Re Metallica, Dialogus. Lib. I. Cum Indicibus diversis, quicquid in Opera tractatum est, pulchrè demonstrantibus. Basileae, Sumptibus & Typis, Emanuelis Konig, 1557.
708p., illus. xq669/Ag8d/1557

33 **Agricola.**
De re metallica Libri XII. Quibus officia, instrumenta, machinae, ac omnia denique ad metallicam spectantia, non modo luculentissimè describuntur, sed & per effigies, suis locis insertas, adjunctis Latinis, Germanisq̀u; appellationibus, ita ab oculos ponuntur, ut clariùs tradi non possint. Ejusdem de animantibus subterraneis Liber, ab Autore recognitus. Cum indicibus diversis, quicquid in opere tractatum est, pulchrè demonstrantibus. Basilae Helvet. Sumptibus itemq́ue typis chalcographicis Ludovici Regis, Ludwig König, 1621.
502p., illus. xq699/Ag8d/1621

34 **Aikin, Arthur,** 1773-1854.
Journal of a tour through North Wales and part of Shropshire; with observations in mineralogy and other branches of natural history. London, 1797.
231p., 1 plate. x914.29/Ai4j

35 **Aikin.**
A dictionary of chemistry and mineralogy, with an account of the processes employed in many of the most important chemical manufactures. To which are added a description of chemical apparatus, and various useful tables of weights and measures, chemical instruments, &c., &c. Illustrated with fifteen engravings. By A. & C. R. Aikin. London, printed for J. and A. Arch, 1807.
2v., 15 plates. q540.3/Ai4d

36 **Aikin.**
An account of the most recent discoveries and improvements in chemistry and mineralogy, to the present time; being an appendix to their Dictionary of Chemistry and Mineralogy. By A. and C. R. Aikin. London, printed for J. and A. Arch, 1814.
176p., plate. q540.3/Ai4d/sup.

37 **Aikin.**
A Manual of Mineralogy. First American Edition. Philadelphia, S. W. Conrad, 1815.
275p. x549/Ai4m/1815

38 **Aimo, Battista,** 16th cent.
De alluvionibus; tractatus fertilissimus. In quo incrementi, & decrementi fluminum, ad fundorum vicinorum commoda, & incommoda, cum naturali acquisitione, & amissione dominiorum, & possessionum; materia ita copiosè, & clarè tractatur, ut nil ampliùs desiderandum sit. Venetiis, 1599.
366p., illus. q uncat.

39 **Ainsworth, William Francis,** 1807-1896.
An account of the caves of Ballybunian, county of Kerry, with some mineralogical details. Dublin, W. Curry, 1834.
96p., illus. 551.44/Ai6a

40 **Alberti, Friedrich August von,** 1795-1878.
Halurgische Geologie. Stuttgart und Tübingen, J. G. Cotta'scher Verlag, 1852.
2v. 549.4/Aℓ14h

41 **Albertus Magnus,** 1193-1280.
De mineralibus libri quinque. Auguste Vindelicorum, 1519.
14ℓ,57p. uncat.

42 **Albertus Magnus.**
De secretis mullierum libellus, scholijs auctus, & à mendis repurgatus. Eiusdem de virtutibus herbarum, lapidum, & animalium quorundam libellus. Item de mirabilibus mundi, ac de quibusdam effectibus causatis à quibusdam animalibus, &c. Aiedicimus & ob materiae similitudinem Michaelis Scoti philosophi, De secretis natura opusculum. Cum indice locupletissimo. Lugduni, Excudebar Ioannes Martinus, 1584.
751p. uncat.

The library owns another edition, with the same title, published at Argentorati, per Lazarum Zetznerum, 1607 (uncat.).

The second book on plants, stones, and animals contains *De virtutibus lapidum quorundam,* a description of metals and minerals. The third book discusses the origin of the world in *De mirabilibus foeliciter incipit.* There seems to be no proof that these works are by Albertus Magnus.

43 **Aldrovandi, Ulisse,** 1522-1607.
Musaeum Metallicum in libros IIII. distributum. Bartholomaeus Ambrosinus . . . labore et studio composuit cum indice copiosissimo Marcus Antonius Bernia propriis impensis in lucem edidit. [Bononiae, J.B. Ferronius, 1648].
979p., illus. q uncat.

V.13 of his [works].
Illustrated by hundreds of woodcuts.

GEORGII AGRICOLAE KEMPNICENSIS, MEDICI AC PHILOSOPHI CLARISS.

DE RE METALLICA LIBRI XII.

QUIBUS OFFICIA, INSTRUMENTA, MACHINÆ, AC OMNIA DENIQUE AD METALLICAM SPECTANTIA, NON MODO LUCULENTISSIMÈ describuntur; sed & per effigies, suis locis insertas, adjunctis Latinis, Germanicisq; appellationibus, ita ob oculos ponuntur, ut clariùs tradi non possint.

Ejusdem

DE ANIMANTIBUS SUBTERRANEIS LIBER, ab Autore recognitus.

Cum INDICIBUS diversis, quicquid in Opere tractatum est, pulchrè demonstrantibus.

BASILEÆ HELVET.

Sumptibus itémque typis chalcographicis Ludovici Regis.

Anno MDCXXI.

Cat. No. 33

44 **Alencé, Joachim d',** d. 1707.
Traitté de l'aiman. Divisé en deux parties. La première contient les Expériences; & la seconde les raisons que l'on en peut rendre. Amsterdam, Henry Wetstein, 1687.
148p., illus. uncat.

45 **Allan, Thomas,** 1777-1833.
An alphabetical list of the names of minerals, at present most familiar in the English, French, and German languages, with tables of analyses. Edinburgh, Printed at the Caledonian Mercury Press, 1808.
72p., tables. 549.03/Aℓ5a

46 **Allan.**
Mineralogical nomenclature, alphabetically arranged; with synoptic tables of chemical analyses of minerals. Edinburgh, Printed for A. Constable [etc.], 1814.
1v., various paging. 549.03/Aℓ5m/1814

Introduction signed: Thomas Allan

47 **Allan.**
Mineralogical nomenclature, alphabetically arranged; with synoptic tables of the chemical analyses of minerals. Edinburgh, Printed for A. Constable and Company, 1819.
1v., various paging. 549.03/Aℓ5m

Introduction signed: Thomas Allan

48 **Alléon Dulac, Jean Louis,** 1723-1788?
Mémoires pour servir à l'histoire naturelle des provinces de Lyonnois, Forez and Beaujolais. Lyon, C. Cizeron, 1765.
2v. 570.9445/Aℓ5m

49 **Allin, James.**
Thunder and earthquake, a loud and awful call to reformation consider'd in a sermon preached at Brooklyn, November the first; upon a special fast occasion'd by the earthquake, which happen'd in the evening after the 29th day of October, 1727. Now published with enlargements, second edition. Boston, N.E., printed by Gamaliel Rogers for Joseph Edwards, 1727.
49p. x252/M699e

Bound with MIX, STEPHEN, Extraordinary displays of the divine majesty & power, are to try men, and impress the fear of God on their hearts, that they sin not . . . London, T. Green, 1728.

Under the heading of "Earthquakes are produced by the agency of second causes," Allin gives a history of theories on earthquakes and mentions some locations such as Sicily, Jamaica, China, Antioch, and New England.

50 **Allioni, Carlo,** 1725-1804.
Oryctographiae Pedemontanae specimen, exhibens corpora fossilia terrae adventitia. Parisiis, C. J. B. Bauche, 1757.
82p. 560.9451/Aℓ5o
Bibliography: p. v-vii.

51 **Altmann, Johann Georg.**
Versuch einer historischen und physischen Beschreibung der helvetischen Eisberge. 2.Aufl. Zürich, Heidegger, 1753.
271p., fold. plates. 551.31/Aℓ7v/1753

52 **Amoretti, Charles,** 1741-1816.
See FORTIS, ALBERTO, Mémoires pour servir à l'histoire naturelle . . . de l'Italie . . . 1802.

53 **Andréossy, Antoine François,** 1761-1828.
Mémoires sur le lac Menzaléh et sur la vallée des lacs de Natron et celle du Fleuve-sans-eau. Paris, Didot, an VIII [1800].
102p., fold. maps. 556.21/An2m
Half-title: Mémoires géologiques sur la Basse Egypte.

54 **Anonymous.**
Tremblement de terre, advenu à Lyon le mardy vingtiesme jour de May mil cinq cens septante-huict, peu avant les quatre heures du soir. Lyon, Par Benoist Rigaud., 1578.
7p. uncat.

55 **Anonymous.**
Dreadfull newes: or, A true relation of the great, violent and late earthquake. Hapned [sic] the 27. day of March, stilo romano last, at Callabria, in the Kingdome of Naples, about the houres of three and foure in the afternoone, to the overthrow and ruine of many cities, townes, and castles, and the death of above fifty thousand persons. With the names of the principall of the sayd Cities, Townes and Castles, destroyed by the said Earthquake, according to the Originall Printed and published by Authority at Venice. London, Printed by I. Okes for R. Mabb, 1638.
17p. x551.22/D811
Attributed to Martin Parker in the Short title catalogue.

56 **Anonymous.**
Relacion del temblor, y terromoto que Dios nuestro Señor fue servido de embiar à la ciudad del Cuzco à 31. de Março este año passado de 1650 . . . Madrid, por Iulian de Paredes, 1651.
4p. q uncat.

57 **Anonymous.**
La verge de Jacob, ou L'art de trouver les tresors, les sources, les limites, les metaux, les mines, les mineraux, & autres choses cachées, par l'usage du Bâton fourché. Par I. N. Lyon, Chez Hilaire Baritel, 1693.
137p. uncat.

58 **Anonymous.**
Mémoire . . . au sujet de l'histoire naturelle de la Province de Languedoc. Montpellier, Jean Martel, 1726.
16p. x554.484/R52m

Extrait Société Royale des Sciences de Montpellier.
Bound with Rivière, Guillaume, Mémoire sur quelques singularités du terroir de Gabian, . . . 1717.

59 **Anonymous.**
Principales merveilles de la nature, où l'on traite de la substance de la terre, de la mer, des fleuves, lacs, rivières, montagnes, rochers, etc. . . . tiré des meilleurs auteurs, anciens & modernes, par M. ***. Amsterdam, Aux dépens de la Compagnie, 1745.
2v. in 1 (332p.). 910/P935

60 **Anonymous.**
Memoirs concerning Herculaneum, the subterranean city, lately discovered at the foot of Mount Vesuvius, giving a particular account of the most remarkable buildings, statues, paintings, medals, and other curiosities found there to the present time. Tr. from the Italian of a memorial drawn by the secretary of the Marquis d'Hospital, . . . and illustrated with notes, by William Fordyce . . . To which are added an extract from Xiphilinus's Epitome of Dion and Pliny the younger's two letters, giving an account of the overthrow of this city, and the eruption of Vesuvius that occasioned it. London, Printed for D. Wilson, 1750.
68p. x913.377/M519

61 **Anonymous.**
Metallischer Baumgarten, in welchem das einzige wahre subjectum philosophiae oder primum ens metallorum bloss und gantz offenbar vor Augen gelegt und beschrieben worden ist; von einem Freunde, deme die wahrheit bewust ist, und der einen jeden vom falschen Weg gern ableiten, hingegen zu dem wahren einzigen Brunnen der Metallen führen und bringen will. Franckfurt und Leipzig, Bey J. F. Fleischer, 1753.
87p. 549/M565/1753

62 **Anonymous.**
Philosophische Ergötzungen, oder... Untersuchung wie die wahrhaften Seemuscheln auf die höchsten Berge und in die festesten Steine gekommen, nebst einer deutlichen Erklärung der Erdbeben und anderer wunderbarer Naturbegebenheiten, auf Veranlassung der von Anton Lazaro Moro herausgegebenen neuen Untersuchung der Veränderung des Erdbodens, ausgestellet von einem fleissigen Erforscher philosophischer Lauterkeit. Bremen, J. H. Cramer, 1765.
564p., 2 plates. 550/P549

63 **Anonymous.**
La terre habitable; ou, Essai sur la structure extérieure et intérieure du globe. Londres; et se trouve chez Valade, Paris, 1782.
136p., 2 fold. plates. x551.1/T277

64 **Anonymous.**
Histoire des découvertes faites par divers savans voyageurs dans plusieurs contrées de la Russie & de la Perse, relativement à l'Histoire civile & naturelle, à l'Economie rurale, au Commerce, & etc. Lausanne, J. P. Heubach, 1779-87.
6v. 914.7/Aℓ34F

v.1, 1784.

65 **Anonymous.**
Elements of natural philosophy, arranged under the following heads: matter and motion, the universe, the solar system, the fixed stars, the earth considered as a planet, the atmosphere, meteors, springs, rivers, and the sea, fossils, plants, animals, the human frame and the human understanding. Philadelphia, J. P. Parke, 1807.
235p. x502/Eℓ26

66 **Anonymous.**
Caesarea. The island of Jersey; its history, constitution, government, laws, peculiar privileges, customs, mineralogy, produce, commerce, and other statistics; with a tour round the coast, and in the interior; to which is added its antiquities, and the biography of eminent men, natives of the island. London, Printed for T. Baker, 1840.
330p., illus., fold. map. 914.234/C11

Chapter on mineralogy (p. 108-117) revised by "Mr. König, of the British Museum."

67 **Ansted, David Thomas,** 1814-1880.
Geology, introductory, descriptive, & practical. London, J. Van Voorst, 1844.
2v., illus. 551/An8g

68 **Ansted.**
The geologist's text-book. Chiefly intended as a book of reference for the geological student. London, J. Van Voorst, 1845.
143p. 550/An8g

69 **Ansted.**
The ancient world; or, Picturesque sketches of Creation. London, Van Voorst, 1847.
408p. 560/An8a

70 **Antisell, Thomas,** 1817-1893.
Irish geology, in a series of chapters, containing an outline of the science of geology, and a description of the various rocks distributed on the surface of the island, with some remarks on the climate. Dublin, J. McGlashan, 1846.
84p. 554.15/An8i

71 **Aquarius** [pseudonym].
Floods, their causes, mitigation & cure. Also a scheme for the efficient drainage of lower Yare, Waveney and Bure Valleys, and the improvement of Yarmouth Harbour by the deepening of the water at the bar, etc., by a simple and comparatively inexpensive method. Norwich, Fletcher, [n.d.].
16p. 627.4/Aq3f

72 **Archiac de Saint Simon, Étienne Jules Adolphe Desmier,** 1802-1868.
Études sur la formation crétacée des versants sud-ouest et nord-ouest du plateau central de la France. 1 partie. Paris, Impr. de Fain et Thunot, 1843.
100p., 2 plates. 554.4/Ar2e

Extrait des Annales des sciences géologiques, deuxième année.

73 **Archiac de Saint Simon.**
Histoire des progrès de la géologie de 1834 à [1859]. Publiée par la Société Géologique de France, sous les auspices de M. le Comte de Salvandy . . . Paris, Au lieu des séances de la Société, 1847-1860.
8v. in 9, col. fold. plates, tables. 550.9/Ar2h

v.2 is published in 2 parts.

BAPTISTAE AYMO, PARMENSIS, I. C. PRAESTANTISS.

De Alluuionibus;

TRACTATVS FERTILISSIMVS.

In quo Incrementi, & Decrementi Fluminum, ad fundorum vicinorum commoda, & incommoda, cum naturali acquiſitione, & amiſſione Dominiorum, & Poſſeſſionum; materia ita copiosè, & clarè tractatur, vt nil ampliùs deſiderandum ſit.

Quod ad Venditionum, Emptionum; pactorum inter Emptorem, & Venditorem; Locationum, Conductionum, & Retractuum, in prædijs Ruſticis Fluminibus adhærentibus, negotia exactiſsimè conferre poſsit.

Opus omnibus vtile, delectabile, & admodum neceſſarium.

Cum deſiderabilibus Summarijs, & Indice copioſiſsimo.

PRIVILEGIIS MVNITVM.

VENETIIS, MDXCIX.

Apud Baretium Baretium, & Socios.

Cat. No. 38

v.2-8 have in title: Publiée . . . sous les auspices de M. le Ministre de l'instruction publique.
Contain bibliographies.

74 **Archiac de Saint Simon.**
Géologie et paléontologie. Paris, F. Savy, 1866.
776p., tables. 550/Ar2g

On cover: Suite et fin de l'histoire de la géologie.
Bound with the author's Explication d'un profil géologique de l'Angleterre. Angers, 1867, 20p., map.

75 **Arduino, Giovanni,** 1714-1795.
Raccolta de memorie chimico-mineralogiche e orittografiche del Signor Giovanni Arduino, e di alcuni suoi amici, tratte dal Giornale d'Italia. Venezia, Benedetto Milocco, 1775.
410p., 1 plate. x549/Ar2r

Contains two letters by J. J. Ferber, on the solfatara of Pozzuoli and on mines in Austria as well as Arduino's theory of the earth: Saggio fisico-mineralogico di lythoconia, e orognosia.

76 **Arduino.**
Esame Chimico, e Considerazioni sopra la Marga, ossia Marna, scoperta nei campi dello Stabilimento a Tabacchi del Sig. Co. Girolamo Manfrino appresso Nona in Dalmazia dal Celebre Sig. Canonico Andrea Zucchini.
[n.p., n.d.]. 30p. 631.1/Ar2e

Riprodotta nel volume secondo del Nuovo Giornale d'Italia del Sig. Gio: Venezia, Antonio Perlini, 1791.
Bibliographical footnotes.

77 **Arkansas. Geological Survey.**
First report of a geological reconnoissance of the northern counties of Arkansas, made during the years 1857 and 1858, by David Dale Owen, principal geologist, assisted by William Elderhorst, chemical assistant; Edward T. Cox, assistant geologist. Little Rock, Johnson & Yerkes, state printers, 1858.
256p., front., illus., plates (partly col.). 557/Ar4o/1858

Contents. — Report [etc.] by David Dale Owen. — Chemical report of the ores, rocks, and mineral waters of Arkansas, by William Elderhorst. — Report of a geological reconnoissance of a part of the state of Arkansas made during the years 1857 and 1858, by Edward T. Cox.

78 **Arkansas. Geological Survey.**
Second report of a geological reconnoissance of the middle and southern counties of Arkansas. Made during the years 1859 and 1860. By David Dale Owen, principal geologist, assisted by Robert Peter, chemical assistant; M. Leo Lesque-

reux, botanist; Edward Cox, assistant geologist. Philadelphia, C. Sherman & son, printers, 1860.

433p., front., illus., plates (partly col.), fold. chart. 557/Ar4o/1859

Ed. by Richard Owen.

Contents. — Report [etc.] from notes of David Dale Owen, ed. by E. T. Cox. — Survey of the Fourche cove in Pulaski County, Arkansas, by Joseph Lesley, topographical assistant. — Chemical analysis of the soils, subsoils, underclays, clays, and nitre-earths of Arkansas. Collected by D. D. Owen and analyzed by Robert Peter. — Chemical analyses made for the survey, by William Elderhorst. — Botanical and palaeontological report on the survey, by Leo Lesquereux. — Second report of a geological reconnoissance of a part of the state of Arkansas, made during the years 1859 and 1860, by Edward T. Cox.

79 **Artis, Edmund Tyrell,** 1789-1847.

Antediluvian phytology, illustrated by a collection of the fossil remains of plants, peculiar to the coal formations of Great Britain, selected for their novelty and interest, from upwards of a thousand specimens now in the possession of the author, and systematically described, with the review of facilitating the study of this important branch of geology. Including remarks on the systems of Count Sternberg, Baron Schlotheim, Professor Martius, and Mons. Brongniart; also communications from Professor Buckland and other eminent geologists. London, Printed for the author by Nichols and Son, 1838.

24 numbered leaves, 24 plates. q561/Ar7a/1838

80 **Artis.**

See MANTELL, GIDEON ALGERNON, A pictorial atlas of fossil remains . . . 1850.

81 **Ashe, Thomas,** 1770-1835.

Memoirs of mammoth, and various other extraordinary and stupendous bones, of incognita, or non-descript animals, found in the vicinity of the Ohio, Wabash, Illinois, Mississippi, Missouri, Osage, and Red rivers, &c. &c. Published for the information of those ladies and gentlemen, whose taste and love of science tempt them to visit the Liverpool museum. Liverpool, Printed by G. F. Harris, 1806.

60p. 569.6/As3m

82 **Association of American Geologists and Naturalists.**

Reports of the first, second, and third meetings of the Association of American geologists and naturalists, at Philadelphia, in 1840 and 1841, and at Boston in 1842. Embracing

its proceedings and transactions. Boston, Gould, Kendall & Lincoln, 1843.
[9]-544p., illus., 21 i.e. 19 plates (part double; incl. maps).
506/ASA 2a

Plate IV is omitted; it may be found in Amer. Jour. Science, vol. XLI. Plates VII and IX are printed on the same leaf.

83 **Atkinson, Christopher William.**
A historical and statistical account of New Brunswick, B.N.A., with advice to emigrants. 3d ed., greatly improved and cor. . . . Edinburgh, Printed for Anderson & Bryce, 1844.
284p., illus., fold. map. 917.15/At5e/1844

84 **Attumoneli, Michele,** 1753-1826.
Della eruzione del Vesuvio accaduta nel mese di agosto dell'anno MDCCLXXIX; ragionamento istorico-fisico. Napoli, Stamperia Abbaziana, 1779.
147p., fold. plate. 551.21/At8d

85 **Aubuisson de Voisins, Jean François d',** 1769-1841.
Des mines de Freiberg en Saxe et de leur exploitation. Leipsic, Pierre Phil. Wolf et Comp., 1802.
3v., 8 fold. plates. 622.09432/Au1d

86 **Aubuisson de Voisins.**
Mémoire sur les basaltes de la Saxe, accompagné d'observations sur l'origine des basaltes en général. Lu à la classe des sciences physiques et mathématiques de l'Institut National, en frimaire an xi. Paris, Courcier, Imprimeur-libraire, 1803.
176p. 552.28/Au1m

87 **Aubuisson de Voisins.**
An account of the basalts of Saxony, with observations on the origin of basalts in general. Tr., with notes, by P. Neill . . . Edinburgh, A. Constable & Co., 1814.
276p., fold. map. 552.28/Au1mEn

88 **Aubuisson de Voisins.**
Traité de géognosie; ou, Exposé des connaissances actuelles sur la constitution physique et minérale du globe terrestre. Strasbourg, F. G. Levrault, 1819.
2v., illus. 550/Au1t

89 **Aubuisson de Voisins.**
Geognosie, oder Darstellung der jetzigen Kentnisse über die physische und mineralische Beschaffenheit der Erdkugel. Deutsch bearb. durch J. G. Wiemann. Dresden, In der Arnoldischen Buchhandlung, 1821-1822.
2v. in 1, illus. 550/Au1tGw

90 **Aubuisson de Voisins.**
Histoire de l'établissement des fontaines à Toulouse. Paris, Carilian-Goeury et V. Dalmont, 1839.
64p., 2 plates, tables. 551.49/Au1h

Extrait des Annales des Ponts et Chaussées.

91 **Auldjo, John,** d. 1857.
Sketches of Vesuvius with short accounts of its principal eruptions, from the commencement of the Christian era to the present time. Naples, G. Glass, 1832.
96p., front., 15 plates. 551.21/Au5s

92 **Babington, William,** 1757-1833.
A new system of mineralogy, in the form of catalogue, after the manner of Baron Born's systematic Catalogue of the collection of fossils of Mlle Éléonore de Raab. London, Printed for the author, by T. Bensley, 1799.
279p. x549/B11n

93 **Bacci, Andrea,** d. 1600.
De thermis Libri septem. Opus locupletissimum, non solum medicis necessarium, verumetiam studiosis variarum rerum naturae perutile. In quo agitur de universa aquarum natura . . . De balneis totius orbis, & de methodo medendi per balneas. Deque lavationum, simul atque exercitationum institutis in admirandis thermis Romanorum . . . Venetiis, Apud V. Valgrisium, 1571.
[64], 509p., illus. xq613.41/B12d

94 **Bacci.**
De Gemmis et Lapidibus pretiosis, eorumque viribus & usu tractatus, Italica lingua conscriptus: Nunc vero non solum in Latinum sermonem conversus, verum etiam utilissimis annotationibus & observationibus auctio redditus, A. Wolfango Gabelchovero. Francofurti, Ex officina Matthiae Beckeri, impensis Nicolai Steinii, 1603.
231p. uncat.

95 **Bacon, Francis,** 1561-1626.
Sylva Sylvarum or Natural Historie. In ten centuries. 3rd ed. London, Printed by J. H., 1631.
258p., port. 500/B13s

The library owns the following other editions of this work: London, 1635 (q uncat.), 1639 (q uncat.); Amsterdam, 1648 (x500/B13sLg/1648); The Hague, 1648 (x500/B13sLg/1648[2]); London, 1658 (q uncat.), 1664 (q uncat.), 1670 (q uncat.), and 1676 (q500/B13s/1676).

The library also owns his *Opera Omnia* in the following editions: Frankfurt, 1665 (uncat.); Leipzig, 1694 (uncat.); and London, 1730 (xq192/B13/1730). Translations of his *Works* into English include London, 1733 (x192/B13/1733), 1737 (x192/B13/1737), 1740 (xq192/B13/1740), 1753 (xq192/B13/1753), and 1765 (xq192/B13/1765).

96 **Bacon.**

Novum organum scientarum. Lugd. Bat., apud Adrianum Wijngaerde et Franciscum Moiardum, 1645.

435p., front. x501.8/B132b/1645

The library owns the following other editions of this work: The Hague, 1650 (x501.8/B132n/1650); and Amsterdam, 1660 (x501.8/B132n/1660).

97 **Baier, Johann Jacob,** 1677-1735.

Oryktographia Norica, sive rerum fossilium et ad minerale regnum pertinentium, in territorio Norimbergensi ejusque vicinia observatarum succincta descriptio. Cum iconibus lapidum figuratorum fere ducentis. Norimbergae, Impensis Wolfgangi Michahellis, Bibliopolae, 1708.

102p., 6 fold. plates, t.p. engraved. x551/B140

Library owns 2 copies. Copy 2 bound with author's Sciagraphia musei sui. Norimbergae, 1730.

This work was translated and edited in Erlanger Geologische Abhandlungen, vol. 29, 1958.

98 **Baier.**

Sciagraphia musei sui accedunt supplementa Oryctographiae Norica. Cum fig. aen. Norimbergae, Apud Haeredes W. M. Endteri et I. A. Engel, 1730.

3 fold. tables. x551/B140

Bound with author's Oryktographia Norica, 1708.

99 **Baier.**

Monumenta rerum petrificatarum praecipua Oryctographiae Noricae supplementi loco iungenda interprete filio Ferdinando Iacobo Baiero; cum tabulis aeneis quindecim. Norimberg, G. Lichtenstegeri, 1757.

20p., 15 leaves of plates (some fold.), illus. xq551/B14o/sup.

100 **Bakewell, J.**

An introduction to the study of mineralogy; or, Student's pocket companion. London, Sherwood, Gilbert & Piper, 1829.

120p. 549/B17i

101 **Bakewell, Robert,** 1768-1843.

An introduction to geology, illustrative of the general struc-

Alberti
magni philosopho
rum maximi de
mineralibus
libri quin
que.

Cat. No. 41

ture of the earth; comprising the elements of the science, and an outline of the geology and mineral geography of England. London, J. Harding, 1813.
362p., 5 plates (4 col.), front. (map). 550/B17i

102 **Bakewell.**
An introduction to geology: comprising the elements of the science in its present advanced state, and all the recent discoveries; with an outline of the geology of England and Wales. 3d ed., entirely recomposed and greatly enl. London, Longman, Rees, Orme, Brown, and Green, 1828.
540p., illus., plates. 550/B17i/1828

103 **Bakewell.**
An introduction to geology; comprising the elements of the science in its present advanced state, and all the recent discoveries; with an outline of the geology of England and Wales. London: 3d ed., entirely recomposed, and greatly enl. With new plates. 1st. American ed., ed. by Professor Silliman, of Yale college, with an appendix containing an outline of his course of lectures on geology. New Haven, H. Howe, 1829.
400, 128p., illus., 7 plates (2 col., 4 fold.) incl. front. uncat.

104 **Bakewell.**
An introduction to geology: intended to convey a practical knowledge of the science, and comprising the most important recent discoveries; with explanations of the facts and phenomena which serve to confirm or invalidate various geological theories. 4th ed., greatly enl. London, Longman, Rees, Orme, Brown, Green & Longman, 1833.
589p., illus., map. 550/B17i/1833a

Copy imperfect: map wanting.

105 **Bakewell.**
An introduction to geology: intended to convey a practical knowledge of the science, and comprising the most important recent discoveries; with explanations of the facts and phenomena which serve to confirm or invalidate various geological theories. 2d American from the 4th London ed., ed. by Prof. B. Silliman. New Haven, H. Howe & Co., 1833.
479p., illus., 8 plates, map. 550/B17i/1833

106 **Bakewell.**
An introduction to geology: intended to convey a practical knowledge of the science, and comprising the most important

recent discoveries; with explanations of the facts and phenomena which serve to confirm or invalidate various geological theories. 3d American from the 5th London ed., ed., with an appendix, by Prof. B. Silliman . . . New Haven, B. & W. Noyes, 1839.
596p., front., illus., 8 plates, map. 550/B17i3

107 **Bakewell.**
Travels; comprising observations made during a residence in the Tarentaise and various parts of the Grecian and Pennine Alps, and in Switzerland and Auvergne, in the years 1820, 1821, and 1822 . . . London, Printed for Longman, Hurst, Rees, Orme and Brown, 1823.
2v., col. fronts., illus., plates (partly col.). 914.94/B17t

Numerous geological observations, many on glacial features.

108 **Balardini, Lodovico,** 1796-1891.
Sulla torba della provincia bresciana; memorie del . . . Lodovico Balardini . . . e di Stefano Grandoni. Brescia, Tipografia della Minerva, 1842.
106p. 553.21/B18s

"Premiata dall'Ateneo di Brescia giusta il programma pubblicato il 16 aprile 1840."

109 **Balbi, Adriano,** 1782-1848.
Delle primarie altitudini del globo; saggio d'ipsometria generale. Milano, Stabilimento di Civelli Giuseppe e C., 1846.
137p. 551.43/B18d

Cavato dal volume secondo delle Memorie dell'I. R. Istituto Lombardo.

110 **Balsamo-Crivelli, Nobile Gabriel.**
Prospetto elementare di una descrizione geologica dell'Italia condotta sullo schizzo di una carta geologica italiana in aggiunta al corso di geologia di Beudant. Milano, Vallardi, 1847.
552p., geol. map. 549/B46mia

111 **Barba, Alvaro Alonso,** b. 1569.
The first book of the art of mettals, in which is declared the manner of their generation; and the concomitants of them. Written in Spanish by Alvaro Alonso Barba . . . in 1640. Translated into English in the year 1669. London, printed for S. Mearne, 1670.
156p. x669/B23aEs/1670

Bound with his The second book of the art of metals . . . 1670.

112 **Barba.**
The second book of the art of metals, wherein is taught the common way of refining silver by quicksilver, with some new rules added for the better performance of the same. Written in Spanish by Alvaro Alonso Barba in 1640. Translated into English. London, printed for S. Mearne, 1670.
91p. x669/B23aEs/1670

Bound with his The first book of the art of mettals . . . 1670.

113 **Barba.**
The art of metals, in which is declared the manner of their generation, and the concomitants of them written in Spanish in 1640 . . . in two books. Translated by the R. H. Edward Earl of Sandwich in 1669. London, Printed for S. Mearne, 1674.
2v in 1. uncat.

114 **Barba.**
Traité de l'art métalique, extrait des oeuvres d'Alvare-Alfonse Barba . . . auquel on a joint un Mémoire concernant les mines de France; avec un tarif qui démontre les opérations qu'il faudroit faire pour tirer de ces mines l'or & l'argent qu'en tiroient les Romains, lorsqu'ils étoient maîtres des Gaules. Ouvrage enrichi de figures en taille-douce. Paris, G. Saugrain, 1730.
264p., 8 fold. plates. 669/B23aF/1730

Translation of Arte de los metales.

115 **Barba.**
A collection of scarce and valuable treatises upon metals, mines, and minerals . . . Being, a translation from the learned Albaro Alonso Barba . . . and the observations of several ingenuous persons of our own country, founded on many years experience. London, Printed for J. Hodges, 1739.
p.1-170. 4 pts. in 1 v., illus. x669/B23aEs/1739

Each of pts. 2-[4] has special t.p.

Bound with Plattes, Gabriel, A discovery of subterranean treasure . . . 1738, and Houghton, Thomas, Rara avis in terris or the compleat miner . . . 1738.

116 **Barba.**
A collection of scarce and valuable treatises upon metals, mines and minerals . . . Being, a translation from the learned Albaro Alonso Barba . . . and the observations of several ingenuous persons of our own country, founded on many years experience. The 2d ed. London, Printed for J. Hodges, 1740.
319p., front., tables, diagr. x669/B23aEs/1740

Preface signed: G[abriel] P[lattes].
Translation by Edward Montagu, 1st Earl of Sandwich.
Bound with Houghton, Thomas, Rara avis in terris . . . [n.d.].

117 **Barba, Antonio,** 1751-1827.
Ragionamento fisico-chimico sull'eruzione ultima del Vesuvio accaduta a'15. giugno 1794 . . . Napoli, 1794.
38p. 551.21/B232r

118 **Barber, Thomas,** artist.
Barber's picturesque guide to the Isle of Wight; illustrated by forty-five views and maps engraved on steel. Including a geological map of the island by Dr. Mantell. New ed., completed to the end of 1850. London, Henry G. Bohn, 1850.
124p., illus., port., 2 fold. col. maps. 914.228/B23p/1850

119 **Barelli, Vincenzo.**
See SARDINIA (KINGDOM). AZIENDA GENERALE ECONOMICA DELL'INTERNO, Cenni de statistica mineralogica degli stati di S. M. il Re di Sardegna . . . 1835.

120 **Barral, Pierre,** 1742-1826.
Mémoire sur les trapps et les roches volcaniques. Dans lequel on donne l'origine des laves en boules, & des articulations des prismes basaltiques. Bastia, L'imprimerie d'E. Batini, 1789.
48p., illus. uncat.

121 **Barral.**
Mémoire sur des roches coquillières trouvées à la cime des Alpes dauphinoises et sur des colonnes d'un temple de Sérapis à Pouzzol, près de Naples, par M. B.** Grenoble, Impr. Peyronard, 1813.
26p., fold. plate. 564/B27m

Lu à la Société des Sciences et des Arts de Grenoble, le 14 Avril 1812.

122 **Barrère, Pierre,** 1690-1755.
Observations sur l'origine et la formation des pierres figurées, et sur celles qui, tant extérieurement qu'intérieurement, ont une figure régulière & déterminée. Paris, d'Houry et fils, 1746.
67p., fold. plates. 560/B27o

123 **Bartholin, Caspar,** 1655-1738.
De fontium fluviorumque origine ex pluviis. Dissertatio Physica. Hafniae, J. P. Bockenhoffer, 1689.
27p. uncat.

124 **Bartholin.**
De fontium fluviorumque origine ex pluviis. Dissertatio Physica. [n.p., n.d.].
26p. uncat.

Bound with his Specimen philosophiae naturalis praecipua physices capita exponens . . . 1698.

125 **Bartholin.**
. . . Specimen philosophiae naturalis praecipua physices capita exponens . . . Accedit De fontium fluviorumque origine ex pluviis, dissertatio physica. Oxoniae, Typis Leon. Lichfield, Impensis Henr. Clements, 1698.
175p. uncat.

Bound with his De fontium fluviorumque in origine ex pluviis . . . n.d.

126 **Bartholinus, Erasmus,** 1625-1698.
. . . Experimenta crystalli Islandici disdiaclastici quibus mira & insolita refractio detegitur. Hafniae, Danielis Paulli, 1670.
60p., illus. uncat.

127 **Bartholinus.**
. . . De naturae mirabilibus quaestiones academicae. Hafniae, Sumptibus Petri Hauboldi, Literis Georgii Gödiani, 1674.
200p. uncat.

128 **Bartolus de Saxoferrato.**
La Tiberiade di Bartole da Sasferrato del modo di dividere l'alluvioni, l'isole, & gl'alvei. Con l'annotationi, et espositioni di Claudio Tobaldutÿ da Montalboddo. Roma, Heredi di Gio: Gigliotto, 1587.
190p., illus. uncat.

129 **Barton, Richard,** b. 1722.
A dialogue, concerning some things of importance to Ireland; particularly to the County of Ardmagh. Being part of a design to write the natural, civil and eclesiastical history of that county. Dublin, Oli Nelson, 1751.
27p. x549.9415/B285ℓ

Bound with his Lectures in natural philosophy . . . 1751.

130 **Barton.**
Lectures in natural philosophy, designed, to be a foundation, for reasoning pertinently, upon the petrifications, gems, crystals, and sanative quality of Lough Neagh in Ireland; and intended to be an introduction, to the natural history of several counties contiguous to that lake, particularly the

VLYSSIS ALDROVANDI
PATRICII BONONIENSIS
MVSAEVM METALLICVM
IN LIBROS IIII DISTRIBVTVM
BARTHOLOMÆVS AMBROSINVS
in patrio Bonon. Archigymnasio Simpl
Med. professor ordinarius, Musęi Illustriss[mi]
Senatus Bonon. et Horti publici Pręfectus
Labore, et Studio composuit
cum Indice copiosissimo
MARCVS ANTONIVS BERNIA
proprys impensis in lucem edidit
AD SERENISSIMVM
RANVTIVM II
FARNESIVM
PARMÆ PLACENTIÆ ET C.
DVCEM VI.

EFFODIVNTVR OPES

Cat. No. 43

county of Ardmagh. Dublin, Printed for the author, by A. Reilly, 1751.

[8]p., 1.ℓ., 209 (i.e. 185)p., 7 plates (part fold., incl. front., map), diagr. x549.9415/B285ℓ

Bound with his A dialogue, concerning some things of importance to Ireland . . . 1751; and some remarks, towards a full description of Upper and Lower Lough Lene . . . 1751.

131 **Barton.**

Some remarks, towards a full description of Upper and Lower Lough Lene, near Killarny, in the County of Kerry. Dublin, S. Powell, 1751.

14p., illus. x549.9415/B285ℓ

Bound with his Lectures in natural philosophy . . . 1751.

132 **Bartram, John,** 1699-1777.

Observations on the inhabitants, climate, soil, rivers, productions, animals, and other matters worthy of notice. Made by Mr. John Bartram, in his travels from Pensilvania to Onondago, Oswego and the lake Ontario, in Canada. To which is annex'd, a curious account of the cataracts at Niagara. By Mr. Peter Kalm . . . London, Printed for J. Whiston and B. White, 1751.

[9]-94p., front. (fold. plan). x917.4/B28

Bartram and Lewis Evans accompanied Conrad Weiser on a mission from the government of Pennsylvania to the Iroquois, to settle a quarrel between the Indians and the colony of Virginia. Weiser's journal is printed in the set known as Colonial records of Pennsylvania, 1851, v.4, p.660-669.

133 **Bartram.**

See STORK, WILLIAM, A description of East Florida, with a Journal, Kept by John Bartram of Philadelphia . . . 1769.

134 **Bartram, William,** 1739-1823.

Travels through North and South Carolina, Georgia, East and West Florida, the Cherokee country, the extensive territories of the Muscogulges or Creek confederacy, and the country of the Chactaws. Containing an account of the soil and natural productions of those regions; together with observations on the manners of the Indians. Philadelphia, printed by James and Johnson, 1791. London, Reprinted for J. Johnson, 1792.

520p., front., 7 plates (1 fold.), fold. map. x917.5/B28t

135 **Bartram.**

Travels through North and South Carolina, Georgia, East and West Florida, . . . Dublin, For J. Moore, W. Jones, R. M'Allister, and J. Rice, 1793.

520p., front. (port.), 7 plates (part fold.), fold. map. x917.5/B28t/1793

First published Philadelphia, 1791.

136 **Bartram.**

Reisen durch Nord- und Süd-Karolina, Georgien, Ost- und West-Florida, das Gebiet der Tscherokesei, Krihks und Tschaktahs, nebst umständlichen Nachrichten von den Einwohnern, dem Boden und den Naturprodukten dieser wenig bekannten grossen Länder. Aus dem Englischen mit erläuternden Anmerkungen von E. A. W. Zimmermann. [Berlin, In der Vossischen Buchhandlung, 1793].

501p. x917.5/B28tGz

Magazin von merkwürdigen neuen Reisebeschreibungen. Berlin, 1793. 10.bd.

Paging irregular: the last 5 pages after p.496 are numbered 465 to 469.

137 **Bartram.**

Reizen door Noord- en Zuid-Carolina, Georgia, Oost- en West-Florida; de landen der Cherokees, der Muscogulges, of het Creek bondgenootschap en het land der Chactaws. Uit het Engelsch vertaald, door J. D. Pasteur. Haarlem, P. Bohn, 1794.

695p., fold. map. x917.5/B28tDp

138 **Bartram.**

Voyage dans les parties sud de l'Amérique septentrionale; savoir: les Carolines septentrionale et méridionale, la Georgie, les Florides orientale et occidentale, le pays des Chérokees, le vaste territoire des Muscogulges ou de la confédération Creek, et le pays des Chactaws. Imprimé à Philadelphie, en 1791, et à Londres, en 1792, et tr. de l'angl. par P. V. Benoist . . . Paris, [1801].

2v., front. (port.), 3 fold. plates, fold. map. x917.5/B28tFb

139 **Basaltic mountains.** [London? 1807?]

1v. of 50 plates. f552.2/B29

Binder's title.

Some of the plates were previously published in papers by John Strange in the Transactions of the Royal Society and in his De'monti colonnari e d'altri fenomeni vulcanici dello stato veneto (Milano, 1778) (q551.21/St8d).

140 **Baudin, Désiré Pierre,** 1809-1870.

Description historique géologique et topographique du bassin houiller de Brassac (Puy-de-Dôme et Haute-Loire), faite en 1843. Paris, Imprimerie Nationale, 1851.

136p., atlas of 25 diagrams. 553.2/B32b

Added title page: [France] Administration des Mines. Études des gîtes minéraux.

Atlas lacking.

141 **Baudrimont, Alexandre Édouard,** 1806-1880.
Traité élémentaire de minéralogie et de géologie. Paris, Amsterdam, H. Cousin, Ve Legras Imbert et Cie, [1835?].
279p., 12 plates. 549/B33t

Published also as v.2 of G. J. Martin Saint-Ange's Traité élémentaire d'histoire naturelle, 1839.

Running title: Traité élémentaire d'histoire naturelle.

142 **Bauhin, Johann,** 1541-1613.
Historia novi et admirabilis fortis balneique Bollensis in Ducatu Wirtembergico ad acidulas geopingenses . . . adijciuntur plurimae figurae novae varium fossilium, stirpium & insectorum, quae in & circa hunc fontem reperiuntur. Montisbeligardi, 1598.
291p., illus. uncat.

The description of mineral waters for medicinal use is based on the springs at Bad Boll, near Groppingen, Württemberg.

143 **Bauhin.**
Historiae fontis et balnei admirabilis bollensis. Liber quartus. De lapidibus metalicis que miro naturae artificio in ipsis terra visceribus figuratis, nec non de stirpibus, insectis, avibus, aliisque animalibus . . . Montibelgardi, apud Iacobum Foilletum, 1598.
222p., illus. uncat.

Bound with Bauhin, Johann, Historia novi et admirabilis . . . 1598.

Chapter I is a description of Bauhin's fossil collection, including drawings of ammonites, belemnites, etc. Chapter II treats fruit and vegetables.

144 **Bauhin, Kaspar,** 1560-1624.
. . . De lapidis bezaaris oriental. & occident. Cervin. & Germanici Ortu, natura, differentiis, veroq́ue usu ex Veterum & recentiorum placitis Liber . . . priore editione auctior. Basileae, Sumptibus Ludovici Regis, 1625.
302p. uncat.

145 **Baumer, Johann Wilhelm,** 1719-1788.
Naturgeschichte aller Edelsteine . . . aus dem Lateinischen übersetzt von Karl Freiherrn und Reichsrittern von Medinger. Wien, in dem Kraussischen Buchladen, 1774.
149p. 553.8/B83a

Bound with Brückmann, U. F. B, . . . Abhandlung von Edelsteinen . . . 1757.

146 **Baumer.**
Fundamenta geographiae et hydrographiae subterraneae, ad naturae ductum. Giessae, Ex Officina Kriegeriana, 1779.
234p. 3 fold. plates. 551/B324f

147 **Bausch, Johann Lorenz,** 1605-1665.
De lapide aetite schediasma, ad modum et mentem Academiae naturae curiosorum congestum. Lipsiae, Impensis V. J. Trescheri, Typis J. E. Hahnii, 1665.
79p., 5 fold. plates. x553.35/B32d

Issued as pt. 2 of the author's Schediasmata bina curiosa de lapide haematite et aetite. Lipsiae, 1665.

148 **Bauzá, Don Felipe.**
Discurso sobre el estado de la Geografía de la América Meridional (Ueber den gegenwärtigen Zustand der Geographie von Süd-Amerika). n.p., 1814.
36p. 549/F59

Bound with Flade, Christian Gottlob, De re Metallica . . . 1791.

149 **Bayle, Émile,** 1819-1895.
See DUFRÉNOY, OURS PIERRE ARMAND PETIT, Explication des planches de fossiles qui accompagnent les cours lithographiés de géologie . . . 1847 [i.e. 1848].

150 **Bayle.**
See FRANCE, SERVICE DE LA CARTE GÉOLOGIQUE DE LA FRANCE, Explication de la carte géologique de la France, v.4, atlas, pt.1, 1878.

151 **Beaujour, Louis August Félix,** 1763-1836.
Aperçu des États-Unis, au commencement du XIXe. siècle, depuis 1800 jusqu'en 1810 avec des tables statistiques. Paris, L. G. Michaud, imprimeur [etc.], 1814.
274p., fold. map, 17 fold. tables. 330.973/B269a

Map shows geological contacts in color, based on Maclure — See Wells, J. W., Wash. Acad. Sci. Jour., v. 49, no. 7, p. 203.

152 **Beaumont, John, Jr.,** d. 1731.
Considerations on a book, entituled *The Theory of the Earth,* publisht some years since by the learned Dr. Burnet. Dedit omnibus Deus, pro virili portione Sapientiam, ut & inaudita investigare possent, & audite perpendere. Lactan. de Orig. Error. c. 8. London, [Ed. Bohun], 1693.
187p., illus. x213/B93tYb

153 **Beaumont.**
Gleanings of antiquities. Containing, I. An essay for explaining the creation and the deluge, according to the sense of the Gentiles. In a letter to a learned person. II. A discourse of oracles, giving an account of the Sibylline oracles; with an uncommon explication of Virgil's fourth eclogue, and some other parts of his works relating to them. To which is added, An account of the oracles delivered at Delphos, and in the other temples of the Gentiles; and of the rise and cessation of oracles, both among the Jews and Gentiles. III. Some notes concerning familiar spirits. London, Printed for W. Taylor, 1724.
206p., plates. 213/B38g

154 **Becchetti, Filippo Angelico,** 1743-1814.
Teoria generale della terra, esposta all'Accademia Volsca di Velletri. Roma, P. Giunchi, 1782.
410p. 525/B38t

Illustrated half title: Teoria della terra.
"Citazioni" at end of each chapter.

155 **Becher, David,** 1725-1792.
Neue Abhandlung vom Karlsbade. Prag, W. Gerle, 1772.
3v. in 1. 3 fold. plates. 553.7/B38n

Bibliographical footnotes.

156 **Becher, Johann Joachim,** 1635-1682.
Actorum laboratorii chymici Monacensis, seu Physica subterranea libri duo, quorum prior profundam subterraneorum genesin, nec non admirandam globi terr-aque-aërei super & subterranei fabricam, posterior specialem subterraneorum naturam, resolutionem in partes partiumq; proprietates exponit, accesserunt sub finem mille hypotheses seu mixtiones chymicae, ante hâc nunquam visae, ominia plusquam mille experimentis stabilita . . . Francofurti, Joh. Davidis Zunner, 1669.
805p. uncat.

The origin of metals, and early metallurgy.

157 **Becher.**
Chymisches laboratorium oder unter-erdische Naturkündigung. Darinnen enthalten wird: I. Die tieffe Zeugung derer unter-erdischen Dinge: wie auch der wunderbare Bau der ober-und unter-erdischen Erd-Wasser-und Lufft-Kugel: und

JOH. JACOBI BAJERI,

PHILOS. ET MED. D. HVJVSQVE IN ACAD.
ALTDORF. PROF. PVBL.
ILLVSTR. REIP. NORIMB. PHYSICI

ΟΡΥΚΤΟΓΡΑΦΙΑ NORICA,

SIVE

RERVM FOSSILIVM

ET

AD MINERALE REGNVM PERTINENTIVM,

IN

TERRITORIO NORIMBERGENSI
EJVSQVE VICINIA OBSERVATARVM

SVCCINCTA

DESCRIPTIO.

CVM ICONIBVS LAPIDVM FIGVRATORVM
FERE DVCENTIS.

NORIMBERGÆ,
IMPENSIS WOLFGANGI MICHAHELLIS, BIBLIOPOLÆ.
ANNO MDCCIIX.

Cat. No. 97

dann die absonderliche Natur der unter-erdischen Dinge Auflöss — und Zulegung in ihre Theile, und derselben Eigenschafft. II. Neue Chymische Proben, einiger künstlichen gleich darstelligen Verwandlung derer Metallen, nach Anleitung der in vorigen Jahren in Druck gegebenen Physica Subterraneae . . . Frankfurt, Joh. Haass, 1680.
732p. x540.1/B38c

This is a translation of the Physica subterranea with additional material on alchemy.

158 **Becher.**
Chymischer Glücks-Hafen, oder grosse chymische Concordanz und Collection, von funffzehen hundert chymischen Processen: Durch viel Mühe und Kosten auss den besten Manuscriptis und Laboratoriis in diese Ordnung, wie hier folgendes Register auswiset, zusammengetragen . . . Frankfurt, Johann Georg Schiele, 1682.
810p. uncat.

159 **Beck, Lewis Caleb,** 1798-1853.
Mineralogy of New York, comprising detailed descriptions of the minerals hitherto found in the State of New-York, and notices of their uses in the arts and agriculture. Albany, White and Visscher, 1842.
536p., illus., 8 plates. q508.742/N48/v.3

Natural History of New York, v.3.

160 **Becker, Wilhelm Gottlieb,** 1753-1813.
Der Plauische Grund bei Dresden, mit Hinsicht auf Naturgeschichte und schöne Gartenkunst. Nürnberg, In der Frauenholzischen Kunsthandlung, 1799.
2v. in 1, illus., plates, map. q914.321/B378p

Contents: Teil I. Der Plauische Grund bei Dresden. Teil II [pt. 1]. Mineralogische Beschreibung des Plauischen Grundes bis Tharand von Andreas Tauber. Teil II [pt. 2]. Verzeichnis der im Plauischen Grunde und den zunächst angrenzenden Gegenden wildwachsenden Pflanzen von Friedrich Traugott Pursch. Teil II [pt. 3]. Verzeichnis der merkwürdigsten Insecten welche im Plauischen Grunde gefunden werden von Ludwig Heinrich Freiherrn von Block.

161 **Belcher, Edward,** 1799-1877.
Narrative of the voyage of H.M.S. Samarang, during the years 1843-46; employed surveying the islands of the Eastern archipelago; accompanied by a brief vocabulary of the principal languages, . . . By Captain Sir Edward Belcher . . . With notes on the natural history of the islands, by Arthur Adams . . . London, Reeve, Benham, and Reeve, 1848.

2v., front., plates, fold. maps, plan. 919.1/B41n

"Vocabulary of languages" (English, Spanish, Malay, Bisyan, Sooloo, Iloco, Batan, Cagayan, Tagala, Chinese, Japanese, Korean).

162 **Belknap, Jeremy,** 1744-1798.
The history of New-Hampshire. Vol. 1. Comprehending the events of one complete century from the discovery of the river Pascataqua. Philadelphia, Printed for the author by R. Aitken, 1784.
361, 84p. x974.2/B41h

No more publ. of this ed.
Vol. 2-3 printed Boston 1791-92; v. 1 reprinted Boston 1792.

163 **Belknap.**
The history of New-Hampshire. Boston, For the author, 1791-92.
3v., front. (v.2: fold. map). x974.2/B41h/1792

Imprint varies: v.1, Boston: re-printed for the author, 1792. — v.2, Printed at Boston, for the author, by Isaiah Thomas and Ebenezer T. Andrews, 1791. — v.3, Printed at Boston, for the author, by Belknap and Young, 1792.
Vol. 1 first published at Philadelphia, in 1784.
Contents. — v.1, The events of one complete century from the discovery of the river Pascataqua. — v.2, The events of seventy five years, from MDCCXV to MDCCXC. — v.3, A geographical description of the state; with sketches of its natural history, productions, improvements, and present state of society and manners, laws and government.
Vol. 1 lacks leaf with copyright notice. Folded map bound in v.1. Title p. of v.2 mutilated.

164 **Bennett, S.**
A new explanation of the ebbing and flowing of the sea, upon the principles of gravitation. New-York, published for the author, 1816.
79p., fold. plate. uncat.

Presentation copy of the author.

165 **Bennigsen-Förder, Rudolf von,** 1824-1902.
Erläuterungen zur geognostischen Karte der Umgegend von Berlin. Berlin, G. Reimer, 1843.
38p., fold. map. 554.3155/B43e

Bound with Hausmann, Johann Friedrich Ludwig, Geologische Bemerkungen über die Gegend von Baden bei Rastadt, 1844; and Roth, J. L. A., Die Kugelformen im Mineralreich und deren Einfluss auf die Absonderungsgestalten der Gesteine, 1844.

166 **Berendt, Georg Carl,** 1790-1850.
Die im Bernstein befindlichen organischen Reste der Vorwelt,

gesammelt, in Verbindung mit Mehreren/bearb. und hrsg. von Dr. Georg Carl Berendt . . . Berlin, In Commission der Nicolaischen Buchhandlung, 1845-56.
2v., plates, part. col. f560.9431/B45d

Each "Abtheilung" has also special t.-p.

Contents. — 1. Bd. I, Abth. Der Bernstein und die in ihm befindlichen Pflanzenreste der Vorwelt, bearbeitet von Professor Dr. H. R. Goeppert . . . und Dr. G. C. Berendt. II. Abth. Die im Bernstein befindlichen Crustaceen, Myriapoden, Arachniden und Apteren der Vorwelt, bearbeitet von C. L. Koch . . . und Dr. G. C. Berendt. — 2. Bd. I. Abth. Die im Bernstein befindlichen Hemipteren und Orthopteren der Vorwelt, bearbeitet von E. F. Germar . . . und Dr. G. C. Berendt. II. Abth. Die im Bernstein befindlichen Neuropteren der Vorwelt, bearbeitet von F. J. Pictet-Baraban . . . und Dr. H. Hagen.

167 **Bérenger, Laurent Pierre,** 1749-1789.
See PICQUET, J. F., Voyage dans les Pyrénées françoises . . . 1789.

168 **Berger, Johann Gottfried.**
De thermis Carolinis commentatio, qua omnium origo fontium calidorum itemq[ue] acidorum ex pyrite ostenditur. Guelferbytti, Impensis G. Freitagii, 1709.
158p. 551.23/B453d

169 **Bergman, Torbern Olof,** 1735-1784.
Physik beskrivelse over Jordkloden . . . Kiobenhavn, 1771.
648p., 2 fold. tables. uncat.

170 **Bergman.**
Physisk beskrifning öfver jord-klotet, på Cosmographiska Sällskapets vägnar författad . . . Upsala, J. Edman, 1773-1774.
2v., illus. front. x910/M288w/v.2,pt.1 & 2

Pt. 1 and 2 of v.2 are in fact v.2 and v.3 of a set of four volumes entitled "Werlds Bedskrifning" (Description of the World).

171 **Bergman.**
See TROIL, UNO VON, Letters on Iceland . . . 1780.

172 **Bergman.**
Physikalische Beschreibung der Erdkugel, auf Veranlassung der Cosmographischen Gesellschaft verfasset . . . Aus dem Schwedischen übersetzt von Lampert Hinrich Röhl. 2. verm. und verb. Aufl. Greifswald, A.F.Röse, 1780.
2v., illus. x551/B454pGr/1780

Vol. 1 has added t.p., engr., with title: Weltbeschreibung. It contains a physical description of the earth while Vol. 2 refers to changes on the surface of the earth followed by a review of theories of the earth including his own. The two volumes represent a translation of Physisk beskrifning öfver jord-klotet . . . 1773-1774.

173 **Bergman.**
Opuscula physica et chemica. Ed. nova emendatior. Lipsiae, I. G. Müller, 1780-90 [v.1, 1788].
6v. in 3, fold. plates. 508/B454o

Imprint varies: v.2, Upsaliae, J. Edman, 1780.
Vol. 4-6: Editionis curam post auctoris mortem gessit Ernestus Benjamin. Gottl. Hebenstreit.

174 **Bergman.**
Kleine physische und chymische Werke. Aus dem Lateinischen übersetzt von Heinrich Tabor. Frankfurt am Mayn, J. G. Garbe, 1782-89.
5v., fold. plates. 508/B45oGt

Vol. 4 published by J. C. Gebhard; v.5, by Gebhard & Koerber.
Translation of Opuscula physica et chemica.

175 **Bergman.**
Traité des affinités chymiques, ou attractions électives; Tr. du latin, sur la dernière éd. de Bergman. Augm. d'un supplément de notes. Paris, Buisson, 1788.
444p., fold. plates. 541.396/B454tE

176 **Bergman.**
Physical and chemical essays; tr. from the original Latin of Sir Torbern Bergman, by Edmund Cullen . . . To which are added notes and illustrations by the translator . . . London, Printed for J. Murray, 1788-1791.
3v. x508/B454oEc/1788

Library owns v.2 only which contains a description of minerals, and ores and the use of the blow-pipe technique.

177 **Bergman.**
Essays, physical and chemical. Translated from the original Latin. To which are added notes and illustrations by the translator. Edinburgh, Printed for G. Mudie and J. & J. Fairbairn, 1791.
446p. x508/B454e/1791

Library owns v.3 only which contains "Thoughts on a Natural System of Fossils."

178 **Bergman.**
Manuel du minéralogiste, ou Sciagraphie du règne minéral, distribué d'après l'analyse chimique. Mise au jour par M. Ferber . . . et traduite et augmentée de notes par M. Mongez le jeune . . . A Paris, Chez Cuchet, 1784.
343p., fold. plate. 549/B452sFm

179 **Bergman.**

Manuel du minéralogiste; ou, Sciagraphie du règne minéral, distribuée d'après l'analyse chimique. Mise au jour par M. Ferber, et traduite et augmentée de notes par M. Mongez le jeune. Nouv. éd., considérablement augm. par J. C. Delamétherie. Paris, Cuchet, 1792.

2v., illus. 549/B452sFm/1792

Translation and revision of Sciagraphia regni mineralis, secundum principia proxima digesti.

180 **Bergman.**

Grundriss des Mineralreichs, in einer Anordnung nach den nächsten Bestandtheilen der Körper. Aus dem Lateinischen mit einigen Zusätzen von D. Joseph Xav. Lippert. Wien, Krauss, 1787.

207p. 549/B452sGℓ

Translation of Sciagraphia regni mineralis.

181 **Bergmann, Fridericus Conradus.**

Salium actione ut causa sanitatis, ex rationibus chymicis: Lipsiae, Langenhemiana, 1756.

12p. 549/F59

Bound with Flade, Christian Gottlob, De re Metallica . . . [1791].

182 **Beringer, Johann Bartholomäus Adam,** d. 1740.

Lithographiae wirceburgensis, ducentis lapidum figuratorum, a potiori insectiformium, prodigiosis imaginibus exornatae specimen primum, quod in dissertatione inaugurali physico-historica, cum annexis corollariis medicis, authoritate et consensu inclytae Facultatis Medicae, in alma Eoo-Francica Wirceburgensium Universitate, praeside . . . publicae litteratorum disquisitioni submittit Georgius Ludovicus Hueber medicinae candidatus . . . Anno M.DCCXXVI. mense Majo . . . Prostat Wirceburgi, Apud Philippum Wilhelmum Fuggart, [1726?].

96 (i.e.98)p., front., 21 plates. xq560/B452ℓ

183 **Beringer.**

. . . Lithographia Wirceburgensis, ducentis lapidum figuratorum, a potiori, insectiformium, prodigiosis imaginibus exornata. Ed. 2. Francofurti et Lipsiae, T. Goebhardt, 1767.

96 [i.e. 98]p., 22 plates. xq560/B452ℓ/1767

ERASMI BARTHOLINI

EXPERIMENTA CRYSTALLI ISLANDICI DISDIACLASTICI

Qvibus mira & insolita

REFRACTIO

detegitur.

ANNO 1670.

HAFNIÆ,

Sumptibus DANIELIS PAULLI Reg. Bibl.

Cat. No. 126

184 **Berkenhout, John, 1730?-1791.**
Synopsis of the natural history of Great Britain and Ireland, containing a systematic arrangement and concise description of all the animals, vegetables and fossils, which have hitherto been discovered in these kingdoms. London, Printed for T. Cadell, 1789.
2v. in 1. x570.942/B45o/1789

185 **Bernard.**
Mémoire sur les avantages et les inconvéniens de l'emploi du charbon de pierre ou du bois dans les fabriques, avec la description des différentes mines de charbon qui sont en Provence, & leurs qualités. Qui a remporté le prix, au jugement de l'Académie des Belles-Lettres, Sciences & Arts de Marseille, dans son Assemblée d'après Pâques, tenue le 5 Avril 1780. Par M. Bernard, Directeur-Adjoint de l'Observatoire Royal de la Marine de Marseille . . . Marseille, chez J. Mossy, 1780.
127p., illus. 549/M62m

Bound with Millin, Aubin Louis, Minéralogie homerique . . . 1790.

186 **Bernhardi, Reinhard,** fl. 1830-1841.
Wie kamen die aus dem Norden stammenden Felsbruchstücke und Geschiebe welche man in Norddeutschland und den benachbarten Ländern findet, an ihre gegenwärtigen Fundorte? *in* Jahrbuch für Mineralogie, Geognosie, Geologie und Petrefactenkunde, 1832.
v.3, p.257-267. 550.5/NJA/1832

187 **Beroaldo, Filippo,** 1453-1505.
Opusculum Phillipi Beroaldi De terremotu & pestilentia, cum annotamentis Galeni. Addita est explicatarum in annotamentis Galeni dictionum tabula. [1510.]
[20], [4] leaves. uncat.

Colophon: Hoc, Philippi Beroaldi de terremotu, opus Argentorati in Mathiae Schurerij Heluetensis officina, staneis calamis, exscriptū est. Anno, IHESV in cuius obitu tota machina portentoso, necnō nature insueto tremore concussa fuit. M.D.X. Mense Maij.

188 **Beroldingen, Franz Cölestin von,** 1740-1798.
Physicalisch Chemische Beschreibung des in dem Bissthum Paderborn gelegenen Gesundbrunnens zu Driburg, nebst angehängten Bemerkungen die Mineralwasser überhaupt betreffend. Hildesheim, Gedruckt bey C. W. Schlegel, 1783.
205p. 613.12/B45p

189 **Beroldingen.**
Bemerkungen auf einer Reise durch die pfälzischen und zweybrückschen Quecksilber-Bergwerke, hrsg. und mit einer Vorrede begleitet von J. D. Brandis. Berlin, F. Nicolai, 1788.
240p., fold. col. map. 553.45/B45b

190 **Bertereau, Martine de.**
La restitution de Pluton . . . Des Mines & Minières de France, cachées et detenües jusques à présent au ventre de la terre, par le moyen desquelles les Finances de sa Majesté seront beaucoup plus grandes que celles de tous les Princes Chrestiens, & ses sujets plus heureux de tous les Peuples. Ensemble la raison pourquoy lesdites Mines & Minières ont esté iusques à présent presque inutiles & sans profit à la Souveraineté & Maiesté Royale. Paris, Chez Hervé du Mesnil, 1640.
uncat.

She was the Baronne de Beausoleil.

191 **Bertrand, Élie,** 1713-1797.
Mémoires sur la structure intérieure de la terre. Zurich, Heidegger, 1752.
152p. x551/B462m

Bound with his Essai sur les usages des montagnes, avec une lettre sur le Nil . . . 1754.

192 **Bertrand.**
Essai sur les usages des montagnes, avec une lettre sur le Nil. Zurich, Heidegger, 1754.
412p. x551/B462m

Bound with his Mémoires sur la structure intérieure de la terre.

193 **Bertrand.**
Dictionnaire universel des fossiles propres, et des fossiles accidentels, contenant une description des terres, des sables, des sels, des soufres, des bitumes, des pierres simples & composées, communes & prétieuses, transparentes & opaques, amorphes & figurées, des minéraux, des métaux, des pétrifications du règne animal, & du règne végétal &c. avec des recherches sur la formation de ces fossiles, sur leur origine, leurs usages &c. La Haye, P. Gosse, Junior, et D. Pinet, 1763.
2v. in 1. 283p., 256p. 550.3/B46d

194 **Bertrand.**
Recueil de divers traités sur l'histoire naturelle de la terre et des fossiles. Avignon, L. Chambeau, 1766.
552p., port. 550/B463r

Contents. — Mémoires sur la structure intérieure de la terre. — Essai sur les usages des montagnes. — Mémoires historiques et physiques sur les tremblemens de terre. — Essai d'une distribution méthodique des fossiles. — Essai de minérographie et d'hydrographie du canton de Berne. — Essai de minérographie du canton de Berne, systématiquement rangée. — Catalogue du Cabinet d'histoire naturelle de M. E. Bertrand. — Lettre sur les inondations et l'usage des montagnes de l'Abissinie. — Lettre sur la diminution des mers, et l'origine des montagnes.

195 **Bertrand.**
Élémens d'oryctologie; ou, Distribution méthodique des fossiles, par M. B. C. P. de la C. de P. Neuchâtel, Impr. de la Société Typographique, 1773.
137p., fold. tables. x560/B457e

M. B. C. P. de la C. de P. translates M. Bertrand, Conseiller Palatin de la Cour de Pologne.

For identification see Bestermann Correspondence of Voltaire Letter D 18599.

196 **Bertrand, Louis,** 1731?-1812?
Renouvellemens périodiques des continens terrestres. 2.éd., corrigée et considérablement augmentée. Genève, J. J. Paschoud, 1803.
406p., plate. 550.1/B47r/1803

197 **Bertrand, Phillippe,** 1730?-1811.
Lettre à M. le Comte de Buffon; Intendant du Jardin et du Cabinet du Roi, de l'Académie Françoise, de celle des Sciences, etc., etc., ou, Critique, et nouvel essai sur la théorie générale de la terre. Avec une notice du dernier discours de M. Pallas . . . sur la formation des montagnes, sur les changemens arrivés au globe, etc. Besançon, 1780.
190p. 550.1/B46ℓ

198 **Bertrand.**
Nouveaux principes de géologie, comparés et opposés à ceux des philosophes anciens et modernes, notamment de J. C. Lamétherie, qui les a tous analysés dans sa Théorie de la Terre: ou manière plus simple d'observer et d'expliquer, l'un par l'autre, les principaux faits naturels. Avec un Abrégé de la géologie nouvelle. Paris, L'auteur, 1797.
538p. 550/B465n

199 **Bertrand.**
Nouveaux principes de géologie, minéralogie, géographie-physique etc., comparés et opposés à ceux des philosophes anciens et modernes, jusqu'à J. C. Delamétherie, qui les a

tous analysés dans sa Théorie de la Terre: avec l'Abrégé d'une géologie toute nouvelle. 2. éd., rev. et augm. Paris, l'Auteur, 1803.
559p. 550/B465n/1803

200 **Bertrand de Doue, Jacques Mathieu.**
Description géognostique des environs du Puy en Velay, et particulièrement du bassin au milieu duquel cette ville est située. Paris, F.-G. Levrault, 1823.
240p., fold. map, profiles. 554.481/D46d

201 **Berzelius, Jöns Jacob,** 1779-1848.
An attempt to establish a pure scientific system of mineralogy, by the application of the electro-chemical theory and the chemical proportions. Tr. from the Swedish original by John Black. London, printed for R. Baldwin; Edinburgh, W. Blackwood, 1814.
144p. 549/B464fEb

Translation of Försök att genom användande af den elektrokemiska theorien, . . . samt lüren om de kemiska proportionerna grundlägga et rent vetens kapligt system for mineralogien.

202 **Berzelius.**
Nouveau système de minéralogie. Traduit du Suédois . . . Paris, Chez Méquignon-Marvis, 1819.
314p., table. 549/B464nFb

203 **Besson, Jacques,** 16th century.
L'art et science de trouver les eaux et fontaines cachées soubs terre, autrement que par les moyens vulgaires des agriculteurs & architectes . . . Orléans, Pour Pierre Trepperel, 1569.
84p. uncat.

204 **Betti, Zaccaria,** 1732-1788.
Descrizione de un meraviglioso ponte naturale nei monti veronesi. Verona, M. Moroni, 1766.
22p., fold. plates. 914.5314/B46d

Title vignette.
Author's name given in dedication.

205 **Beudant, François Sulpice,** 1787-1850.
Voyage minéralogique et géologique, en Hongrie, pendant l'année 1818. Relation historique. Paris, Verdière, 1822.
4v., 7 plates, fold. maps. 554.391/B46v

Vol. 4: atlas.

206 **Beudant.**
Minéralogie. Ouvrage adopté par le Conseil royal de l'Instruction Publique pour l'enseignement dans les collèges et approuvé par Monseigneur l'Archevêque de Paris pour l'enseignement dans les établissements religieux. Paris, Langlois et Leclercq; [etc., etc., 1841].
295p., incl. illus., tables (part double). 549/B46m

(Cours élémentaire d'histoire naturelle par MM. Milne-Edwards, A. de Jussieu et Beudant).

With this is bound the author's Géologie, Paris [1841].

207 **Beudant.**
Minéralogie [Géologie]. Ouvrage adopté par le Conseil de l'Instruction Publique pour l'enseignement dans les lycées et collèges et approuvé par Monseigneur l'Archevêque de Paris . . . 5.éd. Paris, Langlois et Leclercq, 1851.
2v. in 1, illus. 549/B46m/1851

Cours élémentaire d'histoire naturelle . . . par MM. Milne-Edwards, A. de Jussieu, et F.-S. Beudant.

208 **Beudant.**
. . . Mineralogia-geologia . . . Ed. italiana per cura del dottore Giusti Arpesani . . . Milano, F. Vallardi, 1846.
589p. 549/B46mia

Bound with Balsamo-Crivelli, Nobile Gabriel, comp., Prospetto elementare di una descrizione geologica dell' Italia . . . 1847.

209 **Beudant.**
Die Mineralogie und Geologie. Nach der abermals vermehrten Aufl. aus dem Französischen neu übersetzt von J. G. Kurr. Stuttgart, Scheible, Rieger & Sattler, 1848.
720p., illus. 549/B46mGk

Populäre Naturgeschichte der drei Reiche, von F. S. Beudant, Milne-Edwards, A. v. Jussieu, 2. Bd.

210 **Beverley, Robert,** ca. 1673-ca. 1722.
The history of the present state of Virginia, in four parts. I. The history of the first settlement of Virginia, and the government thereof, to the present time. II. . . . The natural productions and conveniences of the country, suited to trade and improvement. III. . . . The native Indians, their religion, laws and customs, in war and peace. IV. . . . The present state of the country, as to the polity of the government, and the improvements of the land. London, printed for R. Parker, 1705.

D. JOANNIS BARTHOLOMÆI ADAMI BERINGER,

PHILOSOPHIÆ ET MEDICINÆ DOCTORIS IN ALMA EOO - FRANCICA WIRCEBURGENSIUM UNIVERSITATE PROFESSORIS PUBL. ORDIN. FACULT. MEDICÆ ASSESSORIS, REVERENDISSIMI ET CELSISSIMI PRINCIPIS WIRCEBURGENSIS CONSILIARII, ET ARCHIATRI, AULÆ, NEC NON PRINCIPALIS SEMINARII DD. NOBILIUM ET CLERICORUM, AC MAGNI HOSPITALIS JULIANAEI PRIMO LOCO MEDICI,

LITHOGRAPHIA WIRCEBURGENSIS,

DUCENTIS LAPIDUM FIGURATORUM,

A

POTIORI, INSECTIFORMIUM,

PRODIGIOSIS IMAGINIBUS

EXORNATA.

EDITIO SECUNDA.

FRANCOFURTI ET LIPSIÆ.

APUD TOBIAM GOEBHARDT, BIBLIOPOLAM BAMBERGENSEM ET WIRCEBURGENSEM. 1767.

Cat. No. 183

16, 104, 40, 64, 83p., front., 14 plates, fold. table. x975.5/B46h

Book II, Chapter III is titled "Of the Earths, and Soil."

211 **Beverley.**
The history of Virginia, in four parts. I. The history of the first settlement of Virginia, and the government thereof, to the year 1706. II. The natural productions and conveniences of the country, suited to trade and improvement. III. The native Indians, their religion, laws, and customs, in war and peace. IV. The present state of the country, as to the polity of the government, and the improvements of the land, the 10th of June 1720. By a native and inhabitant of the place. 2d ed., rev. and enl. by the author. London, Printed for B. and S. Tooke [etc.], 1722.
284, 24p., front., 14 plates, fold. table. x975.5/B46h2

212 **Beverley.**
Histoire de la Virginie, contenant I. L'histoire du premier établissement dans la Virginie, & de son gouvernement jusques à présent. II. Les productions naturelles & les commoditez du païs, avant que les Anglois y negociassent, & améliorassent. III. La religion, les loix, & les coutumes des Indiens naturels, tant dans la guerre, que dans la paix. IV. L'état présent du païs, tant à l'égard de la police, que de l'amélioration du païs. Par un auteur natif & habitant du païs. Tr. de l'anglois . . . Amsterdam, T. Lombrail, 1707.
433p., 14 plates. 975.5/B46hF/1707

213 **Bianconi, Giovanni Giuseppe,** 1809-1878.
De mare olim occupante planities et colles Italiae, Graeciae, Asiae Minoris, etc., et de aetate terreni quod geologi appellant Marnes Bleues dissertatio [I-V]. Bononiae, Ex typographaeo Emygdii, 1846-52.
193p., fold. col. map. 551.781/B47d

Issued in parts.

214 **Bidermann, Io. Gottlob.**
De primis rei Metallicae Inventoribus pauca praefatus; ad Orationes Augusti diei festo consecratas officiose. Freibergae, Barthelii, [1763].
[7p.] 549/F59

Bound with Flade, Christian Gottlob, De re Metallica . . . [1791].

215 **Bigot de Morogues, Pierre Marie Sébastien,** 1776-1840.
Mémoire historique et physique sur les chutes de pierres tombées sur la surface de la terre à diverses époques. Orléans,

Impr. de Jacob Aîne, 1812.
360p. 523.51/B48m

216 **Binet, Étienne.**
Essay des merveilles de nature, et des plus nobles artifices, piece tres necessaire à tous ceux qui font profession d'Eloquence . . . par René François, seconde edition, reveuë, corrigée, augmentée par l'autheur. Besanson, lean Praye, 1622.
547p. uncat.

Chapters XXI to XXIX deal with jewels, precious stones, metallurgy of silver and gold, mines of metals and minerals.

217 **Binninger, Ludwig Reinhard,** 1742-1776.
Oryctographiae agri Buxovillani et viciniae specimen. Argentorati, J. H. Holtzii, 1762.
36p. 554.4383/B51o

218 **Biringucci, Vannuccio,** 1480-1539?
De la pirotechnia libri X. Dove ampiamente se tratta non solo di ogni sorte & diversita de Miniere, . . . ma anchora quanto si ricerca intorno à la prattica di quelle cose di quel che si appartiene à l'arte de la fusione over gítto de metalli come d'ogni altra cosa simile à questa. Venetia, [per Venturnio Roffinello], 1540.
[352p.] uncat.

219 **Biringucci.**
Pirotechnia. Li diece libri della pirotechnia, Nelli quali si tratta non solo la diversità delle minere, ma ancho quanto si ricerca alla prattica di esse: e di quanto s'appartiene all'arte della fusione over getto de metalli, e d'ogni altra cosa à questa somigliante. Venetia, [per Comln da Trino de Monferrato], 1550.
168 leaves, illus. x662/B53p/1550

First edition, Venezia, 1540.
Errors in foliation; leaves 126 and 164 wrongly numbered 124 and 146.

220 **Biringucci.**
Pirotechnia. Li diece libri della pirotechnia, Nelli quali si tratta non solo la diversità delle minere, ma ancho quanto si ricerca alla prattica di esse: e di quanto s'appartiene all'arte della fusione over getto de metalli, e d'ogni altra cosa à questa somigliante. Venetia, [per Comin da Trino de Monferrato], 1558.
168 leaves, illus. x622/B53p/1559

1559 in coloph.

221 **Biringucci.**
Pirotechnia . . . Nella quale si tratta non solo la diversità delle minere, ma ancho quanto si ricerca alla prattica di esse: e di quanto s'appartiene all'arte della fusione over getto de metalli, e d'ogni altra cosa à questa somigliante. Far Campane, Artigliere, fuocchi artificiati, & altre diverse cose utilissime. Bologna, per Gioseffo Longhi, 1678.
630p., illus. x622/B53p/1678

222 **Biringucci.**
La pyrotechnie, ou art du feu, contenant dix livres, ausquels est amplement traicté de toutes sortes & diversités de minieres, fusions & separations de metaux: des formes & moules pour getter artilleries, cloches, & toutes autres figures: des distillations, des mines, contremines, pots, boulets, fusees, lances, & autres feux artificiels, concernans l'art militaire, & autres choses dependantes du feu. Traduite d'Italien en François par feu maistre Iaques Vincent. Paris, C. Fremy, 1556.
231 leaves, illus. x622/B53pFv

223 **Bischof, Gustav,** 1792-1870.
Die Wärmelehre des Innern unsers Erdkörpers, ein Inbegriff aller mit der Wärme in Beziehung stehender Erscheinungen in und auf der Erde. Nach physikalischen, chemischen und geologischen Untersuchungen . . . umgearbeitete und weiter ausgeführte Ausgabe einer gekrönten Preisschrift. Mit Holzschnitten. Leipzig, J. A. Barth, 1837.
512p. 551.12/B54w

224 **Bischof.**
Lehrbuch der chemischen und physikalischen Geologie. Bonn, A. Marcus, 1847-54.
2v. in 5. 550/B54ℓ

v.1 issued in 2 pts.; v.2, in 7 pts.

225 **Blondeau, Charles Félix.**
Manuel de minéralogie; ou, Traité élémentaire de cette science d'après l'état actuel de nos connaissances. 4.éd., entièrement refondue, mise dans un nouvel ordre et rendue plus complète et plus à la portée des gens du monde par un grand nombre d'augmentations, par MM. D** et Julia de Fontenelle. Bruxelles, Société belge de librairie, etc., 1837.
443p., 2 fold. plates (1 col.). 549/B622m/1837

226 **Blondel, François,** 1613-1703.
Thermarum Aquisgranensium, et Porcetanarum elucidatio, & thaumaturgia. Sive, Admirabilis earumdem natura, & admirabi-

liores sanationes; quas producunt in usibus balneationis, potationis. Ed. 3, sincerissima, prioribus auctior, & emendatior. Aquisgrani, Sumptibus authoris, Typis J. H. Clemens, 1688.
160p., illus. x613.38/B62t/1688

227 **Blount, Thomas Pope.**
A natural history: containing many not common observations: extracted out of the best modern writers. London, printed for R. Bentley, 1693.
469p. uncat.

The latter part of the book contains observations on mountains, volcanoes, subterranean fires, earthquakes, islands, the origin of fountains and damps in mines, on diamonds, precious stones, petrifications, gold and silver and generation of metals.

228 **Blum, Johann Reinhard,** 1802-1883.
Lithurgik oder Mineralien und Felsarten nach ihrer Anwendung in ökonomischer, artistischer und technischer Hinsicht systematisch abgehandelt. Mit 3 in den Text eingedruckten Figuren und 3 Stahlstichen. Stuttgart, E. Schweizerbart, 1840.
501p., 3 plates, diagrs. 553/B62ℓ

Naturgeschichte der drei Reiche... Supplement zum mineralogischen Theile.
"Literatur": p. 5-6.

229 **Blum.**
Nachtrag [I-III] zu den Pseudomorphosen des Mineralreichs nebst einem Anhange über die Versteinerungs- und Vererzungs-Mittel organischer Körper. Stuttgart, E. Schweizerbart, 1847-63.
3v. in 1. 549/B62p/sup.1-3

Sup.1-3 in 1v.
Imprint varies.

230 **Blum.**
Grundzüge der Mineralogie und Geognosie. Stuttgart, Franckh'schen Buchhandlung, 1850.
162p. 549/B62g

Aus der "Neuen Encyklopädie für Wissenschaften und Künste," Band III. besonders abgedruckt.

231 **Blum.**
Die Einschlüsse von Mineralien in krystallisirten Mineralien, deren chemische Zusammensetzung und die Art ihrer Entstehung. Von Reinhard Blum [et al. Haarlem, Die Erban Loosjes, 1854].
3 pts. in lv., plates (part col.). q549/B62e

Part 1 by R. Blum; pt. 2 by G. Leonhard; pt. 3 by A. H. Seyfert und E. Söchting.

Bibliographical footnotes.

232 **Blum.**

Lehrbuch der Oryktognosie . . . mit 333 krystallographischen Figuren. 2. verm. und verb. Aufl. Stuttgart, E. Schweizerbart, 1854.

673p. 549/B62ℓ

233 **Blum.**

Handbuch der Lithologie oder Gesteinlehre. Erlangen, F. Enke, 1860.

356p., illus., diagrs. 552/B52h

234 **Blum.**

Die Pseudomorphosen des Mineralreichs. Vierter Nachtrag. Heidelberg, C. Winter's Universitätsbuchhandlung, 1879.

212p. 549/B62p/Sup.4

235 **Boase, Henry Samuel,** 1799-1883.

A treatise on primary geology; being an examination, both practical and theoretical of the older formations. London, Printed for Longman, Rees, Orme, Brown, Green & Longman, 1834.

399p., illus. 550/B629t

236 **Boate, Gerard,** 1604-1650.

Irelands naturall history. Being a true and ample description of its situation, Greateness, Shape and Nature; of its Hills, Woods, Heaths, Bogs; of its Fruitful Parts and profitable Grounds, with the several way of Manuring and improving the same; with its Heads or Promontories, Harbours, Roades and Bayes; of its Springs and Fountaines, Brookes, Rivers, Loghs; of its Metalls, Mineralls, Freestone, Marble, Sea-coal, Turf, and other things that are taken out of the ground. . . . London, for John Wright, 1652.

186p. uncat.

237 **Boate.**

Histoire naturelle d'Irlande contenant une description tres exacte de sa situation, de sa grandeur, de sa figure, de la nature de ses Montagnes, de ses Forests, de ses Bruyeres, de ses Marais & de ses terres labourables avec le desnombrement de ses Caps, de ses Haures, de ses Rades & de ses Bayes, de ses Fontaines, de ses Ruisseaux, de ses Rivieres & de ses Lacs, des Metaux, des Mineraux, etc. Traduit de l'Anglois de Gerard

DICTIONNAIRE UNIVERSEL
DES
FOSSILES PROPRES,
ET DES
FOSSILES ACCIDENTELS,

CONTENANT UNE DESCRIPTION

Des terres, des ſables, des ſels, des ſoufres, des bitumes, des pierres ſimples & compoſées, communes & prétieuſes tranſparentes & opaques, amorphes & figurées, des minéraux, des métaux, des pétrifications du règne animal, & du règne végétal &c. avec des recherches ſur la formation de ces foſſiles, ſur leur origine, leurs uſages &c.

PAR MR. E. BERTRAND,

Prémier Paſteur de l'Egliſe Françoiſe de Berne; Membre des Acad. de Berlin, de Goettinguc, dc Stockholm, de Florence, de Leipſic, de Mayence, de Bavière, de Lyon, de Nanci, de Bâle, de la Société Oeconomique de Berne &c.

TOME PREMIER.

A LA HAYE,

Chez PIERRE GOSSE Junior, ET DANIEL PINET.

M. DCC. LXIII.

Cat. No. 193

Boate. Paris, Chez Robert de Ninville, 1666.
334p. uncat.

238 **Boate.**
See A natural history of Ireland . . . 1726.

239 **Boccone, Paolo,** 1633-1703.
Sylvii Bocconis Curiöse Anmerckungen über ein uñ ander natürliche Dinge. Aus seinem noch nie im Druck gewesenen Museo, experimentaliphysico zusammen gezogen, und im Durchreisen durch Teutschland, zum Andencken seiner in Teutscher Sprach zum Druck hinterlassen. Franckfurt, M. Rohrlachs seel. Wittib, 1697.
501p., illus., port. x550/B63c

Upper right hand corner of port. and upper left hand corner of t.p. mutilated and repaired, with some loss of letters at top of t.p.

240 **Bock, Friedrich Samuel,** 1716-1786.
Versuch einer kurzen Naturgeschichte des preussischen Bernsteins und einer neuen wahrscheinlichen Erklärung seines Ursprunges, . . . Königsberg, Bey J. D. Zeisens Wittwe und J. H. Hartungs Erben, 1767.
146p. 553.29/B63v

"Von den Schicksalen des Bernsteins, in der gelehrten Geschichte: p. [3]-16."

241 **Böckh, August,** 1785-1867.
The public economy of Athens, to which is added a Dissertation on the silver mines of Laurion. Translated by George Cornewall Lewis. 2d ed., rev. London, J. W. Parker, 1842.
688p. 336/B63sEℓ/1842

Bibliographical footnotes.

242 **Boegner, J.**
Das Erdbeben und seine Erscheinungen. Nebst einer chronologischen Uebersicht der Erderschütterungen im mittleren Deutschland vom 8. Jahrhundert bis auf die neueste Zeit und ihres Zusammenhanges mit vulkanischen Erscheinungen in entfernten Ländern. Frankfurt am Main, H. L. Brönner, 1847.
208p., fold map. 551.22/B63e

Mit einer Karte vom Verbreitungsbezirk des Erdbebens vom 29 Juli 1846.

243 **Bonald, Victor,** 1780-1871.
See DEBREYNE, PIERRE JEAN CORNEILLE, Teoria biblica de la cosmogonia . . . 1854.

244 **Bonardo, Giovanni Maria.**
La grandezza, larghezza, e distanza di tutte le sfere, ridotte a nostre miglia, cominciando dall'inferno, fino alla sfera, dove stanno i beati, e la Grandezza delle stelle, con le vere cagioni de piu segnalati effeti naturali, che si generano in ciascuno elemento, e in ciascun cielo . . . Con alcune chiare annotationi, per ciascun Capitolo, di Luigi Grotto Cieco di Hadria. Venetia, Fabio & Agostino Zoppini, 1584.
134 folios, ports. x523/B64g

245 **Bonardo.**
La grandezza, larghezza, e distanza di tutte le sfere, ridotte a nostre miglia, cominciando dall'inferno, fino alla sfera, dove stanno i beati, e la Grandezza delle stelle, con le vere cagioni de piu segnalati effetti naturali, che si generano in ciascuno elemento, e in ciascun cielo . . . Con alcune chiare Annotationi, per ciascun Capitolo, di Luigi Groto Cieca di Hadria. [Venice], Giacomo Zoppini, 1600.
273p., port. x523/B64g/1600

246 **Bonardo.**
La minera del mondo nella qual si tratta delle cose più secrete, e più rare de corpi semplici nel mondo elementare, e de'corpi composti, inanimati, & animati d'anima vegetativa, sensitiva, e ragionevole. Venetia, Appresso Fabio & Agostino Zoppini Fratelli, 1585.
112 leaves, ports. 504/B64m

247 **Bonardo.**
La minera del mondo . . . Nella qual si tratta delle cose più secrete, e più rare de'corpi semplici nel mondo elementare, e de'corpi composti, inanimati, & animati d'anima vegetativa, sensitiva, e ragionevole. Venetia, Fabio & Agostino Zoppini Fratelli, 1589.
112 leaves. uncat.

248 **Bonini, Filippo Maria.**
Il Tevere incatenato overo l'arte di frenar l'acque correnti. Roma, Francesco Moneta, 1663.
422p. uncat.

249 **Bonnard, Auguste Henri de,** 1781-1857.
Aperçu géognostique des terrains. Paris, Chez Deterville, 1819.
266p. 550/B65a

"L'essai géognostique suivant a été rédigé comme article du Nouveau dictionnaire d'histoire naturelle." — Avant-propos.

Note in ms. mounted on t.p.: À Monsieur Freiesleben Hommage de l'auteur.

250 **Bonnard.**
Sur la constance des faits géognostiques qui accompagnent le gisement du terrain d'arkose, à l'est du plateau central de la France. Paris, Impr. de Mme Huzard, 1828.
106p., 3 fold. plates. x551/B642s

Ext. des Annales des mines, 2e série. t.IV, 1828.

251 **Boodt, Anselm Boèce de,** 1550?-1632.
. . . Gemmarum et lapidum historia, qua non solum ortus, natura, vis & precium, sed etiam modus quo exiis, olea, salia, tincturae, essentiae, arcana & magisteria arte chymica confici possint, ostenditur. Opus principibus, medicis, chymicis, physicis, ac liberalioribus ingeniis utilissimum. Cum variis figuris, Indiceq́, duplici & copioso. Hanoviae, Typis Wechelianis apud C. Marnium & heredes J. Aubrii, 1609.
294p., illus., fold. tables, diagrs. x553.8/B64g

UI copy imperfect: fold. tables wanting.
Pages 292 and 293 incorrectly numbered 286 and 287.

252 **Boodt.**
Gemmarum et lapidum historia, quam olim edidit Anselmus Boetius de Boot . . . Nunc vero recensuit, a mendis repurgavit, commentariis, & pluribus, melioribusque figuris illustravit, & multo locupletiore indice auxit, Adrianus Toll . . . Lugduni Batavorum, ex officina Joannis Maire, 1636.
576p., illus., diagrs., 2 fold. tables. x553.8/B64g/1636

253 **Boodt.**
Gemmarum et lapidum historia quam olim edidit Anselmus Boetius de Boot . . . Postea Adrianus Tollius, . . . recensuit, figuris melioribus, & Commentariis pluribus illustravit, & Indice auxit multo locupletiore. Tertia Editio longe purgatissima . . . Lugduni Batavorum, ex officina Ioannis Maire, 1647.
576p., 210p., illus., fold. tables. 553.8/B64g/1647

254 **Boodt.**
Le parfaict ioaillier, ou Histoire des pierreries: où sont amplement descrites leurs naissance, iuste prix, moyen de les cognoistre, & se garder des contrefaites, facultez medecinales, & proprietez curieuses. . . . Et de nouveau enrichi de belles annotations, indices & figures, par André Toll . . . Lyon, I. A. Huguetan, 1644.
746p., illus., 2 fold. tables. x553.8/B64gF

First French translation of Gemmarum et lapidum historia. Translated by Jean Bachou.

255 **Boon Mesch, Antonius Hendrik van der,** 1804-1874.
Disputatio geologica de incendiis montium igni ardentium insulae Javae, eorumdemque lapidibus. Lugduni Batavorum, Haak, 1826.
126p., 3 plates. 551.21/B644d

256 **Borch, Michel Jean,** 1751-1810.
Lythologie sicilienne; ou, Connaissance de la nature des pierres de la Sicile, suivie d'un Discours sur la calcara de Palerme. Rome, B. Francesi, 1778.
228p. 552/B645ℓ/1778

257 **Borch.**
Minéralogie sicilienne docimastique et métallurgique; ou, Connaissance de tous les minéraux que produit l'île de Sicile, avec les détails des mines et des carrières, et l'histoire des travaux anciens et actuels de ce pays. Suivie de la Minérhydrologie sicilienne; ou, La description de toutes les eaux minérales de la Sicile. Par l'auteur de la Lythologie sicilienne. Turin, Frères Reycends, 1780.
264p., fold. tables. 549.9458/B64m

258 **Borelli, Giovanni Alfonso,** 1608-1679.
Historia, et meteorologia incendii Aetnaei anni 1669. Responsio ad censuras Rev. P. Honorati Fabri contra librum auctoris De vi percussionis. Regio lulio, in officina Dominici Ferri, 1670.
162p., fold. plates, diagrs. x551.21/B645h

259 **Born, Ignaz von,** 1742-1791.
Index fossilium quae collegit, et in classes ac ordines disposuit Ignatius S. R. I. eques a Born . . . Pragae, Apud W. Gerle, 1772-75.
2v. in 1, fold. plates. 560/B464i

Half-title: Lithophylacium Bornianum.

260 **Born.**
Schreiben des Herrn Ignatz von Born . . . an Herrn Franz Grafen von Kinsky . . . ueber einen ausgebrannten Vulkan bey der Stadt Eger in Böhmen. Prag, W. Gerle, 1773.
16p. 551.21/B6452s

261 **Born.**
Briefe über mineralogische Gegenstände auf seiner Reise durch das Temeswarer Bannat, Siebenbürgen, Ober- und

Nieder-Hungarn, an den Herausgeber derselben . . . geschrieben. Frankfurt und Leipzig, J. J. Ferber, 1774.
228p., 2 fold. maps, fold. plan. 549/F37b

Bound with Ferber, Johann Jakob, Briefe aus Wälschland über natürliche Merkwürdigkeiten dieses Landes . . . , 1773.

262 **Born.**
Voyage minéralogique fait en Hongrie et en Transylvanie. Traduit de l'allemand, avec quelques notes, par M. Monnet. Paris, 1780.
405p. 549/B645bFm

Translation of Briefe über mineralogische Gegenstände auf seiner Reise durch das Temeswarer Bannat, Siebenbürgen, Ober-und Nieder-Hungarn.

263 **Born.**
See FERBER, JOHANN JACOB, Lettres sur la minéralogie . . . 1776.

264 **Born.**
Baron Inigo Born's new process of amalgamation of gold and silver ores, and other metallic mixtures, as, by his late Imperial Majesty's commands, introduced in Hungary and Bohemia, from the baron's account in German, translated into English by R. E. Raspe. With twenty-two copper-plates. To which are added, a supplement, or a comparative view of the former method of melting and refining; and an address to the subscribers, giving an account of its latest improvements, and of the quick-silver trade. London, Printed for T. Cadell, 1791.
256p., 22 plates (part. fold.). x669.2/B645uEr

265 **Boselli, Gaspare.**
Saggio di litografia ossia descrizione delle qualità e valore delle pietre, coagulazioni e dei metalli preziosi per istruzione dei giojellieri e degli amatori. Milano, G. Pogliani stampatore-librajo, 1820.
112p. 553.8/B65s

"Avvertenza" signed: Gaspare Boselli.
The library has two copies.

266 **Bossut, Charles,** 1730-1814.
Traité élémentaire d'hydrodynamique: ouvrage dans lequel la théorie et l'expérience s'éclairent ou se suppléent mutuellement: avec des notes sur plusieurs endroits qui ont paru mériter d'être approfondis. Paris, C. A. Jombert, 1775.
2v., 16 fold. plates. 532.5/B65t/1775

His Cours de mathématiques. Cinquième partie.

RECUEIL
DE
DIVERS TRAITÉS
SUR
L'HISTOIRE NATURELLE
DE LA TERRE
ET DES FOSSILES.

Par M. E. BERTRAND, *ci-devant premier Pasteur de l'Eglise Françoise de Berne, & Sécretaire de la Société Œconomique, maintenant Conseiller de la Cour du Roi de Pologne; des Académies de Berlin, de Goettingue, de Suede, de Florence, de Leipsic, de Mayence, de Munich, de Lyon, de Nancy, de Bâle; des Sociétés d'Agriculture de Paris, de Lyon, de Rouen, de Dublin;* &c. &c.

A AVIGNON,
Chez *LOUIS CHAMBEAU*, Imprimeur - Libraire, près les RR. PP. Jésuites.

M. DCC. LXVI.

Cat. No. 194

267 **Bossut.**
Lehrbegriff der Hydrodynamik nach Theorie und Erfahrung vorzüglich für solche, welche zur Ausübung dieser Wissenschaft bestimmt sind. Aus dem Französischen übersezt und mit Anmerkungen und Zusätzen hrsg. von Karl Christ Christian Langsdorf. Frankfurt am Main, Andreäischen Buchhandlung.
v.1, 1792. 532.5/B65tGℓ

Translation of Traité élémentaire d'hydrodynamique.
Contents. — 1. Bd. Theorie der Hydrodynamik.
Library owns v.1 only.

268 **Boubée, Nerée,** 1806-1863.
Géologie populaire à la portée de tout le monde, appliquée à l'agriculture et à l'industrie. Paris, Bureau du Nouveau bulletin d'histoire naturelle, 1833.
263p., front. 550/B663g

269 **Bouchette, Joseph,** 1774-1841.
A topographical description of the province of Lower Canada, with remarks upon Upper Canada, and on the relative connexion of both provinces with the United States of America. Embellished by several views, plans of harbours, battles, &c. London, Printed for the author and pub. by W. Faden, 1815.
640p., front. (port.), plates, map, plans, tables. 917.14/B66t

270 **Boué, Ami,** 1794-1881.
Essai géologique sur l'Écosse. Paris, Ve Coucier [1820].
519p., 7 fold. plates, 2 fold. maps. 554.1/B66e

271 **Boué.**
Geognostisches Gemälde von Deutschland. Mit Rücksicht auf die Gebirgsbeschaffenheit nachbarlicher Staaten. Hrsg. von C. C. von Leonhard . . . Frankfurt am Main, J. C. Hermann'sche Buchhandlung, 1829.
623p., 8 plates (incl. maps). 554.3/B66g

272 **Boué.**
Mémoires géologiques et paléontologiques. t.I. Avec quatre planches. Paris, l'Auteur, [etc., etc.] 1832.
362p., 4 fold. plates. 550/B66m

No more published.

273 **Boué.**
Guide du géologue-voyageur, sur le modèle de l'Agenda géognostica de M. Léonhard. Paris, F. G. Levrault, 1835.
2v. in 1, fold. illus. 550/B66g

274 **Boué.**
Essai d'une carte géologique du globe terrestre, présenté le 22 septembre 1843 à la Réunion des naturalistes d'Allemagne, à Gratz. Publié par J. Andriveau-Goujon, sous les auspices et la direction de la Société géologique de France, 1845.
Col. map, 62 x 89 cm. G3200/.C5/1845/.B6

Scale not given.
"Voir le mémoire publié dans le Bulletin de la Société géologique de France, 2ème Serie, Tome 1er, 1844."
Includes "Hauteur des principales sommités du globe, suivant leur longitude."

275 **Boué.**
Der ganze Zweck und der hohe Nutzen der Geologie, in allgemeiner und in specieller Rücksicht auf die Oesterreichischen Staaten und ihre Völker. Eine Erweiterung des am 15. Februar 1850 im Vereine der Freunde der Naturwissenschaften zu Wien gehaltenen Vortrages. Wien, W. Braumüller, 1851.
127p. 550.8/B66g

"Bibliographischer appendix": p. 123-124.

276 **Bouguer, Pierre,** 1698-1758.
La figure de la terre, déterminée par les observations de messieurs Bouguer & de la Condamine, de l'Académie Royale des Sciences, envoyés par ordre du Roy au Pérou, pour observer aux environs de l'équateur... Avec une relation abrégée de ce voyage qui contient la description du pays dans lequel les operations ont été faites. Paris, C. A. Jombert, 1749.
394p., 7 plates (incl. fold. map, diagrs.). 506/PAH/sup./no.8

Académie des sciences, Paris. Mémoires. Suite des mémoires. no. 8.

277 **Bouguer.**
Justification des Mémoires de l'Académie Royale des Sciences de 1744 et du livre de la figure de la terre, déterminée par les observations faites au Pérou, sur plusieurs faits qui concernent les opérations des Académiciens. Paris, C.-A. Jombert, 1752.
44p. 506/PAH/sup./no.9

Académie des sciences, Paris. Mémoires. Suite de mémoires, no. 9.

278 **Bouillet, Jean Baptiste,** 1799-1878.
See DEVÈZE DE CHABRIOL, J.S., Essai géologique et minéralogique sur les environs d'Issoire . . . 1827.

279 **Bouillet,** jt. author.
See LECOQ, HENRI, Vues et coupes . . . Puy-de-Dôme . . . 1830.

280 **Boulanger, Charles Louis,** 1810-1849.
Texte explicatif de la carte géologique du département du Cher, dressée par MM. Boulanger et Bertera . . . Paris, Imprimerie nationale, 1850.
230p., illus. 554.4552/B66t

281 **Bourguet, Louis,** 1678-1742.
Lettres philosophiques sur la formation des sels et des crystaux et sur la génération & le mechanisme organique des plantes et des animaux; à l'occasion de la pierre belemnite et de la pierre lenticulaire. Avec un mémoire sur la théorie de la terre. Amsterdam, F. L'Honoré, 1729.
[44p.], 220p., fold. plate. 500/B66ℓ

282 **Bourguet.**
Traité des pétrifications. Paris, Briasson, 1742.
254p., 60 fold. plates. 560/B66t

283 **Bourguet.**
Traité des pétrifications. Nouvelle éd., corrigée & augmentée. Paris, C.-A. Jombert, 1778.
326p., 60 fold. plates. 560/B66t/1778

284 **Bourrit, Marc Théodore,** 1739-1819.
A relation of a journey to the glaciers, in the dutchy of Savoy. Tr. from the French . . . by C. and F. Davy . . . Norwich, Printed by R. Beatniffe, 1775.
264p., front., 2 plates. 914.94/B66r

Imperfect: lacks front. and 2 plates.

285 **Bourrit.**
Nouvelle description des vallées de glace et des hautes montagnes qui forment la chaîne des Alpes pennines & rhétiennes. Dédiée au roi de France. Ouvrage enrichi de Tableaux, dessinés sur les lieux par l'Auteur, & gravés par M. Moitte. Genève, P. Barde, 1783.
2v., 8 plates, fold. map. 551.31/B66no

286 **Bourrit.**
Nouvelle description des glacières et glaciers de Savoye, Particulièrement de la vallée de Chamouni & du Mont-Blanc, & de la dernière découverte d'une route pour parvenir sur cette haute montagne. Dédiée à M. le Comte de Buffon. Ouvrage qui complète la Description des Alpes Pennines &

Rhétiennes du même Auteur, & de même enrichi de ses tableaux gravés par les meilleurs Artistes. Genève, P. Barde, 1785.
308p., 5 plates. 551.31/B66n

287 **Bowles, William,** 1705-1780.
Introduction à l'histoire naturelle et à la géographie physique de l'Espagne; traduite . . . par le Vicomte de Flavigny. Paris, Chez L. Cellot & Jombert fils, jeune, 1776.
516p. x570.946/B68iFf

Translation of Introducción a la historia natural y la geografía física de España.

288 **Boyle, Robert,** 1627-1691.
Hydrostatical paradoxes, made out by new experiments, (for the most part physical and easie) . . . Oxford, printed by William Hall for Richard Davis, 1666.
347p., 3 fold. plates, illus. x532.2/B697h

289 **Boyle.**
An essay about the origine & virtues of gems. Wherein are propos'd and historically illustrated some conjectures about the consistence of the matter of precious stones, and the subjects wherein their chiefest virtues reside. London, Printed by W. Godbid, 1672.
185p. x553.8/B69e

290 **Boyle.**
Specimen de gemmarum origine & virtutibus. In quo proponuntur & historicè illustrantur quaedam conjecturae circa consistentiam materiae, lapidum praetiosorum, & subjecta, in quibus eorum praecipuae virtutes consistunt. Primum anglicè conscriptum . . . nunc latinè, interprete C.S. Hamburgi, apud Gothofredum Schultz, Amsterodami, apud Joannem Janssonium à Waesberge, 1673.
552p. uncat.

Bound with Boyle's Tractatus . . . and six other anonymous authors.

291 **Boyle.**
Short memoirs for the natural experimental history of mineral waters . . . London, Printed for Samuel Smith, 1684-5.
112p. uncat.

292 **Brackenridge, Henry Marie,** 1786-1871.
Views of Louisiana; together with a Journal of a voyage up the Missouri River, in 1811. Pittsburgh, Printed and published

by Cramer, Spear and Eichbaum, Franklin Head Office, 1814.
304p. x917.63/B72v

293 **Brackenridge.**
Views of Louisiana; containing geographical, statistical and historical notices of that vast and important portion of America. Baltimore, Printed by Schaeffer & Maund, 1817.
323p. 917.63/B72v/1817

294 **Bradbury, John,** fl. 1809.
Travels in the interior of America, in the years 1809, 1810, and 1811; including a description of upper Louisiana, together with the states of Ohio, Kentucky, Indiana, and Tennessee, with the Illinois and western territories . . . Liverpool, Printed for the author; London, Sherwood, Neely, and Jones, 1817.
364p. x917.3/B72t

295 **Bradbury.**
Travels in the interior of America, in the years 1809, 1810, and 1811; including a description of upper Louisiana, together with the states of Ohio, Kentucky, Indiana, and Tennessee, with the Illinois and western territories . . . 2d ed. London, Sherwood, Neely, and Jones, 1819.
[17]-346p., front. (fold. map). x917.3/B72t/1819

296 **Braithwaite, John,** 1797-1870.
Supplement to Captain Sir John Ross's Narrative of a second voyage in the Victory, in search of a northwest passage containing the suppressed facts necessary to a proper understanding of the causes of the failure of the steam machinery of the Victory, and a just appreciation of Captain Sir John Ross's character as an officer and a man of science. London, Chapman and Hall, 1835.
18p. x919.8/R73n

Bound with Ross, Sir John, Narrative of a second voyage . . . 1835, v.1.

297 **Brande, William Thomas,** 1788-1866.
A descriptive catalogue of the British specimens deposited in the geological collection of the Royal Institution. London, Printed for Longman, Hurst, Rees, Orme and Brown, 1816.
212p. 550.74/B73d

298 **Brande.**
Outlines of geology; being the substance of a course of lectures delivered in the theatre of the Royal Institution in the year 1816. London, J. Murray, 1817.
144p., front. 550.4/B73c

L'ART ET SCIENCE DE TROVVER LES EAVX ET FONTAINES CACHEES SOVBS terre, autrement que par les moyens vulgaires des Agriculteurs & Architectes, par Iaques Beſſon Dauphinois, Mathematicien.

A ORLEANS,

Pour Pierre Trepperel, Libraire, en la rue de l'Eſcriuinerie, deuant la porte du cloiſtre ſaincte Croix.

M. D. LXIX.

Auec priuilege du Roy pour dix ans.

Cat. No. 203

299 **Brande.**
Outlines of geology. London, J. Murray, 1829.
234p., illus. 550.4/B73o/1829

"The following pages present an outline of the different courses of lectures which I have delivered in the Royal Institution upon the subject of geology."

300 **Brande.**
Elementi di geologia di T. G. [i.e. G. T.] Brande. Prima traduzione italiana dall'Inglese con note e coll'aggiunta di un Dizionario dei termini geologici del Carlo Ormea. Torino, F. Reycend, 1836.
200p., plates (1 fold. col.). 550.4/B73olo

Translation of: Outlines of geology.

301 **Brander, Gustavus,** 1720-1787.
Fossilia hantoniensia collecta, et in Musaeo britannico deposita. Londini, 1766.
43p., 9 plates. x560.94227/B733f

The descriptions are by D. Solander.
Interleaved.

302 **Brandis, Joachim Dietrich,** 1762-1845.
Anleitung zum Gebrauche des Driburger Bades und Brunnens, nebst einer kurzen Beschreibung der dortigen Anlagen und Gegend. Münster, F. Theissing, 1792.
225p., illus. 613.12/B73a

303 **Brard, Cyprien Prosper,** 1786-1838.
Manuel du minéralogiste et du géologue voyageur. Paris, Impr. de Feugueray. L'auteur, 1805.
470p., 1 fold. plate. 549/B733m

304 **Brard.**
Nouveaux élémens de minéralogie, ou Manuel du minéralogiste voyageur. 2.éd., rev. et considérablement augm. Paris, Méquignon-Marvis, 1824.
686p., fold. plate. 549/B733m/1824

First ed. published under title: Manuel du minéralogiste et du géologue voyageur.

305 **Brard.**
Traité des pierres précieuses, des porphyres, granites, marbres, albâtres, et autres roches propres à recevoir le poli et à orner les monumens publics et les édifices particuliers; suivi de la description des machines dont on se sert pour tailler, polir et travailler ces pierres; et d'un coup d'oeil général sur l'art

du marbrier; ouvrage utile aux joailliers, lapidaires, bijoutiers; aux architectes, décorateurs, etc., etc. Orné de planches. Paris, F. Schoell, 1808.
2v. in 1, plates, tables. 553.8/B73t

Paged continuously.

306 **Brard.**
Minéralogie appliquée aux arts; ou, Histoire des minéraux qui sont employés dans l'agriculture, l'économie domestique; la fabrication des sels, des combustibles et des métaux; l'architecture et la décoration; la peinture et le dessin; les arts mécaniques; la bijouterie et la joaillerie. Ouvrage destiné aux artistes, fabricans et entrepreneurs. Paris, F. G. Levrault, 1821.
3v., illus. (part. fold.). 549/B733mi

307 **Breihaupt, August.**
Die Paragenesis der Mineralien. Mineralogisch, geognostisch und chemisch beleuchtet, mit besonderer Rücksicht auf Bergbau. Freiberg, J. G. Engelhardt, 1849.
276p., fold. plate. 553/B74p

308 **Breislak, Scipione,** 1748-1826.
Essais minéralogiques sur la Solfatare de Pouzzole. Tr. du manuscrit italien par François de Pommereuil . . . Naples, J. Giaccio, 1792.
240p. 551.21/B74e

309 **Breislak.**
Voyages physiques et lythologiques dans la Campanie; suivis d'un mémoire sur la constitution physique de Rome; avec la carte générale de la Campanie, d'après Zannoni; celle des cratères éteints entre Naples et Cumes; celle du Vésuve, du plan physique de Rome, etc., etc. Tr. du manuscrit italien, et accompagnés de notes, par le Général Pommereuil . . . Paris, Dentu, Imprimeur-libraire, 1801.
2v., fold. plates, fold. maps. 554.5/B75vFp

310 **Breislak.**
Introduzione alla geologia. Milano, Stamperia Reale, 1811.
2v. in 1. 550/B74i

311 **Breislak.**
Atlas géologique; ou, Vues d'amas de colonnes basaltiques faisant suite aux Institutions géologiques de Scipion Breislak. Milan, 1818.
55 plates, illus. 551/B74a

312 **Breislak.**
Descrizione geologica della provincia di Milano, pubblicata per ordine dell'I. R. governo di Lombardia. Milano, Imperiale regia stamperia, 1822.
260p., fold. map. 550/B74d

Bibliographical footnotes.

313 **Breislak.**
Descrizione geologica della provincia di Milano pubblicata per ordine dell'I. R. governo di Lombardia. 2.ed., con le notizie su la vita e su le opere dell'autore. Milano, G. Silvestri, 1845.
311p. 550/B74d/1845

Added t.-p.: Biblioteca scelta di opere italiane, antiche e moderne. v.491.

Notizie su la vita e su le opere dell'autore scritte dal sig. G. Gley: p. [v]-viii.

Bibliographical footnotes.

314 **Brewster, David,** 1781-1868.
See MILLER, HUGH, The foot-prints of the Creator... [with] Hugh Miller: sketches of his life and writings by Sir D. Brewster... 1851.

315 **Breyn, Johann Philipp,** 1680-1764.
Epistola de melonibus petrefactis montis Carmel vulgo creditis ad D. Johannem Anderson. Cui annexe binae epistolae ad authorem datae, altera De aquis matrialibus Olonicensibus, nos ex Africa in Belgium deferri coepit. Lipsiae, Literis I. Titii, 1722.
48p., plate. 550/B759e

The first letter was written by Georg R. Remus, the second by Nathanael Sendel.

316 **Brisson, Mathurin Jacques,** 1723-1806.
Principes élémentaires de l'histoire naturelle et chymique des substances minérales; ouvrage utile aux écoles centrales. Paris, F. Didot, 1797.
245p. uncat.

317 **Brocchi, Giovanni Battista,** 1772-1826.
Trattato mineralogico e chimico sulle miniere di ferro del departimento del Mella, con l'esposizione della costituzione fisica delle montagne metallifere della Val-Trompia. Brescia, N. Bettoni, 1807-08 [v.1, 1808].
2v., fold. table. x553.35/B78t

Manuscript notes in margins, on last flyleaf of v.1, and on 5p. tipped in at end of v.2.

318 **Brocchi.**
Memoria mineralogica sulle Valle di Fassa in Tirolo. Milano, G. Silvestri, 1811.
233p. 549.945/B78m

319 **Brocchi.**
Conchiologia fossile subappennina, con osservazioni geologiche sugli Appennini e sul suolo adiacente. Milano, Stamperia reale, 1814.
2v. 564/B78c

320 **Brocchi.**
Conchiologia fossile subappennina, con osservazioni geologiche sugli Appennini e sul suolo adiacente. Con atlante de sedici tavole. Milano, G. Silvestri, 1843.
2v. 564/B78c/1843

Added t.p.: Biblioteca scelta de opere italiane, antiche e moderne, p.452-454.

321 **Brocchi.**
Dello stato fisico del suolo di Roma, memoria per servire d'illustrazione alla carta geognostica di questa città. Roma, Nella stamperia de Romanis, 1820.
281p., fold. col. plates. 554.5632/B68d

322 **Brochant de Villiers, André Jean François Marie,** 1772-1840.
Traité élémentaire de minéralogie, suivant les principes du professeur Werner . . . rédigé d'après plusieurs ouvrages allemands, augmenté des découvertes les plus modernes, et accompagné de notes pour accorder sa nomenclature avec celle des autres minéralogistes français et étrangers. Avec 18 tableaux et une planche. Paris, Chez Villier, [1801-1803].
2v., tables. 549/B781t

Plates separately bound are missing from this set.

323 **Brochant de Villiers.**
Traité élémentaire de minéralogie, suivant les principes du Professeur Werner, conseiller des mines de Saxe; rédigé d'après plusieurs ouvrages allemands, augmenté des découvertes les plus modernes, et accompagné de notes pour accorder sa nomenclature avec celle des autres minéralogistes français et étrangers. 2.éd. Avec 18 tableaux et une planche. Paris, Mme. veuve Villier, 1808.
2v., fold. plate, tables. 549/B781t/1808

324 **Brochant de Villiers.**
De la cristallisation considérée géométriquement et physiquement; ou, Traité abrégé de cristallographie. Suivi d'un précis de nos connaissances actuelles sur les phénomènes physiques de la cristallisation. Strasbourg, Impr. F. G. Levrault, 1819.
206p., 16 plates. 548/B68d

325 **Broderip, William John,** 1789-1859.
See DE LA BECHE, HENRY THOMAS, Researches in theoretical geology . . . 1834.

326 **Brodie, Peter Bellinger,** 1815-1897.
A history of the fossil insects in the secondary rocks of England. Accompanied by a particular account of the strata in which they occur, and of the circumstances connected with their preservation. By the Rev. Peter Bellinger Brodie. London, J. van Voorst, 1845.
130p., 11 plates. 565.7/B779h

327 **Broeck, Victor van den.**
Traité abrégé de docimasie ou Résumé des leçons données à l'École des Mines du Hainaut. Mons, Impr. de L. F. Moureaux, 1841.
238 (i.e. 338)p., illus., tables. 669.92/B78t

328 **Bromell, Magnus von,** 1679-1731.
. . . Mineralogia et lithographica Svecana; das ist, Abhandlung derer in dem Königreich Schweden befindlichen Mineralien und Steinen; ehemahls in Schwedischer Sprache abgefasst, nunmehro aber ihrer besondern Merckwürdigkeit halben ins Teutsche übers., mit einem Vorbericht von dem vor kurtzer Zeit in Schweden entblössten Gold-Erz begleitet, und mit darzu dienlichen Kupfern ans Licht gestellt, von Mikrandern. Stockholm und Leipzig, G. Kiesewetter, 1740.
[26], 148p., illus., fold. plate. uncat.

329 **Brongniart, Adolphe Théodore,** 1801-1876.
Prodrome d'une histoire des végétaux fossiles. Paris [etc.], F. G. Levrault, 1828.
223p. 561/B78p

330 **Brongniart.**
Histoire des végétaux fossiles; ou Recherches botaniques et géologiques sur les végétaux renfermés dans les diverses couches du globe. Paris, Fortin, Masson, 1828-36.
2v., plates. q561/B78h

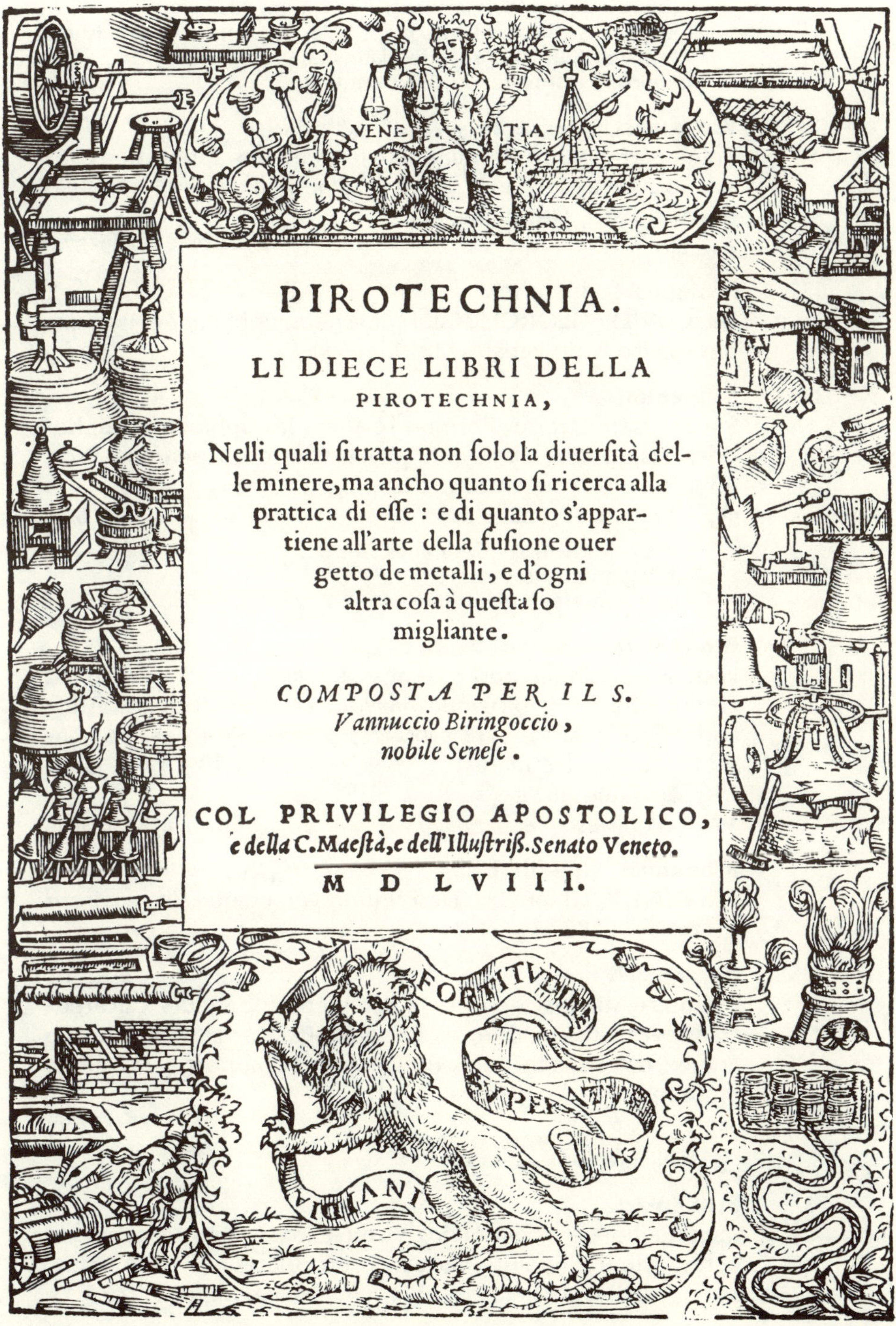

PIROTECHNIA.

LI DIECE LIBRI DELLA PIROTECHNIA,

Nelli quali ſi tratta non ſolo la diuerſità delle minere, ma ancho quanto ſi ricerca alla prattica di eſſe: e di quanto s'appartiene all'arte della fuſione ouer getto de metalli, e d'ogni altra coſa à queſta ſomigliante.

COMPOSTA PER IL S. Vannuccio Biringoccio, nobile Seneſe.

COL PRIVILEGIO APOSTOLICO, e della C. Maeſtà, e dell'Illuſtriß. Senato Veneto.

M D L V I I I.

Cat. No. 220

v.II has imprint: Paris, Crochard et comp., 1837.
Of v.II only the first livraison (p. 1-72, 30 plates) was published.
v.I, plates 128, 130 precede plate 119.
Imperfect: v.II plates 27 & 29 wanting.

331 **Brongniart, Alexandre,** 1770-1847.
Traité élémentaire de minéralogie, avec des applications aux arts; ouvrage destiné à l'enseignement dans les lycées nationaux. Paris, Deterville, 1807.
2v., fold. plates. 549/B78t

332 **Brongniart,** jt. author.
See CUVIER, GEORGES, Essai sur la géographie minéralogique des environs de Paris . . . 1811.

333 **Brongniart.**
Sur le gisement; ou, Position relative des ophiolites, euphotides, jaspes, etc., dans quelques parties des Apennins. Paris, Impr. de Madame Huzard, 1821.
64p., 2 fold. col. plates 554.5/B78s

Extrait des Annales des mines, 1821.
Bibliographical footnotes.
Copy imperfect: plate 1, figures 1-2 missing.

334 **Brongniart.**
Histoire naturelle des crustacés fossiles, sous les rapports zoologiques et géologiques. Savoir: Les trilobites, par Alexandre Brongniart . . . Les crustacés proprement dits, par Anselme-Gaëtan Desmarest . . . Avec onze planches. Paris, Chez F.-G. Levrault, 1822.
154p., 11 plates. q565.3/B86h

335 **Brongniart,** jt. author.
See CUVIER, GEORGES, Description géologique des environs de Paris . . . 1822.

336 **Brongniart.**
Mémoire sur les terrains de sédiment supérieurs calcaréo-trappéens du Vicentin, et sur quelques terrains d'Italie, de France, d'Allemagne, etc., qui peuvent se rapporter à la même époque. Paris, F. G. Levrault, 1823.
86p., illus., 6 plates. 554/B78m

Bibliographical footnotes.

337 **Brongniart.**
Introduction à la minéralogie, ou exposé des principes de cette science et de certaines propriétés des minéraux. Réimpression de l'éd. de 1824, sans changemens. Paris, F. G.

Levrault, 1825.
2 p.l., 159p., 2 plates. 549/B789i/1825

"Article inséré dans le tome XXXI du Dictionnaire des sciences naturelles, au mot minéralogie."

338 **Brongniart.**
Classification et caractères minéralogiques des roches homogènes et hétérogènes. Paris, F. G., Levrault, 1827.
144p. 552/C659

Bound with Classificazione delle rocce secondo i più celebri autori per servire allo studio della geologia. Milano, Tip. Sonzogna e comp., 1814.

339 **Brongniart.**
Tableau des terrains qui composent l'écorce du globe: ou, Essai sur la structure de la partie connue de la terre. Paris, F. G. Levrault, 1829.
435p. 551.13/B78t/1829

340 **Brongniart.**
Tableau des terrains qui composent l'écorce du globe; ou, Essai sur la structure de la partie connue de la terre. Bruxelles, Société Belge de Librairie, etc., 1838.
422p. 551.13/B78t

341 **Brongniart.**
Tableau de la distribution méthodique des espèces minérales suivi dans le cours de minéralogie fait au Muséum royal d'histoire naturelle en 1833. Présentant leur nomenclature univoque ou Linnéenne, et leur nomenclature caractéristique ou définition, tirée de leur composition définie et de leur forme fondamentale. Paris, Roret, 1883.
46p. 549/B78ta

342 **Bronn, Heinrich Georg,** 1800-1862.
System der urweltlichen Konchylien: durch Diagnose, Analyse und Abbildung der Geschlechter erläutert. Zum Gebrauche bey Vorlesungen über Petrefactenkunde und zur Erleichterung des Selbststudiums derselben. Heidelberg: J. C. B. Mohr, 1824.
56p., 7 fold. plates. q564/B789s

343 **Bronn.**
System der urweltlichen Pflanzenthiere durch Diagnose, Analyse und Abbildung der Geschlechter erläutert. Zum Gebrauche bey Vorlesungen über Petrefactenkunde und zur Erleichterung des Selbstudiums derselben. Heidelberg, J. C.

B. Mohr, 1825.
47p., 7 fold. plates. q563/B78s

344 **Bronn.**
Paläontologische Collectaneen, hauptsächlich als beliebiges Ergänzungs-Heft zum Neuen Jahrbuch für Mineralogie, Geologie, Geognosie und Petrefakten-Kunde, Jahrgänge 1840-1843, dienend. Stuttgart, E. Schweizerbart'sche Verlagshandlung, 1843.
156p. 016.56/B78p

345 **Bronn.**
Index palaeontologicus: oder Übersicht der bis jetzt bekannten fossilien Organismen. Unter Mitwirkung der Prof. H. R. Goppert und Herm. v. Meyer. Stuttgart, E. Schweitzerbart, 1848-1849.
2v. in 3. 560/B78i

346 **Bronn.**
Lethaea geognostica; oder, Abbildung und Beschreibung der für die Gebirgs-Formationen bezeichnendsten Versteinerungen. 3. stark verm. Aufl., bearb. von H. G. Bronn & F. Roemer. Mit einem Atlas von 124 Tafeln . . . Stuttgart, E. Schweizerbart, 1850-56.
3v., and Atlas of 123 plates, col. tables. q560/B78h/1850
q560/B78h/1850/atlas

347 **Brooke, Henry James,** 1771-1857.
A familiar introduction to crystallography; including an explanation of the principle and use of the goniometer. With an appendix, containing the mathematical relations of crystals; rules for drawing their figures; and an alphabetical arrangement of minerals, their synonymes, and primary forms. Illustrated by nearly 400 engravings on wood. London, W. Phillips, 1823.
508p., diagrs. 548/B79f

348 **Brooke.**
See PHILLIPS, WILLIAM, An elementary introduction to mineralogy . . . 1852.

349 **Brooks, William Alexander.**
Treatise on the improvement of the navigation of rivers; with a new theory on the cause of the existence of bars. London, J. Weale, 1841.
154p. 627.12/B79t

350 **Browallius, Johan,** 1707-1755.
Historische und physikalische Untersuchung der vorgegebenen Verminderung des Wassers und Vergrösserung der Erde, bey welcher sowohl der Ursprung und Fortgang dieses Lehrsatzes, inner- und ausserhalb Schweden, als auch dessen eigentliche Beschaffenheit umständlich und reiflich geprüfet worden. Aus dem Schwedischen übersetzet, und mit einigen Erläuterungen vermehret von Karl Ernst Klein. Stockholm, G. Kiesewetter, 1756.
280p., map. x550.1/B81hGk

351 **Brown, Samuel R.,** 1775-1817.
The western gazetteer; or Emigrant's directory, containing a geographical description of the western states and territories, viz., the states of Kentucky, Indiana, Louisiana, Ohio, Tenessee [sic] and Mississippi; and the territories of Illinois, Missouri, Alabama, Michigan, and North-Western. With an appendix, containing sketches of some of the western counties of New-York, Pennsylvania and Virginia; a description of the great northern lakes; Indian annuities, and directions to emigrants. Auburn, N.Y., Printed by H. C. Southwick, 1817.
[7]-360p. 917.7/B815w

352 **Brown, Thomas,** 1785-1862.
Illustrations of the fossil conchology of Great Britain and Ireland, with descriptions and localities of all the species. London, Smith, Elder, 1849.
273p., 98 plates. q564/B81i

353 **Browne, Patrick,** 1720-1790.
The civil and natural history of Jamaica. In three parts. Containing, I. An accurate description of that island . . . with a brief account of its former and present state, government, revenues, produce, and trade. II. A history of the natural productions . . . native fossils . . . III. An account of the nature of climates in general, and their different effects upon the human body. London, printed for the author, 1756.
503p., illus. xq917.292/B81c

354 **Brückmann, Franz Ernst,** 1697-1753.
Magnalia Dei in locis subterraneis, oder Unterirdische Schatz-Cammer aller Königreiche und Länder, in ausführlicher Beschreibung aller, mehr als MDC. Bergwercke durch alle vier Welt-Theile . . . nebst Anmerckung aller derjenigen Länder und Oerter, wo Edelgesteine zu finden, in geographischer Ordnung und einigen Kupffer-Figuren zu besichtigen . . . Braunschweig, 1727-30.

2v., front. (port.), 51 plates (partly fold.). xq622/B83

Vol. 2 published at Wolffenbüttel.

355 **Brückmann.**
See HERMANN, LEONARD DAVID, Monumentum gratitudinis marmoreum . . . 1729.

356 **Brückmann, Urban Friedrich Benedict,** 1728-1812.
Abhandlung von Edelsteinen, nebst einer Beschreibung des so genannten salzthalischen Steins. Braunschweig, verlegts die Fürstl. Waysenhausbuchhandlung, 1757.
143p. 553.8/B83a

Bound with Baumer, Johann Wilhelm. Naturgeschichte aller Edelsteine . . . aus dem Lateinischen übersetzt von Karl Freiherrn und Reichsrittern von Medinger, 1774.

357 **Brugmans, Sebald Justinus,** 1763-1819.
Lithologia Groningana, iusta ordinem Wallerii digesta, cum synonimis aliorum, imprimis Linnaei et Cronstedii. Publice defensa, cum summos in philosophia honores, in Academia Groningo-omlandica consequeretur. Groningae, Apud P. Doekema & R. Muller, 1781.
120p., fold. plate. 552/B833ℓ

358 **Brullé, Gaspard Auguste,** 1809-1873.
Sur le gisement des insectes fossiles, et sur les secours que l'étude de ces animaux peut fournir à la géologie. Paris, Imprimerie de Terzuolo [etc.], 1839.
30p. 565.7/B83s

Thèse — Faculté des sciences de Paris.
Bibliographical footnotes.

359 **Bry, Theodor de,** 1528-1598.
[America pt. 1 Latin] Admiranda narratio, fida tamen, de commodis et incolarum ritibus Virginiae . . . Anglico scripta sermone à Thoma Hariot . . . nunc autem primum Latio donata à C.C.A. Francoforti ad Moentum, Typis Ioannis Wecheli, Sumptibus vero Theodori de Bry, venales reperiuntur in officina Sigismundi Feirabendii, 1590.
34 [84]p., 28 col. illus., map. xq910.8/B84iL

Illus. 1-23 have title; Vivae imagines et ritus incolarum eius provinciae in America, quae Virginia appelate est . . . Omnia diligenter observata, et ad vivum expressa a Joanne With . . . Illus. 24-28 have title: Pictorum, Brittaniae partem olim incolentium aliquot icones . . .

Imperfect: t.p. wanting, supplied by facsim.; extra copy of leaf a^2 laid in; illus. IV (2nd group) bound before illus. III.

I L

TEVERE INCATENATO

OVERO

L'Arte di frenar l'acque correnti

ALLA SANTITÀ DI N. S.

PAPA ALESSANDRO VII.

DELL'ABBATE

FILIPPO MARIA BONINI,

VICARIO GENERALE DI PELESTRINA.

Con le Tauole de' Capitoli, delle figure, e delle cose più notabili.

IN ROMA,

Nella Stampa di Francesco Moneta. MDCLXIII.

CON LICENZA DE' SUPERIORI.

A spese dell'Autore.

Cat. No. 248

First ed., first issue. Cf. Church, Cat. of books relating to . . . America, no. 140.

Text consists of a Latin translation of Harriot's A brief and true report of the new found land of Virginia, London, 1588. Cf. Church, 135.

360 **Buch, Christian Leopold, von,** 1774-1853.

Beobachtungen über den Kreuzstein. Der Linneischen Societät zu Leipzig mitgetheilt von Leopold von Buch. Leipzig, Wagner & Co., 1794.

28p., fold. plate. 549.6/B85b

Bibliographical footnotes.

361 **Buch.**

Geognostiche Beobachtungen auf Reisen durch Deutschland und Italien. Berlin, Haude und Spener, 1802-09.

2v., fold. plates, fold. maps. 554.3/B85g

2. bd. has additional title: Mit einem Anhange von mineralogischen Briefen aus Auvergne an den Geh. Ober-Bergrath Karsten von demselben Verfasser.

362 **Buch.**

A mineralogical description of the environs of Landeck, in the county of Glatz. With a mineralogical map. Tr. from the French, with notes, by Charles Anderson . . . Edinburgh, A. Constable & Co.; [etc., etc.], 1810.

140p., front. (fold. map). 554.314/B85vEa

Translation of: Versuch einer mineralogischen Beschreibung von Landeck.

363 **Buch.**

Reise durch Norwegen und Lappland. Berlin, G. C. Nauck, 1810.

2v., fold. maps. 914.81/B85r

Added t.-p., engr., with vignettes.

364 **Buch.**

Travels through Norway and Lapland, during the years 1806, 1807 and 1808, by Leopold von Buch . . . Tr. from the original German by John Black. With notes and illustrations, chiefly mineralogical, and some account of the author, by Robert Jameson . . . Illustrated with maps and physical sections . . . London, Printed for H. Colburn, 1813.

460 (i.e. 440)p., fold. maps. 914.81/B85rEb

p. 433-440 numbered 453-460.

365 **Buch.**
Voyage en Norvège et en Laponie, fait dans les années 1806, 1807 et 1808. Traduit de l'allemand, par J. B. B. Eyriés; précédé d'une introduction de M. A. de Humboldt; suivi d'un Mémoire de M. de Buch, sur la limite des neiges perpétuelles dans le nord, et enrichi de cartes et de coupes de terrein. Paris, Gide Fils, 1816.
2v., maps. 914.81/B85rFe

366 **Buch.**
Description physique des Iles Canaries, suivie d'une indication des principaux volcans du globe. Tr. de l'allemand par C. Boulanger... Rév. et augm. par l'auteur. Paris et Strasbourg, F. G. Levrault, 1836.
525p., illus., tables. 916.49/B85pFb

367 **Buckland, William,** 1784-1856.
Vindiciae geologicae: or, The connexion of geology with religion, explained in an inaugural lecture delivered before the University of Oxford, May 15, 1819, on the endowment of a readership in geology by His Royal Highness the Prince Regent. Oxford, At the University press for the author, 1820.
38p. 215/B85v

368 **Buckland.**
See LYON, GEORGE FRANCIS, A narrative of travels in northern Africa ... [with an account of the rocks and minerals by Prof. Buckland]. 1821.

369 **Buckland.**
Reliquiae diluvianae; or Observations on the organic remains contained in caves, fissures, and diluvial gravel, and on other geological phenomena, attesting the action of an universal deluge. By the Rev. William Buckland. London, J. Murray, 1823.
303p., 23 plates, 4 maps, fold. table. 560/B85r

Handwritten humorous epitaph to Prof. Buckland by A. Wateley appended.

370 **Buckland.**
Reliquiae diluvianae; or, Observations on the organic remains contained in caves, fissures, and diluvial gravel, and on other geological phenomena, attesting the action of an universal deluge. 2d ed. London, J. Murray, 1824.
303p., 23 plates, 4 maps (2 col.), fold. table. 560/B85r/1824

371 **Buckland.**
Geology and mineralogy considered with reference to natural theology. London, William Pickering, 1836.
2v. 550.1/B85

372 **Buckland.**
Geology and mineralogy considered with reference to natural theology. [2d ed.] London, W. Pickering, 1837.
2v., 69 (i.e. 87) plates. 560/B85g/1837

Half-title: The Bridgewater treatises on the power, wisdom and goodness of God, as manifested in the creation. VI.
v.2 (the plates) lacking.

373 **Buckland.**
Geology and mineralogy considered with reference to natural theology. By the Rev. William Buckland . . . Philadelphia, Carey, Lea and Blanchard, 1841.
2v., 69 (i.e. 87) plates. 560/B85g/1841

Half-title: The Bridgewater treatises on the power, wisdom, and goodness of God, as manifested in the creation.

374 **Buckland.**
La géologie et la minéralogie dans leurs rapports avec la théologie naturelle. Traduit de l'anglais par M. L. Doyère. Paris, Crochard, 1838.
2v., 69 [i.e. 87] plates. 560/B85gFd

Translation of Geology and mineralogy considered with reference to natural theology.

375 **Buckland.**
Die Urwelt und ihre Wunder, oder allgemeine Darstellung der Geschichte des Erdkörpers. Nach der zweiten Aufl. . . . Anhange von W. P. Schimper. Stuttgart, Paul Neff, 1838.
406p., plates. uncat.

Translation of his Geology and mineralogy . . .

376 **Buckland.**
Geologie und Mineralogie in Beziehung zur natürlichen Theologie. Aus dem Englischen, nach der 2. Ausg. des Originals, übers. und mit Anmerkungen und Zusätzen versehen, von L. Agassiz. Neuchâtel, Gedruckt bei Petit-Pierre, 1838-39 [v.1, 1839].
2v, 69 [i.e. 87] plates. 560/B85gGa

Translation of Geology and mineralogy considered with reference to natural theology.

377 **Buckman, James,** 1816-1884.
See MURCHISON, RODERICK IMPEY, Outline of the geology... of Cheltenham... 1845.

378 **Bucquet, Jean Baptiste Michel,** 1746-1780.
Introduction à l'étude des corps naturels, tirés du règne minéral. Paris, J. T. Herrissant, 1771.
2v., illus. 549/B858i

379 **Buettner, David Sigismund,** 1660-1719.
Coralliographia subterranea. Seu, Dissertatio de corallis fossilibus, in specie, de lapide corneo, Horn- oder gemeinem Feuer-Stein. Cum tabulis aeneis. Lipsiae, F. Croschuffii, 1714.
68p., 4 fold. tables. 563.6/B86c

380 **Buffon, George Louis Leclerc,** 1707-1788.
Oeuvres complètes de Buffon, avec les descriptions anatomiques de Daubenton, son collaborateur, nouvelle édition. Dirigée par M. Lamouroux. Paris, Verdière et Ladrange, 1824-32.
44v. 500/B86o/1824

Library has v. 7-40 only.

381 **Buffon.**
Oeuvres complètes de Buffon, avec les suites. Accompagnées de 300 vignettes représentant plus de 800 animaux, d'un beau portrait de Buffon, et d'un frontispice représentant le génie de la nature, dessinés par Victor Adam, et gravés sur acier par MM. Muller, Lalaisse, Durand... etc. Nouvelle éd.... Paris, E.-L.-C. Mauprivez, 1835-36.
6v. in 4, front., port., plates. 500/B86o/1835

382 **Buffon.**
Büffon's sämmtliche Werke, sammt den Ergänzungen, nach der Klassifikation von G. Cuvier. Einzige Ausgabe in deutscher Uebersetzung von H. Schaltenbrand... Köln, 1837-40.
9v. 500/B86oGs

383 **Buffon.**
Histoire naturelle, générale et particulière, avec la description du Cabinet du Roy... Paris, Imprimerie Royale, 1749-1804.
44v. 500/B86h

Library owns v. 1-24, 27-31, Suppl. 1-7.

384 **Buffon.**
Histoire naturelle, générale et particulière... Deux-Ponts, Sanson, 1785-91, [v.54, 1790].

54v., plates (part col.), port., fold. maps, tables. x500/B86h/1785

Vols.46-65 have title: Histoire naturelle des minéraux.
Dedication signed by Buffon and his collaborator Daubenton.
Contents. — v.1-13. [Histoire générale]. — v.14-27. Quadrupèdes. — v.28-45. Oiseaux. — v.46-54. Histoire naturelle des minéraux.

385 **Buffon.**

Histoire naturelle, générale et particulière. Nouvelle éd., accompagnée de Notes, et dans laquelle les Supplémens sont insérés dans le premier textes, à la place qui leur convient. L'on y a ajouté l'histoire naturelle des Quadrupèdes et des Oiseaux découverts depuis la mort de Buffon, celle des Reptiles, des Poissons, des Insectes et Vers; enfin, l'histoire des Plantes dont ce grand Naturaliste n'a pas eu le tems de s'occuper. Redigé par C. S. Sonnini. Paris, F. Dufart, [1799-1805].

500/B86h/1799

Library owns v.5-16.

386 **Buffon.**

Allgemeine Naturgeschichte. Eine frëye mit einigen Zuzätzen vermehrte Übersetzung nach der neuesten französ: Ausg. von 1769: in 8° . . . Berlin, Joachim Pauli, 1771-74.

7v. in 4, fronts. (v.1, 3, 5), plates, 2 fold. maps, tables. B500/B86hiG

The library owns vols. 1-2 in one vol. (2 editions); vols. 3-4 in one vol. (2 editions); and vol. 7 in one vol. (1 edition).

387 **Buffon.**

The natural history of animals, vegetables, and minerals, with the theory of the earth in general. Translated from the French of Count de Buffon by W. Kenrick. London, Printed for, and sold by T. Bell, 1775-76.

6v., illus., plates, port. x500/B86hEk

Vol.2 translated by W. Kenrick and J. Murdoch, v.3-5 by W. Kenrick and others.
Clippings of pictures and text from newspapers have been inserted in v.1 and 2.

388 **Buffon.**

Natural history, general and particular. Translated into English, illustrated with above 300 copper-plates, and occasional notes and observations, by William Smellie . . . 2nd ed. . . . London, Printed for W. Strahan and T. Cadell, 1785.

9v., front. (port.). v.1, plates (partly fold.). x500/B86hEs/1785

Translation of Histoire naturelle, générale et particulière.

ANSELMI

BOETII DE BOODT

BRVGENSIS BELGÆ,

RVDOLPHI SECVNDI,

IMPERATORIS ROMANORVM,

Perſonæ Medici,

GEMMARVM ET LAPIDVM Hiſtoria,

Qua non ſolum ortus, natura, vis & precium, ſed etiam modus quo ex iis, olea, ſalia, tincturæ, eſſentiæ, arcana & magiſteria arte chymica confici poßint, oſtenditur.

OPVS PRINCIPIBVS, MEDICIS, CHYmicis, Phyſicis, ac liberalioribus ingeniis vtiliſſimum.

Cum variis figuris, Indiceq; duplici & copioſo.

HANOVIÆ,

Typis Wechelianis apud Claudium Marnium & heredes Ioannis Aubrii.

M. DC. IX.

Cat. No. 251

Vol. 9 does not have edition statement.
v.1, plates.

389 **Buffon.**

Natural history. Containing a theory of the earth, a general history of man, of the brute creation, and of vegetables, minerals, &c. &c. &c. From the French. With notes by the translator. London, Printed for the proprietor, and sold by H. D. Symonds, 1797.

10v., plates. x500/B86hEb/1797

At head of title: Barr's Buffon.

390 **Buffon.**

Barr's Buffon. Buffon's natural history. Containing a theory of the earth, a general history of man, of the brute creation, and of vegetables, minerals, &c. &c. &c. From the French. With notes by the translator . . . London, 1807-11.

500/B86hEb

Library owns vols.1-10.

391 **Buffon.**

A natural history, general and particular; containing the history and theory of the earth, a general history of man, the brute creation, vegetables, minerals, &c. &c. Translated from the French by William Smellie. A new edition corr. and enl. To which is added A history of birds, fishes, reptiles, and insects, embracing the recent discoveries of eminent naturalists, together with an account of the most curious foreign plants, by Henry Augustus Chambers. Vol. I [-II]. London, Printed for T. Kelly, by W. Clowes, [1816?].

2v., 52 col. plates, ports. 500/B86hEs/1816

No more published in this edition.

392 **Buffon.**

Histoire naturelle des minéraux. Paris, Imprimerie royale, 1783-1787.

8v. 549/B864h

393 **Buffon.**

See LELARGE DE LIGNAC, JOSEPH ADRIEN, Lettres à un Amériquain sur l'Histoire naturelle . . . de monsieur de Buffon . . . 1751.

394 **Buffon.**

See BERTRAND, PHILIPPE, Lettre à M. le comte de Buffon . . . 1780.

395 **Burat, Amédée,** 1809-1883.
Description des terrains volcaniques de la France centrale. Paris, F. G. Levrault, 1833.
342p., 10 fold. plates. 554.4/B89d

396 **Burat.**
Description de quelques gîtes métallifères de l'Algérie, de l'Andalousie (Espagne), de Taunus et du Westerwald (Prusse) et de la Toscane. Paris, Langlois et Leclercq, 1846.
163p., illus. 553.3/B89d

397 **Burat.**
De la houille. Traité théorique et pratique des combustibles minéraux (houille, anthracite, lignite, etc.). Paris, Langlois et Leclercq, 1851.
482p., illus., plates, diagrs. 553.2/B888d

398 **Burat.**
Géologie appliquée; traité du gisement et de l'exploitation des minéraux utiles. 3. éd. Paris, Langlois et Leclercq, 1855-56.
2v., illus., plates, maps, plans, diagrs. 622/B89g/1855

399 **Burnet, Thomas,** 1635-1715.
Telluris . . . Theoria Sacra: orbis nostri originem & mutationes generales, quas aut jam subiit, aut olim subiturus est, complectens. Libri duo priores de diluvio & paradiso. Londini, Gualt. Kettilby, 1681.
306p., illus., maps. uncat.

400 **Burnet.**
Telluris theoria sacra: orbis nostri originem & mutationes generales, quas aut jam subiit, aut olim subiturus est, complectens. Libri duo posteriores de conflagratione mundi, et de futuro rerum statu. Londini, Gualt. Kettilby, 1689.
262p., front. uncat.

401 **Burnet.**
Telluris, theoria sacra, originem & mutationes generales orbis nostri quas aut jam subiit, aut olim subiturus est, complectens. Accedunt Archaeologiae Philosophicae, sive doctrina antiqua de rerum originibus. Amstelaedami, Apud Joannem Wolters, 1699.
558p., illus., folded map. x213/B93t/1699

402 **Burnet.**
The theory of the earth: containing an account of the original of the earth, and of all the general changes which it hath

already undergone, or is to undergo, till the consummation of all things. The two first books concerning the Deluge, and concerning Paradise. London, printed by R. Norton, for W. Kettilby, 1684.
327p., illus. q uncat.

403 **Burnet.**
The theory of the earth: containing an account of the original of the earth and of all the general changes which it hath already undergone, or is to undergo, till the consummation of all things. The two last books, concerning the burning of the world, and concerning the new heavens and new earth. London, Printed by R. Norton, for Walter Kettilby, 1690.
224p., illus., front. uncat.

404 **Burnet.**
The theory of the earth: containing an account of the original of the earth, and of all the general changes which it hath already undergone, or is to undergo, till the consummation of all things. The two last books, concerning the burning of the world, and concerning the new heavens and new earth. London, printed by R. Norton, for W. Kettilby, 1690.
224p., front. q uncat.

405 **Burnet.**
The sacred theory of the earth, in which are set forth the wisdom of God displayed in the works of the creation, salvation, and the consummation of all things, until the destruction of the world by fire, including the blessed millennium, or, the reign of Christ with his saints upon earth. London, T. Kinnersley, 1816.
716p., illus., maps. 213/B93tE/1816

406 **Burnet.**
A review of the theory of the earth, and of its proofs: especially in reference to scripture. London, printed for Walter Kettilby, 1690.
52p. q uncat.

407 **Burnet.**
An answer to the late exceptions made by Mr. Erasmus Warren against the theory of the earth. London, printed by R. Norton, for Walter Kettilby, 1690.
86p. q uncat.

408 **Burnet.**
A short consideration of Mr. Erasmus Warren's defence of his exceptions against the theory of the earth: in a letter to

a friend. London, printed by R. Norton, for W. Kettilby, 1691.
42p. q uncat.

Bound with Burnet, Thomas, An answer to the late exceptions . . . London, 1690.

409 **Burnet.**
See KEILL, JOHN, An examination of Dr. Burnet's theory of the earth . . . 1734.

410 **Burtin, François Xavier de,** 1748-1818.
Oryctographie de Bruxelles; ou, Description des fossiles tant naturels qu'accidentels découverts jusqu'à ce jour dans les environs de cette ville. Bruxelles, Impr. de le Maire, se vend chez l'auteur, 1784.
152p., 32 col. plates. q560.9493/B95o

Bound with Rumpf, Georg Eberhard, Thesaurus imaginum piscium testaceorum, 1739. Also bound with 85p. handwritten text plus illustrations, including some in color, pasted on sheets, all with "title" extracts from D'Acosta's Elements of Conchology. 1776, dated Cr. Rec. 1777, signed Stillingfleet.

411 **Cadet, Jean Marcel,** 1751-1835.
Mémoire sur les jaspes et autres pierres précieuses de l'isle de Corse, suivi de Notes sur l'histoire naturelle, de la traduction du Critias & de divers morceaux du Timée de Platon. A Bastia, De l'Imprimerie de la veuve Batini, 1785.
239p. 553.8/C11m

412 **Caels, Theodoric Pierre,** 1739-1819.
Ratio occurrendi morbis a mineralium abusu produci solitis. Amstelodami, Apud J. van Harrevelt, 1781.
117p. 615/C11r

413 **Caldwell, Joseph,** 1773-1835, supposed author.
A tour through part of Virginia, in the summer of 1808. In a series of letters, including an account of Harper's Ferry, the Natural Bridge, the new discovery called Weir's Cave, Monticello, and the different medicinal springs, hot and cold baths, visited by the author. New-York, Printed for the author, R. C. Southwick, printer, 1809.
31p. x918.55/T644

Variously attributed to S. L. Mitchill, Joseph Caldwell (Cushing, Anonyms, 1889, p.677), and to T. Caldwell. Cf. Sabin, Bibl. amer., no. 9917.

414 **Calzolari, Francesco,** 1521-1600.
Musaeum Franç. Calceolari iun. Veronensis a Benedicto Ceruto medico incaeptum, et ab Andrea Chiocco med. physico

excellentiss. collegii luculenter descriptum & perfectum, in quo multa ad naturalem, moralemq́. philosophia spectantia, non pauca ad rem medicam pertinentia erudité proponuntur, & explicantur, non sine magna rerum exoticarum supellectile, quae artifica plané manu in aes incisae, studiosis exhibentur. [Veronae, apud Angelum Tamum, 1622.]
746p., illus., front. q uncat.

415 **Camerarius, Elias,** 1673-1734.
See WOODWARD, JOHN, Géographie physique . . . 1735.

416 **Camper, Petrus,** 1722-1789.
Lettres sur quelques objets de minéralogie, à Mr. le professeur Petrus Camper. La Haye, Impr. chez de Groot, 1789-[90].
2 pts. in 1v., 7 plates. 549/C15ℓ

Pt. 2 has title: Lettres à Mr. G. Forster.

417 **Capaccio, Giulio Cesare,** ca. 1550-1631.
Balneorum quae Neapoli, Puteolis, Baiis Pithecusis extant, virtutes, Thermarum, et Balnearum apud antiquos structurae, usus, ministeria. Ad aegrorum commoda. Neapoli, Apud Constantinum Vitalem, 1604.
88p. uncat.

Bound with his Puteolana historia . . . 1604.

418 **Capaccio.**
Puteolana Historia . . . Accessit eiusdem De Balneis libellus, Neapoli, Excudebat Constantinus Vitalis, 1604.
208p., illus. uncat.

Bound with his Balneorum quae Neapoli . . . 1604.

419 **Capper, Benjamin Pitts.**
A topographical dictionary of the United Kingdom containing every city, town, village, hamlet, parish, district, object, and place in England, Wales, Scotland, Ireland, and the small islands dependent. With forty-seven maps. London, Printed for G. B. Whittaker, 1825.
865p., 152p., 39p., maps. 914.2/C172t

420 **Cardano, Girolamo,** 1501-1576.
De subtilitate libri XXI. Ad illustriss. Principem Ferrandum Gonzagam, Mediolanensis provinciae Praefectum. Nurimbergae, apud Ioh. Petreium, 1550.
371p., illus. q uncat.

The library owns another edition, published at Basel, 1582 (q500/C17d) as well as the complete works, *Opera omnia,* 1663, 10 v. (xq875/C178/1663).

LE PARFAICT IOAILLIER, OV HISTOIRE DES PIERRERIES:

OV SONT AMPLEMENT DESCRITES leur naissance, iuste prix, moyen de les cognoistre, & se garder des contrefaites, Facultez medecinales, & proprietez curieuses.

Composé par ANSELME BOECE DE BOOT, Medecin de l'Empereur Rodolphe II.

Et de nouueau enrichi de belles Annotations, Indices & Figures.

Par ANDRE TOLL, Doct. Med, de Leide.

A LYON,

Chez IEAN-ANTOINE HVGVETAN Marchand Libraire, en ruë Merciere, à l'Enseigne de la Sphere

M. DC. XLIV.

AVEC PRIVILEGE DV ROY.

Cat. No. 254

421 **Cardano.**
Les livres de Hierosme Cardanus medecin milannois, intitulez De la Subtilité, & subtiles inventions, ensemble les causes occultes, & raisons d'icelles. Traduits de Latin en François par Richard le Blanc. Paris, Robert le Fizellier, 1584.
479p. x500/C17dFl/1584

422 **Cardano.**
De rerum varietate libri XVII. Basilea, 1557.
707p. q500/C17dv/1557

The library owns two other editions of 1580 (uncat.) and 1581 (uncat.).

423 **Cardano.**
Offenbarung der Natur und natürlicher Dingen auch mancherlei subtiler Würckingen. Trans. Heinrich Pantaleon. Basel, Heinrich Petri, 1559.
1334p., front. q500/C17dGp

Translation of De rerum varietate libri XVII.

424 **Carlisle, Nicholas,** 1771-1847.
A topographical dictionary of Ireland; exhibiting the names of the several cities, towns, parishes, and villages . . . arranged in alphabetical order. Being a continuation of the topography of the United Kingdom of Great Britain and Ireland. London, Printed for W. Miller, by Savage, 1810.
[705]p. q914.15/C19t

Bibliography: [xvii]-xix.

425 **Carosi, Jan Filip,** 1744-1801.
Essai d'une lithographie de Mloćin, écrit à Varsovie en 1777. Dresde, 1777.
96p. 560.9438/C22e

Dedication signed: Jean Philippe de Carosi.

426 **Carosi.**
Sur la génération du silex et du quarz en partie. Observations faites en Pologne. Cracovie, I. Grebel, 1783.
94p., illus. 549/C222s

427 **Cartheuser, Friedrich August.**
Mineralogische Abhandlungen. Giessen, J. P. Krieger, 1771-73.
2v. in 1. 549/C24m

428 **Carver, Jonathan,** 1710-1780.
Travels through the interior parts of North-America in the years 1766, 1767, and 1768. Illustrated with copper plates.

London, Printed for the author, sold by J. Walter [etc.], 1778.
10, [17]-543p., plates. x917/C25t

Carver's Travels as printed was probably the work of Dr. John Coakley Lettsom. An article on Carver and his book may be found in American historical review, January 1906, v. 11, p.278-302.

429 **Carver.**
Travels through the interior parts of North America, in the years 1766, 1767, and 1768. Illustrated with copper plates. The 2d ed. London, Printed for the author, by W. Richardson, 1779.
12, [17]-543p., plates, 2 fold. maps (incl. front.). x917/C25t/1779

430 **Carver.**
Travels through the interior parts of North America, in the years 1766, 1767, and 1768. Illustrated with copper plates, coloured. The 3d ed. To which is added, some account of the author, and a copious index. London, C. Dilly [etc.], 1781.
22, [22], [17]-543, [21]p., front. (port.), illus., plates (part. col.), fold. maps. x917/C25t/1781

431 **Carver.**
Three years' travels through the interior parts of North-America for more than five thousand miles, containing an account of the Great Lakes and all the lakes, islands, and rivers . . . of the north west regions of that vast continent . . . with a description of the birds, beasts, reptiles, insects, and fishes peculiar to the country. Together with a concise history of the genius, manners, and customs of the Indians inhabiting the lands that lie adjacent to the heads and to the westward of the great river Mississippi; and an appendix describing the uncultivated parts of America that are the most proper for forming settlements. Philadelphia, J. Crukshank, 1784.
217p. x917/C25t/1784

First published in 1778 under title: Travels through the interior parts of North America in the years 1766, 1767, and 1768.

432 **Carver.**
Three Years' Travels Through the Interior Parts of North-America. Philadelphia, Key & Simpson, 1796.
360p. uncat.

433 **Carver.**
Three years' travels throughout the interior parts of North America . . . 4th American, from the last London ed. Charles-

town, Printed by Samuel Etheridge, for West and Greenleaf, no. 56, Cornhill, Boston, 1802.
312p. x917/C25t/1802

434 **Carver.**
Three years' travels throughout the interior parts of North America, . . . Walpole, N.H., Published by Isaiah Thomas & co., 1813.
[17]-280p. x917/C25t/1813

435 **Carver.**
Reisen durch die innern Gegenden von Nord-Amerika in den Jahren 1766, 1767 und 1768, mit einer Landkarte. Aus dem Englischen. Hamburg, C. E. Bohn, 1780.
456p., fold. map. x917/C25tG

Neue Sammlung von Reisebeschreibungen. 1. th.

436 **Carver.**
Voyage dans les parties intérieures de l'Amérique septentrionale, pendant les années 1766, 1767 & 1768. Ouvrage traduit sur la troisième édition angloise . . . avec des remarques & quelques additions du traducteur. Yverdon, 1784.
436p. 917/C26t.Fm

Translation of Travels through the interior parts of North-America.

437 **Casey, William R.,** civil engineer, jt. author.
See JOHNSON, EDWIN F., Reports on the Bear Mountain Railroad . . . 1845.

438 **Casselberry, Isaac.**
A description of certain fossil bones found near Evansville, Ia. [Indiana]: with speculations concerning the animal, of which they are remains. Printed at the Courier office, Evansville, 1845.
8p., fold. plate. uncat.

439 **Castelnau, Francis,** 1812-1880.
Essai sur the système silurien de l'Amérique septentrionale. Paris, Chez P. Bertrand, [etc., etc.] 1843.
56p., 27 plates (incl. map). 560.97/C25e

Extrait des Comptes rendus des séances de l'Académie des sciences, séance du 6 mars 1843.

440 **Castillon, Jean Louis,** ca. 1720-1793.
Les dernières révolutions du globe, ou, Conjectures physiques sur les causes de la dégradation actuelle des tremblemens de

terre, & sur la vraisemblance de leur cessatio[n] prochaine. Bouillon, Impr. de la Société typographique, 1771.

269p. 551.2/C27d

Running title: Conjectures physiques.

441 **Catalogo mineralogico Werneriano.**

See ITALY. DIREZIONE GENERALE DI PUBBLICA INSTRUZIONE. Catalogo di una collezione di minerali disposta seconda il sistema del celebre Werner . . . 1813.

442 **Catesby, Mark,** 1683-1749.

The natural history of Carolina, Florida and the Bahama Islands; containing the figures of birds, beasts, fishes, serpents, insects, and plants; particularly the forest-trees, shrubs, and other plants, not hitherto described, or very incorrectly figured by authors. Together with their descriptions in English and French. To which, are added observations on the air, soil, and waters; with remarks upon agriculture, grain, pulse, roots, &c. To the whole, is prefixed a new and correct map of the countries treated of. London, Printed at the expence of the author, 1731-43.

2v., 100, 100, 20 col. plates, col. map. xf570.975/C28n

Vol.2, p.59 numbered 56.

Vol.2, plate 61, unnumbered; plate 62 numbered 30, plate 80 numbered 62.

App., plate 3, 7, 9 unnumbered.

"The figures were etched by himself from his own paintings, and the colored copies executed under his inspection." — Appleton's cyclopedia of Amer. biog. v.1, p.557.

Vol. 1 contains supplement: An account of Carolina and the Bahama Islands. Relation de la Caroline, et des Isles de Bahama. xliv p.; v.2: Appendix. 1748? 20p., 20 col. plates.

443 **Catesby.**

The natural history of Carolina, Florida, and The Bahama Islands; containing the figures of birds, beasts, fishes, serpents, insects and plants; particularly the forest-trees, shrubs and other plants, not hitherto described, or very incorrectly figured by authors. Together with their descriptions in English and French. To which are added, observations on the air, soil, and waters; with remarks upon agriculture, grain, pulse, roots &c. To the whole is prefixed a new and correct map of the countries treated of. Rev. by Mr. Edwards . . . Histoire naturelle de la Caroline, de la Floride, & des isles de Bahama . . . London, Printed for C. Marsh [etc.], 1754.

2v., 220 col. plates, fold. map. xf570.975/C28n/1754

English and French in parallel columns.

444 **Catesby.**

The natural history of Carolina, Florida, and the Bahama Islands; containing the figures of birds, beasts, fishes, serpents, insects and plants; particularly, those not hitherto described, or incorrectly figured by former authors, with their descriptions in English and French. To which is prefixed, a new and correct map of the countries, with observations on their natural state, inhabitants, and productions. Revised by Mr. Edwards. To the whole is now added a Linnaean index of the animals and plants . . . Histoire naturelle de la Caroline, de Floride, et des isles de Bahama; contenant les dessins des oiseaux, des quadrupèdes, des poissons, des serpens, des insectes, et des plantes, qui se trouvent dans ces pays-là; et en particulier, de ceux qui n'ont point été décrits jusqu'à présent par les auteurs, ou peu exactement dessinés. Avec leurs descriptions en françois & en anglois. On trouve au commencement une carte de ces pays, avec des remarques sur leur état naturel, leurs habitans, & leurs productions. Revue par Monsieur Edwards . . . On y a ajouté une table selon le système de Linnaeus . . . London, Printed for B. White, 1771.

2v., 220 col. plates, fold. map. xf570.975/C28n/1771

English and French in parallel columns.
t.p. of v.1 repaired.
Imprint missing.

445 **Catullo, Tommaso Antonio,** 1782-1869.

Osservazioni sopra i monti che circoscrivono il distretto di Belluno . . . divise in due parti. Verona, Tipografia Mainardi, 1818.

147p. 551.43/C29o

"Lettera del professore T. A. Catullo al signore Giovanni de Brignoli . . ."; p.[129]-147.
Bibliographical footnotes.

446 **Catullo.**

Saggio di zoologia fossile. Padova, Dalla tipografia del Seminario, 1827.

348p., 8 plates. q560/C29s

447 **Catullo.**

Elementi de mineralogia applicati alla medicina e alla farmacia, di T. Antonio Catullo. Padova, Coi tipi della Minerva, 1833.

2v. in 1, [512p.], illus., tables. 549.1/C29e

LETTRES PHILOSOPHIQUES

SUR LA FORMATION
DES SELS ET DES CRYSTAUX.
ET SUR
La Génération & le Mechanisme Organique
DES PLANTES ET DES ANIMAUX;
A L'OCCASION
DE LA PIERRE BELEMNITE ET DE LA PIERRE LENTICULAIRE.
AVEC UN MEMOIRE
SUR LA THEORIE DE LA TERRE.

PAR Mr. BOURGUET.

A AMSTERDAM,
Chez FRANCOIS L'HONORE.
MDCCXXIX.

Cat. No. 281

448 **Catullo.**
Sopra le acque termali del territorio padovano; memoria geologica letta nella I. R. Accademia de scienze, lettere ed arti di Padova il dí 15 marzo 1836 . . . Padova, Tipografia della Minerva, 1836.
21p. q553.7/C29s

"Estratta dal volume IV. dei Nuovi saggi dell'Accademia suddetta."
Bibliographical footnotes.

449 **Catullo.**
Trattato sopra la costituzione geognostico-fisica dei terreni alluviali o postdiluviani delle province venete. Ed. notevolmente accresciuta. Padova, Presso A. Zambeccari, 1844.
464p. 554.531/C29t/1844

"Scritti publicati da Tomaso Antonio Catullo": p. [457]-464.

450 **Catullo.**
Trattato sopra la costituzione geognostico-fisica dei terreni alluviali o postdiluviani delle province venete. Padova, Tip. Cartallier e Sicca, 1858.
512p. 554.531/C29t/1858

451 **Caumont, Arcisse de,** 1802-1873.
Mémoire géologique sur quelques terrains de la Normandie occidentale . . . Paris, Treuttel et Wurtz [etc.], 1825.
15 (i.e.151)p., 5 fold. plates. 554.42/C31m

Extrait des Mémoires de la Société linnéenne du Calvados, 1825.

452 **Cautley, Proby Thomas,** 1802-1871, jt. author.
See FALCONER, HUGH, Fauna antiqua Sivalensis . . . 1846-49.

453 **Celsius, Anders,** 1701-1744, praeses.
De luna non habitabili. Upsaliae, 1740.
24p. 523.3/C33d

Diss.-Uppsala (Ericus Engman, respondent).

454 **Cerini, Giuseppe.**
Della pressione idrostatica a cui sono soggette le acque sotterranee anche in località non dominate da' monti superiori; ossia, Cause idrostatico-fisiche della forza saliente nei pozzi trivellati detti artesiani. Milano, Presso L. Nervetti, 1831.
99p. 551.49/C335d

455 **Cermelli, Pier Maria.**
Carte corografiche, e memorie riguardanti le pietre, miniere, e i fossili per servire alla storia naturale delle provincie del

Patrimonio, Sabina, Lazio, Marittima, Campagna, e dell'Agro Romano abbozzate e raccolte dal prefetto degli studii del Reale collegio Fernandiano alla Nunziatella. Napoli, V. Flauto, 1782.

46p., fold. maps. q554.5/C33c

456 **Cesalpino, Andrea,** 1519-1603.

De metallicis libri tres . . . Noribergae, curante Conrado Agricola, 1602.

222p. uncat.

457 **Cesi, Bernardo,** 1581-1630.

Mineralogia, sive naturalis philosophiae thesauri, in quibus metallicae concretionis medicatorúmque fossilium miracula, terrarum pretium, colorum & pigmentorum apparatus, concretorum succorum virtus, lapidum atque gemmarum dignitas continentur. Hos publici iuris fecit R. P. Bernardus Caesius . . . Lugduni, I. &. P. Prost, 1636.

626p. xq549/C33m

In double columns.

458 **Ceva, Giovanni,** 1647 or 1648-1734.

Opus hydrostaticum. Mantuae, Ex typographía S. Benedicti, apud A. Pazzoni impressorem, [1728].

235p., illus. 551.483/C33o

459 **Chambers, Robert,** 1802-1871.

Vestiges of the natural history of creation. 5th ed. London, J. Churchill, 1846.

423p. 550/C352v5

First edition, London, 1844; an American edition was published New York, Wiley and Putnam, 1845.

460 **Chambers.**

Vestiges of the natural history of creation. With a sequel. New York, Harper & Brothers, [1847?].

303p. 550/C352v/1847

461 **Chambers.**

Vestiges of the natural history of creation. 10th ed., with extensive additions and emendations. London, John Churchill, 1853.

325p. 550/C352v/1853

462 **Chambers.**

Vestiges of the natural history of creation, with a sequel. New York, Harper & Brothers, 1862.

303p. 550/C352v/1862

463 **Chambers.**
Vestiges of the natural history of creation. With an introd. by Henry Morley. London, New York, G. Routledge, 1887.
286p. 550/C352v/1887

Morley's universal library.

464 **Chambers.**
Sporen van de natuurlijke geschiedenis der schepping, of, Schepping en voortgaande ontwikkeling van planten en dieren, onder den invloed en het beheer der natuurwetten. Uit het engelsch vertaald door dr. J. H. van den Broek . . . Met een voorwoord van prof. G. J. Mulder. . . Utrecht, J. G. Broese, 1849.
2v. in 1. 550/C352vDb

With this are bound, as issued, the author's Vervolg der Sporen van de natuurlijke geschiedenis der schepping . . . uit het engelsch vertaald door dr. J. H. van den Broek. Utrecht, 1849; and Mason, Thomas Monck. Schepping door de regtstreeksche tusschenkomst van God, in tegenstelling van eene schepping door natuurwetten. Eene wederlegging van de Sporen van de natuurlijke geschiedenis der schepping. Utrecht, 1849.

465 **Chambers.**
Atlas van de belangrijkste plant en diervormen voorkomende in de Sporen van de natuurlijke geschiedenis der schepping, uit de beste werken bijeenverzameld door dr. J. H. van den Broek . . . Utrecht, J. G. Broese, 1850.
v.1, 51 plates. 550/C352vDb/Atlas

466 **Chambers.**
Natürliche Geschichte der Schöpfung des Weltalls, der Erde und der auf ihr befindlichen Organismen, begründet auf die durch die Wissenschaft errungenen Thatsachen. Aus dem Englischen nach der 6. Aufl. von Carl Vogt . . . Braunschweig, F. Vieweg und Sohn, 1858.
330p., illus. 550/C352vGv/1858

Zweite verb. Aufl.
Translation of Vestiges of the natural history of creation.

467 **Chambers.**
Explanations; a sequel to "Vestiges of the natural history of creation." By the author of that work. London, J. Churchill, 1845.
198p. 550/C352e/1845

468 **Chambers.**
Explanations: a sequel to "Vestiges of the natural history of

creation." By the author of that work. 2d ed. London, J. Churchill, 1846.
205p. 550/C352e/1846

With his Vestiges of the natural history of creation. 5th ed. London, 1846.

469 **Chambers.**
Ancient sea-margins, as memorials of changes in the relative level of sea and land. Edinburgh, W. & R. Chambers; [etc., etc], 1848.
337p., front., illus., fold. map. 551.13/C35a

470 **Chambers.**
Tracings of Iceland and the Faröe Islands. London and Edinburgh, W. & R. Chambers, 1856.
85p., maps, front. uncat.

471 **Chappe d'Auteroche, Jean.**
A voyage to California, to observe the transit of Venus . . . with an historical description of the author's route through Mexico, and the natural history of that province. Also a voyage to Newfoundland and Sallee, to make experiments on Mr. Le Roy's time keepers by Monsieur de Cassini. London, printed for Edward and Charles Dilly, 1778.
215p., fold. map. uncat.

The chapter on the natural history of Mexico is actually an extract from a letter from Mexico addressed to the Royal Academy of Sciences at Paris by Don Joseph Anthony de Alzate y Ramyrez. He mentions mines, basalt, phonolite, and some petrifications.

472 **Charleton, Walter,** 1619-1707.
Physiologia, Epicuro-Gassendo-Charltoniana: or a fabrick of science natural upon the hypothesis of atoms. London, Thos. Newcomb, 1654.
475p. xq500/C38p

473 **Charpentier, Jean G. F. de,** 1786-1855.
Essai sur la constitution géognostique des Pyrénées . . . Ouvrage couronné par l'Institut Royal de France. Paris, F.-G. Levrault, 1823.
633p., fold. map. 554.489/C38e

474 **Charpentier.**
Essai sur les glaciers et sur le terrain erratique du bassin du Rhône. Lausanne, M. Ducloux, 1841.
363p., illus., plates, map. 551.31/C38e

475 **Charpentier, Johann Friedrich Wilhelm Toussaint von,** 1728-1805.
Mineralogische Geographie der Chursächsischen Lände. Leipzig, S. L. Crusius, 1778.
432p., front., 7 fold. plates (incl. col. map). 554.321/C38m

476 **Charpentier.**
Beobachtungen über die Lagerstätte der Erze, hauptsächlich aus den Sächsischen Gebirgen. Leipzig, Georg Joachim Göschen, 1799.
206p., 7 plates. uncat.

477 **Chastellux, François Jean de,** 1734-1788.
Voyage de Mr. le chevalier de Chastellux en Amérique. [Cassel], 1785.
228p. x917/C38v/1785

A surreptitious edition, being a reprint of the extracts from his ms. journal first published in a Gotha periodical. cf. the Paris edition of 1786, p. 7.

478 **Chastellux.**
Voyages de M. le marquis de Chastellux dans l'Amérique Septentrionale dans les années 1780, 1781 & 1782 . . . Paris, Prault, 1786.
2v. in 1, fold. maps. 917/C38v

479 **Chastellux.**
Voyages de m. le marquis de Chastellux dans l'Amérique Septentrionale, dans les années 1780, 1781 & 1782 . . . 2.éd. Paris, Prault, 1788-91.
2v., 3 fold. plates, 2 fold. maps. 917/C38v/1788

480 **Chastellux.**
Remarks on the travels . . . in North America. London, printed for G. & T. Wilkie, 1787.
80p. x917/C38vYs

By [Simcoe, John Graves, 1752-1806].

481 **Chaubard, Louis Anastase,** 1785-1854.
Éléments de géologie mis à la portée de tout le monde, et offrant la concordance des faits géologiques avec les faits historiques, tels qu'ils se trouvent dans la Bible, les traditions égyptiennes et les fables de la Grèce. 2.éd. Paris, Chez l'auteur, 1838.
516p., 4 plates, tables. 550.1/C39e/1838

RELIQUIÆ DILUVIANÆ;

OR,

OBSERVATIONS

ON THE

ORGANIC REMAINS

CONTAINED IN

CAVES, FISSURES, AND DILUVIAL GRAVEL,

AND ON

OTHER GEOLOGICAL PHENOMENA,

ATTESTING THE ACTION OF AN

UNIVERSAL DELUGE.

BY THE REV. WILLIAM BUCKLAND, B. D. F. R. S. F. L. S.

MEMBER OF THE GEOLOGICAL SOCIETY OF LONDON; OF THE IMPERIAL SOCIETIES OF MINERALOGY AND NATURAL HISTORY AT PETERSBURG AND MOSCOW; AND OF THE NATURAL HISTORY SOCIETY AT HALLE; HONORARY MEMBER OF THE AMERICAN GEOLOGICAL SOCIETY; CORRESPONDENT OF THE MUSEUM OF NATURAL HISTORY OF FRANCE; FELLOW OF C. C. C. AND PROFESSOR OF MINERALOGY AND GEOLOGY IN THE UNIVERSITY OF OXFORD.

LONDON:

JOHN MURRAY, ALBEMARLE-STREET.

1823.

Cat. No. 369

482 **Chikhachev, Petr Aleksandrovich,** 1808-1890.
Coup d'oeil sur la constitution géologique des provinces méridionales du Royaume de Naples, suivi de quelques notions sur Nice et ses environs, par Pierre de Tchihatchoff. Berlin, S. Schropp, 1842.
248p., 2 fold. col. maps. 554.57/C43c

483 **Childrey, Joshua,** 1623-1670.
Britannia Baconia: or, The natural rarities of England, Scotland, & Wales. According as they are to be found in every shire. Historically related, according to the precepts of the Lord Bacon; Methodically digested; and the causes of many of them philosophically attempted. With observations upon them, and deductions from them, whereby divers Secrets in Nature are discovered, and some things hitherto reckoned Prodigies, are fain to confess the cause whence they proceed. Usefull for all ingenious men of what Profession or quality soever. London, printed for the author, 1661.
185p. 554.2/C43b

484 **Christ, Johann Friedrich,** 1700-1756.
See HEBENSTREIT, JOHANN ERNST, Museum Richterianum . . . 1743.

485 **Churchman, John,** 1753-1805.
The magnetic atlas, or Variation charts of the whole terraqueous globe; comprising a system of the variation and dip of the needle, by which the observations being truly made, the longitude may be ascertained. London, Printed for the author, by Darton and Harvey, 1794.
xxix, [31]-76p., 3 fold. charts. 538.7/C47m

486 **Clarke, Edward Daniel,** 1769-1822.
See OTTER, WILLIAM, The life and remains of Edward Daniel Clarke, professor of mineralogy in the University of Cambridge . . . 1825.

487 **Cleaveland, Parker,** 1780-1858.
Elementary treatise on mineralogy and geology, being an introduction to the study of these sciences, and designed for the use of pupils,—for persons attending lectures on these subjects,—and as a companion for travellers in the United States of America. Illustrated by six plates. Boston, Cummings and Hilliard, 1816.
668p., fold. map, diagrs. x549/C58e

488 **Cleaveland.**
An elementary treatise on mineralogy and geology, designed for the use of pupils, — for persons, attending lectures on those subjects, — and as a companion for travellers in the United States of America. 2d. ed. . . . Boston, Cummings and Hilliard, 1822.
2v., 6 plates, illus., front. 549/C58e2

Paged continuously.

489 **Cleaveland.**
Notebooks on mineralogy, geology, natural history, etc. 1822-57.
5v. MS/500/C58n

Holograph.

American scientist, professor of mathematics and natural history at Bowdoin College, 1805-58; author of the first American work on mineralogy and geology.

490 **Clemens, Ioan. Fridericus Rudolphus.**
Pericula nonnvilla microscopico chemica circa sal seri. Erlangae, Waltheri, 1766.
[30p.], plate. 549/F59

Bound with Flade, Christian Gottlob, De re Metallica . . . [1791].

491 **Clevé, Étienne de.**
Paradoxes, ou traittez philosophiques des pierres et pierreries, contre l'opinion vulgaire. Ausquels sont demonstrez la matiere, la cause efficiente externe, la semence, la generation, la definition, & la nutrition d'icelles. Ensemble la generation de tous les mixtes, sçavoir est des animaux, vegetaux, & mineraux, ou fossiles. Paris, chez la veufue Pierre Chevalier, 1635.
492p. uncat.

492 **Clinton, de Witt,** 1769-1828.
An introductory discourse, delivered before the Literary and Philosophical Society of New-York, on the fourth of May, 1814. New-York, Published by David Longworth, At the Shakespeare-Gallery. N. Van Riper, print., 1815.
148p. x500/C61i

"Notes and illustrations": p.[39]-148.

Includes history of geology in America and mention of Maclure, Mitchill and Kalm.

493 **Cole, Henry,** fl. 1823-1846.
Popular geology subversive of divine revelation! A letter to the Rev. Adam Sedgwick . . . being a Scriptural refutation of

the geological positions and doctrines promulgated in his lately published commencement sermon, preached in the University of Cambridge, 1832. London, Hatchard and son; [etc.], 1834.
136p. 213/C67p

494 **Collette, Cárlos.**
Reconocimiento geológico del señorío de Vizcaya, hecho de órden de su Disputación General. Bilbao, Libreria de Delmas, 1848.
159p., fold. col. map. 554.66/C68r

495 **Collin, Alexandre,** 1808-1890.
Recherches expérimentales sur les glissements spontanés des terrains argileux, accompagnées de considérations sur quelques principes de la mécanique terrestre. Paris, Carilian-Goeury et V. Dalmont, 1846.
2v. in 1. 21 plates. 624.1/C69r

Bibliographical footnotes.
Contents. — [1] Texte. — [2] Atlas.

496 **Collini, Cosimo Alessandro,** 1727-1806.
Journal d'un voyage, qui contient differentes observations mineralogiques; particulieremfnt[sic] sur les agates, et le basalte. Avec un detail sur la maniere de travailler les Agates. Mannheim, C. F. Schwan, 1776.
384p., 15 plates. x552/C69j

497 **Collini.**
Considérations sur les montagnes volcaniques. Mémoire lu dans une séance de l'Academie électorale des sciences et belles lettres de Mannheim le v. novembre MDCCLXXXI. À l'occasion de la St. Charles, fête de S. A. S. E. Palatine et de Bavière Charles Théodore, fondateur et protecteur de cette académie, . . . avec une table et une carte qui concernent les montagnes. Mannheim, C. F. Schwan et M. Fontaine; [etc., etc.], 1781.
64p., illus., table. 551.21/C69c

498 **Collini.**
Remarques sur la pierre élastique du Brésil et notices sur les marbres flexibles de Monseigneur le Prince Borghese. Mannheim, Schwan et Goetz, 1805.
23p. 552/C69r

499 **Collot, Georges Henri Victor,** 1752?-1805.
Voyage dans l'Amérique Septentrionale, ou Description des pays arrosés par le Mississippi, l'Ohio, le Missouri et autres

rivières affluentes; observations exactes sur le cours et les sondes de ces rivières, sur les villes, villages, hameaux et fermes . . . Avec un atlas de 36 cartes, plans, vues et figures. Paris, A. Bertrand, 1826.

2v., and atlas of 36 plates (incl. 11 fold. maps, 15 plans). x917.7/C69v

Atlas is in English, with added t.-p. in French.

500 **Colonna, Fabio,** 1567-1650.

. . . Purpura. Hoc est de purpura ab animali testaceo fusa de hoc ipso Animali alijsa rarioribus testaceis quibusdam. Romae, Apud Jacobum Mascardum, 1616.

42p., illus. uncat.

Issued with the Purpura are his Minus cognitarum Plantarum, first and second parts. The first part is entitled Minus cognitarum rariorumque . . . Roma, Apud J. Mascardum, 1616. 340p., illus., and includes the Aquatilibum, et terrestrium aliquot animalium . . . Roma, Apud J. Mascardum, 1616. 73p. The second part is entitled Minus cognitarum stirpium . . . Roma, Apud J. Mascardum, 1616. 99p., illus.

501 **Colonna,** De Glossopetris Dissertatio, 1616.

See SCILLA, AGOSTINO, De corporibus marinis . . . 1752 and 1759.

502 **Comstock, John Lee,** 1789-1858.

An introduction to mineralogy; adapted to the use of schools, and private students. Illustrated by nearly two hundred wood cuts. 2d ed., improved. Hartford, B. B. Barber, 1832.

143 (i.e.343)p., illus. x549/C73i/1832

Page 343 numbered 143.

503 **Comstock.**

An introduction to mineralogy: adapted to the use of schools and private students. Illustrated by nearly two hundred woodcuts. 3d ed., improved. New York, Robinson, Pratt & co., 1841.

369p., diagrs. 549/C73i/1841

504 **Comstock.**

Outlines of geology: intended as a popular treatise on the most interesting parts of the science. Together with an examination of the question, whether the days of creation were indefinite periods. Designed for the use of schools and general readers. 3d ed. New York, Robinson, Pratt, & co., 1839.

384p., illus. 550/C73o/1839

505 **Connecticut. Geological Survey.**
Report on the geology of the state of Connecticut. By James G. Percival. Published under the direction of the commissioners appointed by the legislature. New Haven, Osborn & Baldwin, printers, 1842.
495p., front. (fold. map). x557/C761r

506 **Conybeare, William Daniel,** 1787-1857.
Outlines of the geology of England and Wales, with an introductory compendium of the general principles of that science, and comparative views of the structure of foreign countries, by the Rev. W. D. Conybeare and William Phillips. Part I. London, W. Phillips, 1822.
470p., illus. (fold. map), col. plates, fold. tables, diagrs. x554.2/C76o

507 **Conybeare.**
Ten plates comprising a plan, sections, and views, representing the changes produced on the coast of East Devon, between Axmouth and Lyme Regis by the subsidence of the land and elevation of the bottom of the sea, on the 26th December, 1839 and 3rd of February, 1840. From drawings by W. Dawson, W. D. Conybeare and Mrs. Buckland; with a geological memoir, and sections, descriptive of these and similar phaenomena by W. D. Conybeare. The whole revised by Professor Buckland. London, J. Murray, 1840.
14p., illus., 10 col. plates. 551.22/C76t

508 **Cooper, Thomas,** 1759-1840.
On the connection between geology and the Pentateuch, in a letter to Professor Silliman, from Thomas Cooper, M. D., to which is added an appendix. Boston, Abner Kneeland; London, Watson, 1845.
72p. uncat.

509 **Cordier, Pierre Louis Antoine,** 1777-1861.
An essay on the temperature of the interior of the earth . . . Read to the Academy of sciences at their sessions of June 4th, June 9th and July 23d, 1827. Tr. from the French by the Junior class in Amherst College. Amherst, Printed by J. S. & C. Adams, 1828.
94p. 551.12/C81eE

510 **Cornalia, Emilio,** 1824-1882.
Notizie geo-mineralogiche sopra alcune valli meridionali del Tirolo . . . Milano, Tipografia di V. Gugliemini, 1848.

TH. BURNETII TELLURIS THEORIA SACRA,

Originem & Mutationes Generales

ORBIS NOSTRI

Quas aut jam ſubiit, aut olim ſubiturus eſt, complectens.

Accedunt

ARCHÆOLOGIÆ PHILOSOPHICÆ,

Sive

Doctrina Antiqua de Rerum Originibus.

Editio ultima, priori longe correctior, & copioſo rerum indice auctior.

AMSTELÆDAMI,

Apud JOANNEM WOLTERS, M.D.C. XCIX.

Cat. No. 401

49p., front. (fold. map), 2 fold. plates, tables. q549/C814n

Tesi de laurea — Pavia.

511 **Corniani Degli Algarotti, Marco Antonio,** 1768-1845.
Dello Stabilimento delle miniere a relative fabbriche nel distretto di Agordo. Trattato storico, mineralogico, disciplinare. Venezia, F. Andreola, 1823.
477p. 549/C81d

Bibliographical footnotes.

512 **Corrozet, Gilles,** 1510-1568.
Le catalogue des antiques erections des Villes & cites, Fleuves, & Fontaines, assisses es troys Gaules, cestassavoir Celticque Belgicque & Aquitaine, contenant deulx livres. Le premier faict & compose par Gilles Corrozet Parisien. Le second par Claude Champier Lyonnois, Avec ung petit traicte des Fleuves & Fontaines admirables, estans esdictes Gaules. Lyon, Chez Francoys Iuste, [ca. 1535].
68p., illus. uncat.

513 **Cortesi, Giuseppe.**
Saggi geologici degli stati di Parma e Piacenza . . . Piacenza, Torchi del Maino, 1819.
166p., 7 fold. plates (incl. map). q554.5/C81s

Bibliographical footnotes.

514 **Cossali, Pietro,** 1748-1815.
Sull' opinione delle pioggie de' sassi dai vulcani lunari; disquisizione matematica. Modena, Presso la Società tip., 1806.
59p. q523.51/C82s

Inserita nel tomo XIII della Società italiana delle scienze.

515 **Costa, Emanuel Mendes da.**
See CRONSTEDT, AXEL FREDRIK, An essay towards a system of mineralogy . . . 1772.

516 **Cotta, Bernhard von,** 1808-1879.
Anleitung zum Studium der Geognosie und Geologie, besonders für deutsche Forstwirthe, Landwirthe und Techniker. Dresden, Arnold, 1842.
584p., diagrs., tables. 550/C82a

517 **Cotta.**
Grundriss der Geognosie und Geologie, als zweite Auflage der Anleitung zum Studium der Geognosie und Geologie. Dresden, Arnold, 1846.
428p., illus., diagrs., tables. 550/C82a/1846

518 **Cotta.**
Leitfaden und Vademecum der Geognosie als 3. Aufl. des Grundrisses der Geognosie und Geologie. Dresden und Leipzig, Arnoldische Buchhandlung, 1849.
291p., col. front., 3 fold. tables, diagrs. uncat.

519 **Cotta.**
Geologische Briefe aus den Alpen. Leipzig, T. O. Weigel, 1850.
328p., front., col. plates, diagrs. 554.94/C82g

Literatur: p. [vii]-viii.

520 **Cotta.**
Die Alpen. 2.Ausg. Leipzig, T. O. Weigel, 1851.
328p., illus., 5 col. plates, front. 554.94/C82g/1851

First ed. published 1850 under title: Geologische Briefe aus den Alpen.
Bibliography: p. [vii]-viii.

521 **Cotta.**
Gangstudien; oder, Beitrage zur Kenntniss der Erzgänge. Hrsg. von Bernhard Cotta . . . Freiberg, J. G. Engelhardt, 1850-62.
4v., plates, maps, tables, diagrs. 553.1/C82g

Bd. 3-4, Hft. 1, hrsg. von Cotta and Herm. Müller.
No more published after Bd. 4, Hft. 1.
Bd. 1 lacking.

522 **Cotta.**
Die Lehre von den Erzlagerstätten. 2. verb. und verm. Aufl. Freiberg, J. G. Engelhardt, 1859-61.
2v. in 1, illus. 553.1/C82ℓ/1859

523 **Cotta.**
Geologische Bilder. Leipzig, J. J. Weber, 1852.
242p., illus., port. 550/C82g

A collection of articles originally published in the Illustrirte Zeitung, Leipzig.

524 **Cotta.**
Geologische Bilder. 3.verb. und verm. Aufl. Leipzig, J. J. Weber, 1856.
262p., illus., port. 550/C82g/1856

525 **Cotta.**
Deutschlands Boden, sein geologischer Bau und dessen Einwirkungen auf das Leben der Menschen. Leipzig, F. A. Brockhaus, 1854.
2v., fold. col. map, diagrs. (1 fold.), tables. 554.3/C82d

Includes bibliographies.
Contents. — 1. Abth. [Text] — 2. Abth. Beilagen.

526 **Cotta.**
Die Gesteinslehre. Freiberg, J. G. Engelhardt, 1855.
255p. 552/C82g

527 **Cotta.**
Die Gesteinslehre. 2. umgearb. Aufl. Freiberg, J. G. Engelhardt, 1862.
333p. 552/C82g/1862

528 **Cotta.**
Rocks classified and described. A treatise on lithology. An English ed. by Philip Henry Lawrence. With English, German, and French synonyms. Rev. by the author. London, Longmans, Green, and co., 1866.
425p. 552/C82

529 **Cotta.**
Rocks classified and described. A treatise on lithology. An English edition, by Philip Henry Lawrence; with English, German, and French synonyms. Rev. by the author. New ed. London, Longmans, Green & Co., 1878.
420p. 552/C82gEℓ2

530 **Cotta.**
Die Lehre von den Flötzformationen . . . Freiberg, J. G. Engelhardt, 1856.
285p., front. (col. plate), diagrs. 552.5/C82ℓ

531 **Cotta.**
Geologische Fragen. Freiberg, J. G. Engelhardt, 1858.
344p., illus., diagrs., tables. 550/C82geo

532 **Cotta.**
Ungarische und siebenbürgische Bergorte. Ansichten nach Federzeichnungen von Heinrich von Jossa; beschrieben und geologisch erlaütert von Bernhard von Cotta. Leipzig, J. J. Weber, 1862.
48p., 14 plates. 914.39/C82u

533 **Cotta.**
Der Altai. Sein geologischer Bau und seine Erzlagerstätten. Leipzig, J. J. Weber, 1871.
325p., illus., 8 plates (partly fold., partly col., incl. maps), diagrs.

"Petrographische Bemerkungen über Gesteine des Altai . . . von Alfred Stelzner": p. [110]-166.
"Über fossile Pflanzen aus der Steinkohlenformation am Altai, von Dr. H. B. Geinitz": p. [167]-179.

"Bemerkungen über Klima und Vegetation im Altai, von Th. Teplouchow": p. [267]-297.

534 **Cotta.**
Die Geologie der Gegenwart dargestellt und beleuchtet. 4., umgearb. Aufl. mit dem Portrait des Verfassers in Stahlstich. Leipzig, J. J. Weber, 1874.
450p., front. (port.), illus., fold. table. 550/C82ge/1874

535 **Cotta.**
Geologisches Repertorium. Mit einem vollständigen Index. Leipzig, J. J. Weber, 1877.
400p. 016.55/C82g

Beitrage zur Geschichte der Geologie. 1.Abth.

536 **Coxe, William,** 1747-1828.
Travels in Switzerland, and in the country of the Grisons: in a series of letters to William Melmoth, esq., from William Coxe . . . With an historical sketch and notes on the late revolution. The 4th ed. . . . London, T. Cadell, Jun. and W. Davies, 1801.
3v., plates, fold. maps. 914.94/C83t/1801

537 **Coxe.**
Lettres à M. W. Melmoth sur l'état politique, civil et naturel de la Suisse. Traduites de l'anglois et augm. des observations faites dans le même pays par le traducteur. Paris, Belin, 1781.
2v. 914.94/C83tFr

538 **Coxe.**
Lettres à M. W. Melmoth sur l'état politique, civil et naturel de la Suisse. Traduites de l'anglois et augm. des observations faites dans le même pays par le traducteur. Paris, Belin, 1787.
2v., fold. col. map. 914.94/C83tFr/1787

539 **Coxe.**
See RAYMOND DE CARBONNIÈRES, LOUIS FRANÇOIS ELIZABETH, Observations faites dans les Pyrénées . . . 1789.

540 **Cozzens, Issachar,** 1780-1865.
A geological history of Manhattan or New York island together with a map of the island, and a suite of sections, tables and columns, for the study of geology, particularly adapted for the American student. New York, W. E. Dean, 1843.
114p., tables, 9 col. plates (incl. fold. map). 557/N48co

541 **Cranz, David,** 1723-1777.
Historie von Grönland, enthaltend die Beschreibung des Landes und der Einwohner &c., insbesondere die Geschichte

der dortigen Mission der Evangelischen Brüder zu Neu-Herrnhut und Lichtenfels. Barby, H. D. Ebers, [etc., etc.], 1765.
1132, [26]p., fold. plates, 2 fold. maps. 998/C85h

542 **Cranz.**
Historie von Grönland, enthaltend die Beschreibung des Landes und der Einwohner &c., insbesondere die Geschichte der dortigen Mission der Evangelischen Brüder zu Neu Herrnhut und Lichtenfels . . . Barby, H. D. Ebers, 1770.
2v. (1132p.), plates, 2 fold. maps. 998/C85h/1770

"Zweyte Auflage."
Vol.2 has half title only.

543 **Cranz.**
Historie von Grönland, enthaltend die Beschreibung des Landes und der Einwohner &c. Nebst beträchtlichen Zusatzen und Anmerkungen zur naturlichen Geschichte bis auf das Jahr 1779. Nürnberg, 1782.
392p., fold. plates, fold. maps. 919.8/C85h

544 **Cranz.**
The history of Greenland; containing a description of the country, and its inhabitants; and particularly, a relation of the mission, carried on for above these thirty years by the Unitas fratrum, at New Herrnhuth and Lichtenfels, in that country. Tr. from the High Dutch . . . London, Printed for the Brethren's society for the furtherance of the gospel among the heathen and sold by J. Dodsley [etc.], 1767.
x998/C85hE and uncat.

545 **Cranz.**
The history of Greenland; including an account of the mission carried on by the United brethren in that country, from the German of David Crantz. With a continuation to the present time . . . and an appendix, containing a sketch of the mission of the Brethren in Labrador . . . London, Printed for Longman, Hurst, Bees, Orme, and Brown, 1820.
2v., fronts., plates, 2 maps (1 fold.). 998/C85hE/1820

546 **Cranz.**
Historia om Grönland, deruti landet och dess inbyggare, etc., i synnerhet Evangeliska Brödra-Församlingens där warande mission . . . på swenska öfwersatt . . . Stockholm, J. G. Lange, 1769-70.

MVSAEVM
FRANC. CALCEOLARI
IVN. VERONENSIS
A BENEDICTO CERVTO MEDICO
INCÆPTVM,
ET AB ANDREA CHIOCCO
MED. PHYSICO EXCELLENTISS.
COLLEGII
Luculenter Descriptum, & Perfectum.
In quo multa ad naturalem, moralemq; Philosophiã
Spectantia, non pauca ad rem Medicam
pertinentia erudité proponuntur;
& explicantur;
Non sine magna rerum exoticarum
supellectile, quæ artifici plané
manu in æs incisa, studiosis
exhibentur.

Cat. No. 414

3v., plates, 2 fold. maps. 998/C85hS

Vol.2 has title: Grönlandska historiens sednare del om Brödra-Församlingens missioner ifrän 1740 ärs början til 1762 ärs slut.

Vol.3 has title: Fortsättning af Grönland, i synnerhet af Evangeliska Brödra-Församlingens där warande missioner förättningar i Ny-Herrnhut och Lichtenfels ifrän är 1763 til 1768 . . .

Translation of Historie von Grönland.

547 **Crèvecoeur, Michel Guillaume St. Jean de,** 1735-1813.
Voyage dans la haute Pénsylvanie et dans l'état de New York, par un membre adoptif de la nation Onéida. Tr. et pub. par l'auteur des Lettres d'un cultivateur américain . . . Paris, Maradan, 1801.
3v., 11 plates (incl. front., ports., fold. maps, fold. plans), fold. tables. 917.3/C866v

Howes C884 "Original work of this author, presented under the guise of a translation."

Early description of Niagara Falls.

548 **Croizet, Jean Baptiste.**
Recherches sur les ossemens fossiles du département du Puy-de-Dôme, par l'abbé Croizet et Jobert ainé . . . Paris, Chez les principaux libraires, 1828.
224p., plates. q560.944591/C87r

549 **Cronstedt, Axel Fredrik,** 1722-1765.
Försök til mineralogie, eller mineral-rikets upställning. Stockholm, Wildiska tryckeriet, 1758.
251p. 549/C88f

550 **Cronstedt.**
Försök till en mineralogie eller mineral rikets upställning. 2.upl. förbatträd. Stockholm, Truckt hos J. A. Carlbohm, 1781.
302p. 549/C88f/1781

Interleaved, with manuscript notes.

551 **Cronstedt.**
Versuch einer neuen Mineralogie, aus dem Schwedischen übersetzt. Kopenhagen, Auf Kosten der Rothenschen Buchhandlung, 1760.
264p. 549/C88fGwi

Translation of Försök til mineralogie, eller mineral-rikets upställning.

"Vorbericht" signed: G. Wiedeman.

552 **Cronstedt.**
Versuch einer Mineralogie. Aufs neue aus dem Schwedischen übersetzt und nächst verschiedenen Anmerkungen vorzüg-

lich mit äussern Beschreibungen der Fossilien vermehrt von Abraham Gottlob Werner. 1.Bd., 1.Th. Leipzig, S. L. Crusius, 1780.

[72], 254p. 549/C88fGw

No more published?

Translation of Försök till en mineralogie eller mineral rikets upställning.

553 **Cronstedt.**

An essay towards a system of mineralogy. Translated from the original Swedish, with notes, by Gustav von Engestrom. To which is added, a treatise on the pocket-laboratory, containing an easy method, used by the author, for trying mineral bodies, written by the translator. The whole revised and corrected, with some additional notes, by Emanuel Mendes da Costa. London, Printed for E. and C. Dilly, 1770.

329p., 2 fold. plates. x549/C88fEe

554 **Cronstedt.**

An essay towards a system of mineralogy. Translated from the original Swedish, with notes, by Gustav von Engestrom. To which is added, a treatise on the pocket-laboratory, containing an easy method, used by the author, for trying mineral bodies, written by the translator. The whole rev. and corr., with some additional notes, by Emanuel Mendes da Costa. 2d ed.; with an appendix, containing additions and notes, by M. T. Brunnich. London, Printed for E. and C. Dilly, 1772.

329p., 2 fold. plates. x549/C88fEe/1772

555 **Cronstedt.**

An essay towards a system of mineralogy. Tr. from the original Swedish, with annotations, and an additional treatise on the blowpipe. By Gustav von Engestrom . . . The 2d ed., greatly enl. and improved, by the addition of the modern discoveries, and by a new arrangement of the articles, by John Hyacinth de Magellan . . . London, Printed for C. Dilly, 1788.

2v., 2 fold. plates. x549/C88fEe/1788

Paged continuously.

556 **Cronstedt.**

Essai d'une nouvelle minéralogie. Tr. du suédois & de l'allemand de M. Wiedman, &c. &c., par M. Dreux fils. Paris, P. F. Didot le jeune, 1771.

389p. 549/C88fFd

557 **Cronstedt.**
See RINMAN, SVEN, Åminnelse-tal öfver . . . herr Axel Fred. Cronstedt . . . 1766.

558 **Crouch, Edmund A.**
An illustrated introduction to Lamarck's conchology; contained in his Histoire naturelle des animaux sans vertèbres: being a literal translation of the descriptions of the recent and fossil genera accompanied by twenty-two highly finished lithographic plates. London, Longman, Rees, Orme, Brown & Green [etc.], 1827.
47p., 22 col. plates. q594/C88i

559 **Crouch, Nathaniel,** 1632?-1725?
The general history of earthquakes: being an account of the most remarkable and tremendous earthquakes that have happened in divers parts of the world, from the Creation to this time; as they are recorded by sacred and common authors; and particularly those lately in Naples, Smyrna, Jamaica and Sicily. With a description of the famous burning mount, Aetna, in that Island; and the relation of the several dreadful conflagrations and Fiery Irruptions thereof for many ages. Likewise the natural and material causes of earthquakes with the usual signs and prognosticks of their approach; and the consequences and effects that have followed several of them. London, Printed for A. Bettesworth, 1734.
177p., front. x551.22/C884g/1734

560 **Cumberland, George,** 1752-1848.
Reliquiae conservatae from the primitive materials of our present globe, with popular descriptions of the prominent characters of some remarkable fossil encrinites, and their connecting links . . . Illustrated by sixty-six figures . . . Bristol, Printed and Published for J. M. Gutch, 1826.
45p., plates. 563.91/C91r

561 **Cunningham, Robert James Hay.**
. . . On the geology of the Lothians. [Edinburgh, 1838].
160p., front. (fold. map), xvi (i.e. 17) plates. 554.1/C91o

Caption title.
From Memoirs of the Wernerian Natural History Society, vol. vii, 1838.
Title page missing.

562 **Cutler, Jervis,** 1768-1844.
A topographical description of the state of Ohio, Indiana Territory, and Louisiana. Comprehending the Ohio and Mis-

sissippi Rivers, and their principal tributary streams: the face of the country, soils, waters, natural productions, animal, vegetable, and mineral; towns, villages, settlements and improvements: and a concise account of the Indian tribes west of the Mississippi. To which is added an interesting journal of Mr. Chas. Le Rave, while a captive with the Sioux nation, on the waters of the Missouri River. Boston, C. Williams, 1812.
219p., illus. x917.7/C976t

563 **Cuvier, Georges,** 1769-1832.
... Éloges historiques, précédés de l'éloge de l'auteur, par M. Flourens, ... Paris, E. Ducrocq [18-].
356p. 509.4/C98e

Half-title: Bibliothèque classique des célébrités contemporaines.
At head of title: G. Cuvier.
Bibliographical footnotes.
Contents. — Éloge historique de G. Cuvier, par [P.] Flourens. — Notes. — Daubenton. — Joseph Priestley. — Michel Adanson. — A.-F. de Fourcroy. — Henri Cavendish. — Pierre-Simon Pallas. — A.-A. Parmentier et du Comte de Rumford. — René-Just Haüy. — Comte Berthollet. — Sir Humphry Davy.

564 **Cuvier.**
Essai sur la géographie minéralogique des environs de Paris, avec une carte géognostique, et des coupes de terrain; par G. Cuvier... et Alexandre Brongniart... Paris, Potey, 1811.
287p., 2 fold. plates (incl. col. profiles), fold. map. 554.4/C98e

Originally published in Annales du muséum d'histoire naturelle de Paris, tom. XI, 1808 (p. 293-326) — and again, with additions, in Mémoires de la France, année 1810, published 1811 — of which the present edition is a re-issue with addition of a preface.
A new edition, 1822, has title: Description géologique des environs de Paris.

565 **Cuvier.**
Description géologique des environs de Paris, par MM. G. Cuvier et Alex. Brongniart. Nouv. éd., dans laquelle on a inséré la description d'un grand nombre de lieux de l'Allemagne, de la Suisse, de l'Italie, etc., qui présentent des terrains analogues à ceux du bassin de Paris, par M. Alex. Brongniart... Paris et Amsterdam, G. Dufour et E. d'Ocagne, 1822.
428p., 11 (i.e. 17) plates (incl. col. map), fold. col. map.
q554.436/C98d/1822

"Description des végétaux fossiles du terrain de sédiment supérieur, cités dans la description géologique du bassin de Paris, par M. Adolphe Brongniart": p. [353]-371.

566 **Cuvier.**
Recherches sur les ossemens fossiles de quadrupèdes, où l'on rétablit les caractères de plusieurs espèces d'animaux que les révolutions du globe paroissent avoir détruites. Paris, Deterville, 1812.
4v., plates, fold. col. map. 560/C98r/1812

567 **Cuvier.**
Recherches sur les ossemens fossiles, où l'on rétablit les caractères de plusieurs animaux dont les révolutions du globe ont détruit les espèces. Nouv. éd. Paris, G. Dufour et E. d'Ocagne, 1821-24.
5v. in 7, illus., maps. q560/C98r/1821

568 **Cuvier.**
Recherches sur les ossemens fossiles où l'on rétablit les caractères de plusieurs animaux dont les révolutions du globe ont détruit les espèces. 4. éd. . . . Paris, E. D'Ocagne [etc.], 1834-36.
10v., fold. tables, and an atlas of 2v. 560/C98r4
560/C98r4/atlas

569 **Cuvier.**
Le règne animal distribué d'après son organisation, pour servir de base à l'histoire naturelle des animaux et d'introduction à l'anatomie comparée, avec figures, dessinées d'après nature. Paris, Deterville, 1817.
4v. 590/C98r/1817

Plates and index are in v.4.

570 **Cuvier.**
Discours sur les révolutions de la surface du globe, et sur les changements qu'elles ont produits dans le règne animal. 5.ed. Paris G. Dufour, [etc., etc.], 1828.
400p., plates, tables. 560/C98d

571 **Cuvier.**
Discours sur les révolutions de la surface du globe, et sur les changements qu'elles ont produits dans le règne animal; . . . 6. éd. française, rev. et augm. Paris, E. d'Ocagne; [etc., etc.], 1830.
408p., plates, tables. 560/C98d/1830

572 **Cuvier.**
Discours sur les révolutions du globe, avec des notes et un appendice, d'après les travaux récents de MM. Humboldt, Flourens, Lyell, Lindley, etc. Rédigés par le Dr. Hoefer. Paris, F. Didot frères, 1850.

DE METAL-
LICIS LIBRI
TRES.

ANDREA CAESAL-
PINO ARETINO.

Medico & Philosopho
auctore.

NORIBERGÆ,

Recusi, curante Conrado Agricola.

M D CII.

Cat. No. 456

276p., 4 plates, 2 tables. 551.4/C98d

Flourens, [M. J. P.]. Éloge historique de Georges Cuvier, p. i-xlvi.

573 **Cuvier.**
Discours sur les révolutions du globe, avec des notes et un appendice d'après les travaux récents de MM. de Humboldt, Flourens, Lyell, Lindley, etc. rédigés par le Dr. Hoefer. Paris, Firmin-Didot, 1875.
356p., illus. 551.4/C98d/1875

574 **Cuvier.**
Essay on the theory of the earth. Translated from the French of M. Cuvier, by Robert Kerr; with mineralogical notes, and an account of Cuvier's geological discoveries, by Professor Jameson. Edinburgh, Printed for W. Blackwood, 1813.
265p., illus. 551.1/C98eEk

575 **Cuvier.**
Essay on the theory of the earth. Tr. from the French of M. Cuvier, by Robert Kerr; with mineralogical notes, and an account of Cuvier's geological discoveries, by Professor Jameson. 2d ed., with additions. Edinburgh, 1815.
332p., plates. 551.1/C98eEk2

576 **Cuvier.**
Essay on the theory of the earth. With mineralogical notes and an account of Cuvier's geological discoveries, by Professor Jameson. 3d ed., with additions. Edinburgh, Printed for W. Blackwood, 1817.
348p., illus. 551.1/C98eEj/1817

577 **Cuvier.**
Essay on the theory of the earth. With mineralogical notes, and an account of Cuvier's Geological Discoveries, by Professor Jameson. To which are now added, observations on the geology of North America; illustrated by the description of various organic remains, found in that part of the world by Samuel L. Mitchell. New York, Kirk & Mercein, 1818.
431p., 8 plates, front. 551.1/C98e

578 **Cuvier.**
Essay on the theory of the earth. With geological illus., by Professor Jameson. 5th ed., tr. from the last French ed., with numerous additions by the author and translator. Edinburgh, W. Blackwood, 1827.
550p., front., plates. 551.1/C98eEj/1827

579 **Cuvier.**
A discourse on the revolutions of the surface of the globe, and the changes thereby produced in the animal kingdom. Tr. from the French with illustrations and a glossary. Philadelphia: Carey & Lea, 1831.
252p., illus. x560/C98dE

580 **Cuvier.**
Rozprawa o přewratech kůry zemnj, a o proměnách w žiwočistwu gimi způsobených, w ohledu přjrodopisném a dějepisném. Podle 5. wyd. přeložil a rozmnožil Jan Swatopluk Presl. W Praze, W Knjžecj arcibiskupské khihtisk., 1834.
318p., illus. 560/C98dCZp

Translation of Discours sur les révolutions de la surface du globe et sur les changements qu'elles ont produits dans le règne animal.

581 **Cuvier.**
Recueil des éloges historiques lus dans les séances publiques de l'Institut Royal de France. Strasbourg, F. G. Levrault, 1819.
2v. 925/C98r

Bibliographical footnotes.

582 **Cuvier.**
See FLOURENS, MARIE JEAN PIERRE, Cuvier... 1845.

583 **Czechura, Honorius Martin,** 1688-1726.
Mare philosophicum, thalassophilis ad perscrutandum propositum, seu Universa philosophia Aristotelico-neutristica, universi anatomia maris, interjectis quibusdam philosophico-politicis axiomatibus adornata; ac in thesibus triginta quinque, totidemque axiomatibus, & discursibus thalassologis complexa. Pragae, Typis W. Wickhart, 1724.
[20], 564p., plate, 2 maps (1 fold.). q551.46/C99m

Disputatio — Prague (J. W. Schaffgotsch, respondent).

584 **Dale, Samuel,** 1659-1739.
See TAYLOR, SILAS, The history and antiquities of Harwich and Dovercourt . . . 1730.

585 **Dalman, Johan Wilhelm,** 1787-1828.
Über die Palaeaden oder die sogenannten Trilobiten; aus dem schwedischen übersetzt von Friedrich Engelhart. Nürnberg, J. L. Schrag, 1828.
84p., 6 leaves of plates, illus. x565.393/D16oGe

586 **Dana, James Dwight,** 1813-1895.
A system of mineralogy; including an extended treatise on crystallography; with an appendix, containing the application of mathematics to crystallographic investigation, and a mineralogical bibliography. New Haven, Durrie & Peck and Herrick & Noyes, 1837.
1-452, 119p., 181p., illus., 4 plates (fronts.). x549/D19

Bibliography: Appendix p. 89-106.

587 **Dana.**
Geology. With a folio atlas of twenty-one plates. New York, G. P. Putnam, [1849?].
756p., illus., 2 maps, profiles. q559.9/D19g

United States Exploring Expedition. During the years 1838, 1839, 1840, 1841, 1842. Under the command of Charles Wilkes, U.S.N. [v. 10].
Atlas wanting.

588 **Dana.**
On coral reefs and islands. By James D. Dana, geologist of the exploring expedition around the world during the years 1838-1842, C. Wilkes, U.S.N., commander... New York, G. P. Putnam & co., 1853.
143p., illus., 2 fold. maps. 551.96/D19o

From the author's exploring expedition report on geology, with additions.

589 **Dana, James Freeman,** 1793-1827.
Outlines of the mineralogy and geology of Boston and its vicinity, with a geological map. By J. Freeman Dana, M.D., and Samuel L. Dana, M.D. Boston, Cummings and Hilliard, 1818.
108p., front. (col. geol. map). x557.4461/D190

590 **Dana, Samuel Luther,** 1795-1868, jt. author.
See DANA, JAMES FREEMAN, Outlines of the mineralogy and geology of Boston ... 1818.

591 **Darby, William,** 1775-1854.
A geographical description of the state of Louisiana, the southern part of the state of Mississippi, and territory of Alabama ... Together with a map, from actual survey and observation ... of the state of Louisiana, and adjacent countries. 2d ed., enl. and improved. New-York, James Olmstead; New-Orleans, J. Seymour, printer, 1817.
356p., 3 maps (2 fold., incl. front.). x917.6/D24

592 **Darby.**
The emigrant's guide to the western and southwestern states and territories comprising a geographical and statistical description of the states . . . Accompanied by a map of the United States . . . New York, Kirk & Mercein, 1818.
311p., 2 fold. maps (incl. front.), plan, tables. 917.7/D24e

593 **Darby.**
A tour from the city of New-York, to Detroit, in the Michigan territory, made between the 2d of May and the 22d of September, 1818 . . . The tour is accompanied with a map upon which the route will be designated; a particular map of the falls and river of Niagara, and the environs of the city of Detroit. New-York, Published for the author, by Kirk & Mercein, 1819. Printed by E. Worthington, Brooklyn.
228, 63p., 3 fold. maps (incl. front.). 917.3/D24t
Errata slip preceding p.9 wanting.

594 **Darby.**
View of the United States, historical, geographical, and statistical; exhibiting, in a convenient form, the natural and artificial features of the several states, and embracing those leading branches of history and statistics best adapted to develop the present condition of the North American union . . . Philadelphia, H. S. Tanner, 1828.
654p., fold. maps. 917.3/D24

595 **Darcet, Jean,** 1795-1801.
Discours en forme de dissertation sur l'état actuel des montagnes des Pyrénées, et sur les causes de leur dégradation. On y a joint des expériences et des observations sur les variations du baromètre, sur le thermomètre, et autres morceaux de Physique, d'Histoire naturelle & de Chimie, avec une note de M. le Monnier, sur l'Aiguille aimantée. Paris, P. G. Cavelier, 1776.
134p., illus. x551.43/D24d

596 **Darwin, Charles Robert,** 1809-1882.
See FITZROY, ROBERT, 1805-1865, ed., Narrative of . . . the Beagle . . . 1839.

597 **Darwin.**
Journal of researches into the natural history and geology of the countries visited during the voyage of H.M.S. Beagle round the world, under the command of Capt. Fitz Roy. 2d ed., corr., with additions. London, J. Murray, 1845.
519p., illus. 508.3/D25j/1845

598 **Darwin.**
Geological observations on South America. Being the third part of the Geology of the voyage of the Beagle, under the command of Capt. Fitzroy, R.N. during the years 1832 to 1836. London, Smith, Elder and co., 1846.
279p., illus., fold. plates (1 col.), fold. map. 558/D25g

599 **Darwin.**
Journal of researches into the natural history and geology of the countries visited during the voyage of H.M.S. Beagle round the world, under the command of Capt. Fitz Roy, R.N. New York, Harper & Brothers, 1846.
2v., illus. 508.3/D25j/1846

600 **Daubeny, Charles Giles Bridle,** 1795-1867.
A description of active and extinct volcanos; with remarks on their origin, their chemical phaenomena, and the character of their products, as determined by the condition of the earth during the period of their formation. Being the substance of some lectures delivered before the University of Oxford, with much additional matter. London, W. Phillips; [etc., etc.], 1826.
466p., front., illus., 2 maps, diagr. 551.21/D26d

Bibliography: p. [453]-466.

601 **Daubeny.**
A description of active and extinct volcanos, of earthquakes, and of thermal springs; with remarks on the causes of these phaenomena, the character of their respective products, and their influence on the past and present condition of the globe. 2d ed., greatly enl. London, R. & J. E. Taylor, 1848.
743p., illus., 4 plates, 11 (i.e. 7) maps, diagr. 551.21/D26d/1848

Bibliography: p. 729-743.

602 **Daubeny.**
Sketch of the geology of North America, being the substance of a memoir read before the Ashmolean society Nov. 26, 1838. Oxford, Printed by T. Combe, printer to the university, for the Ashmolean society, 1839.
73p., front. (fold. map). 557/Xd

603 **Daubrée, Gabriel Auguste,** 1814-1896.
Scandinaviens Erzlagerstätten . . . Bearb. von Gustav Leonhard . . . Mit fünf Tafeln, Stuttgart, E. Schweizerbart, 1846.
62p., 3 fold. plates, 2 maps. 553.1/D26s

MINERALOGIA,

SIVE

NATVRALIS PHILOSOPHIÆ THESAVRI,

IN QVIBVS METALLICÆ CONCRETIONIS medicatorúmque fossilium miracula, terrarum pretium, colorum & pigmentorum apparatus, concretorum succorum virtus, lapidum atque gemmarum dignitas continentur.

Hos publici iuris fecit R. P. BERNARDVS CÆSIVS *Mutinensis, è Societate* IESV.

PRODERIT HÆC PRETIOSA SVPELLEX NON Philosophiæ modò, ac Medicinæ, verùm etiam sacræ & humanioris literaturæ studiosis.

LVGDVNI,

Sumptib. IACOBI & PETRI PROST.

M. DC. XXXVI.

CVM PRIVILEGIO REGIS.

Cat. No. 457

604 **Daubrée.**
Rapport sur les progrès de la géologie expérimentale . . . Publication faite sous les auspices du Ministère de l'instruction publique. Paris, Imprimerie impériale, 1867.
142p. 551/D26r

Recueil des rapports sur les progrès des lettres et des sciences en France.

605 **Daubrée.**
Études synthétiques de géologie expérimentale . . . Paris, Dunod, 1879.
828p., 7 plates (part. fold., incl. maps), diagrs. 551/D26e

"Indication des publications originales de M. Daubrée relatives aux sujets traités dans cet ouvrage", p. [719]-730.

606 **Daubrée.**
Synthetische Studien zur Experimental-Geologie. Autorisirte Deutsche Ausg. Braunschweig, F. Vieweg und Sohn, 1880.
596p., illus., plates, maps. 551/D26eGg

Translation of Études synthétiques de géologie expérimentale.

607 **Daumas, Melchior Joseph Eugène,** 1803-1871.
Le Sahara algérien; études géographiques, statistiques et historiques sur la région au sud des établissements français en Algérie; ouvrage rédigé sur les documents recueillis par les soins de M. le lieutenant-colonel Daumas . . . et pub. avec l'autorisation de M. le maréchal, duc de Dalmatie . . . Paris, Fortin, Masson et Cie; [etc., etc.], 1845.
339p., fold. map. 916.611/D26s

"Exploration scientifique de l'Algérie pendant les années 1840, 1841, 1842 . . . Prospectus": 4 p. at end.

608 **Dausque, Claude,** 1566-1644.
. . . Terra, et aqua, seu terrae flutantes. Tornaci Nerviorum, Ex typographio Adriani Quinque, 1633.
260p. uncat.

609 **Dausque.**
Terra, et aqua, seu terrae flutantes. Parisiis, apud Fredericum Leonard, 1677.
260p. uncat.

610 **Davies, Thomas Alfred,** 1809-1899.
Answer to Hugh Miller and theoretic geologists. New York, Rudd & Carlton, 1860.
302p. 550.1/D28a

611 **Davreux, Charles Joseph,** 1806-1863.
Essai sur la constitution géognostique de la province de Liège, en réponse à la question proposée par l'Académie royale des sciences, arts et belles-lettres de Bruxelles, pour le concours de 1830, savoir: "Faire la description géologique de la province de Liège; indiquer les espèces minérales et les fossiles accidentels que l'on y rencontre, avec l'indication des localités et la synonymie des noms sous lesquels les substances déja connues ont été décrites." . . . Bruxelles, M. Hayez, imprimeur de l'Académie royale, 1833.
297p., 9 fold. plates. 554.93/D31e

Académie royale des sciences et belles-lettres de Bruxelles. Mémoires couronnés, t.9.

612 **Debreyne, Pierre Jean Corneille,** 1786-1867.
Teoría bíblica de la cosmogonía y de la geología; nueva doctrina fundada sobre un princípio único y universal sacade de la Biblia. Obra destinada especialmente al clero y a los seminarios. . . . traducida del francés por el D. D. Pedro Parcet y D. Juan Cascante . . . Seguida de un tratado titulado: Moisés y los geólogos modernos, por Mr. Victor de Bonald. Barcelona: Libreria religiosa, 1854.
371p. 213/D35t

Bibliographical footnotes.

613 **Defrance, Marin Jacques Louis,** 1758-1850.
Tableau des corps organisées fossiles, précédé de remarques sur leur pétrification. Paris, F. G. Levrault; [etc., etc.], 1824.
136p. 560/D36t

614 **Degner, Johann Hartmann,** 1687-1756.
Dissertatio physica de turfis, sistens historiam naturalem cespitum combustilium qui in multis Europae regionibus, & praecipue in Hollandia reperiuntur ac ligni loco usurpantur. Trajecti ad Rhenum, Apud G. Kroon; G. Stouw, 1729.
190p. 553.21/D36d

615 **De la Beche, Henry Thomas,** 1796-1855.
A selection of the geological memoirs contained in the Annales des mines, together with a synoptical table of equivalent formations, and M. Brongniart's table of the classification of mixed rocks. Tr., with notes, by H. T. De la Beche . . . London, W. Phillips, 1824.
335p., front., 10 fold. plates (incl. maps), fold. tables. 550/D37s and 550.8/D37s/1824

616 **De la Beche.**
A selection of geological memoirs contained in the Annales des mines, written by Brongniart, Humboldt, von Buch, and others. Tr., with notes and additions, by H. T. De la Beche . . . Illustrated with eleven folding plates, including a geological map of France. London, W. Phillips, 1836.
335p., front. (map), 10 fold. plates (incl. 2 maps), fold. table.
550.8/D37s

617 **De la Beche.**
Sections & views, illustrative of geological phaenomena. London, Truettel & Würtz, Truettel, Jr. & Richter, 1830.
71p., 40 plates [part. col., incl. col. maps (part. fold.)]. 550/D37se

618 **De la Beche.**
Coupes et vues pour servir à l'explication des phénomènes géologiques, . . . avec un texte traduit de l'anglais, par H. de Collegno . . . 40 planches coloriées. Paris, Pitois-Levrault et Cie, 1839.
77p., 40 plates (part. col., part. fold., incl. maps, profiles).
550/D37seFp

Translation of Sections & views, illustrative of geological phaenomena.

619 **De la Beche.**
A geological manual. Philadelphia, Carey & Lea, 1832.
535p., illus. 550/D37ge

620 **De la Beche.**
A geological manual. 3d ed., considerably enl. London, C. Knight, 1833.
629p., illus. 550/D37ge3

621 **De la Beche.**
Manuel géologique. 2.éd., publiée à Londres en 1832. Traduction française, rev. et publiée par A.-J.-M. Brochant de Villiers. Bruxelles, Meline, Cans et Cie, 1837.
506p., illus. 550/D37geFb

Translation of A geological manual.
Bibliographical footnotes.

622 **De la Beche.**
Researches in theoretical geology. London, C. Knight, 1834.
408p., illus., 2 plates (incl. front.). 550/D374

"Appendix. Table of the situations and depths at which recent genera of marine and estuary shells have been observed. By W. J. Broderip": pp. [399]-408.

VOYAGES

DE M. LE MARQUIS

DE CHASTELLUX

DANS L'AMÉRIQUE

SEPTENTRIONALE

Dans les années 1780, 1781 & 1782.

Πολλῶν δ' ἀνθρώπων ἴδεν ἄστεα, καὶ νόον ἔγνω.

Multorumque hominum vidit urbes, & mores cognovit.

ODISSÉE, Liv. I.

TOME PREMIER.

A PARIS,

CHEZ PRAULT, IMPRIMEUR DU ROI,

Quai des Augustins, à l'Immortalité.

1786.

Cat. No. 478

623 **De la Beche.**
Researches in theoretical geology . . . with a preface and notes by Prof. Edward Hitchcock . . . New York, F. J. Huntington & Co., Philadelphia, Desilver, Thomas & Co., 1837.
342p., front. (map), diagrs. 550/D37r/1837

624 **De la Beche.**
Recherches sur la partie théorique de la géologie. Tr. de l'anglais par H. de Collegno . . . Paris, Strasbourg, F. G. Levrault, 1838.
289p., front., map, diagrs. 550/D37rFp
Translation of Researches in theoretical geology.

625 **De la Beche.**
L'art d'observer en géologie. Traduit de l'anglais par D. de Collegno. Paris, F. G. Levrault, 1838.
230p., illus. 550/D37hFc
Translation of How to Observe Geology.

626 **De la Beche.**
The geological observer. London, Longman [etc.], 1851.
846p., illus. 550/D37g/1851

627 **De la Beche.**
The geological observer. 2d ed., rev. London, Longman [etc.], 1853.
740p., illus. 550/D37g2

628 **Delanges, Paolo,** d. 1810.
Esperienze ed osservazioni intorno alla pressione delle terre, ed alla resistenza de' muri, che le medesime terre sostengono, la di loro composizione naturale impedendo. Verona, Per gli Eredi di M. Moroni, 1779.
119p., 3 plates. 624.16/D37e

629 **Delius, Christoph Traugott,** 1728-1779.
Abhandlung von dem Ursprunge der Gebürge und der darinne befindlichen Erzadern, oder der sogenannten Gänge und Klüfte; ingleichen von der Vererzung der Metalle und insonderheit des Goldes. Leipzig, Bey C. G. Hilschern, 1770.
156p., front. x549/D37a

630 **De Luc, Jean André,** 1727-1817.
Relation de différents voyages dans les Alpes du Faucigny, par Messieurs D. & D. [J. A. Deluc et P. G. Dentand]. Maestricht, J. E. Dufour, 1776.
138p. 914.947/D38r

631 **De Luc.**
Lettres physiques et morales sur l'histoire de la terre et de l'homme. Adressées à la reine de la Grande Bretagne, Paris, V. Duchesne, 1779-80.
5v. in 6. 550/D38ℓ

Vol. 2 has imprint: La Haye, De Tune, 1780.
Library has only vols. 2 and 3.

632 **De Luc.**
Lettres sur l'histoire physique de la terre, adressées à M. le professeur Blumenbach, renfermant de nouvelles preuves géologiques et historiques de la mission divine de Moyse. Paris, Nyon, 1798.
406p. 550/D38ℓe

633 **De Luc.**
Traité élémentaire de géologie. Paris, Courcier, 1089 [i.e. 1809].
395p. 550/D38t

634 **De Luc.**
An elementary treatise on geology: determining fundamental points in that science, and containing an examination of some modern geological systems, and particularly of the Huttonian theory of the earth. Tr. from the French manuscript by the Rev. Henry De La Fite . . . London, Printed for F. C. and J. Rivington, 1809.
415p. x550/D38Ed

635 **De Luc.**
Geological travels. Tr. from the French manuscript. London, Printed for F. C. and J. Rivington, 1810-11.
3v., plates, fold. map. 554.2/D38e

636 **De Luc.**
Geological travels in some parts of France, Switzerland, and Germany, London, F. C. and J. Rivington, 1813.
2v., 2 fold. maps. 554.4/D38ge

637 **Dentand, Pierre Gédéon,** 1750-1780, jt. author.
See DE LUC, JEAN ANDRÉ, Relation de différents voyages dans les Alpes du Faucigny . . . 1776.

638 **Descartes, René,** 1596-1650.
Principia philosophiae. Amstelodami, apud Ludovicum Elzevirium, 1644.
310p., illus. uncat.

The library owns the following other editions published at Amsterdam: 1656 (uncat.); 1664 (2 copies uncat.); 1672 and 1677 (x194/D45pr); 1685 (uncat.). All volumes are illustrated and contain Descartes' portrait as frontispiece.

The work was translated into French, Paris, 1659 (x194/D45prFp) and Rouen, 1679 (uncat.). It was also translated into Italian, Torino, 1722 (x194/D45prIb).

The work is included in *Opere philosophica,* Amsterdam, 1685, of which the library owns two copies (x194/D45/1685 and x194/D45/1685a).

639 **Descartes.**
Discours de la methode pour bien conduire sa raison, & chercher la verité dans les sciences. Plus la Dioptrique et les Meteores qui sont des essais de cette methode. Paris, Henry le Gras, 1658.
367p., illus. uncat.

640 **Descartes.**
Discours de la methode pour bien conduire, & chercher la verité dans les sciences. Plus la Dioptrique et les Meteores qui sont des essais de cette methode. Revue & corrigée . . . Paris, Michel Bobin & Nicolas le Gras, 1688.
501p., illus. uncat.

641 **Descartes.**
Le monde . . . ou le traité de la lumière et les autres principaux objets des sens. Paris, Michel Bobin & Nicolas le Gras, 1664.
290p., illus. uncat.

642 **Deshayes, Gérard Paul,** 1797-1875.
Description des coquilles fossiles des environs de Paris. Paris, L'auteur, chez Béchet jeune [etc.], 1824.
2v. 564/D45de

Contents. — I. Conchifères. — II. Mollusques.

643 **Deshayes.**
Description des coquilles fossiles des environs de Paris. Atlas. Paris, F. G. Levrault, 1837.
2v. of plates. 564/D45de/atlas

Contents. — I. Conchifères. — II. Mollusques.

644 **Deshayes.**
Description de coquilles caractéristiques des terrains. Paris, F. G. Levrault; [etc., etc.], 1831.
264p., 14 plates. 564/D45d

PARADOXES, OV TRAITTEZ PHILOSOPHIQVES DES PIERRES ET PIERRERIES, contre l'opinion vulgaire.

Aufquels font demonftrez la matiere, la caufe efficiente externe, la femence, la generation, la definition, & la nutrition d'icelles.

Enfemble la generation de tous les mixtes, fçauoir eft des animaux, vegetaux, & mineraux, ou foßiles.

Par ESTIENNE DE CLAVE, Docteur en Medecine.

A PARIS,
Chez la veufue PIERRE CHEVALIER, ruë S. Iacques, à l'Image S. Pierre, proche les Mathurins.

M. DC. XXXV.

Auec Approbation, & Priuilege.

Cat. No. 491

645 **Desmarest, Anselme Gaëtan,** 1784-1838.
See BRONGNIART, ALEXANDRE, Histoire naturelle des crustacés . . . Les crustacés proprement dits. 1822.

646 **Desmarest, Nicolas,** 1725-1815.
Mémoire sur l'origine & la nature du basalte à grandes colonnes polygones, déterminées par l'histoire naturelle de cette pierre, observée en Auvergne, *in* Mémoires de l'Académie Royale des Sciences, Paris, 1771.
p.705-775, fold. map. 506 PAH/1771

The map is entitled: Carte d'une partie de l'Auvergne où sont figurés les courants de laves, dans lesquels se trouve le basalte en prismes, en boules, &c . . . levée par MM. Pasumot et Dailley.

647 **Desmarest.**
Mémoire sur le basalte. Troisième partie. Où l'on traite du basalte des anciens, & où l'on expose l'histoire naturelle des différentes espèces de pierres auxquelles on a donné, en différens temps, le nom de basalte, *in* Mémoires de l'Académie Royale des Sciences, Paris, 1773.
p.599-670. 506 PAH/1773

648 **Desmarest.**
Extrait d'un mémoire sur la détermination de quelques époques de la nature par les produits des volcans, & sur l'usage de ces époques dans l'étude des volcans, *in* Observations sur la physique, sur l'histoire naturelle et sur les arts, 1779.
v.13, p.115-126. 505 JOUP/1779

649 **Desmarest.**
Mémoire sur la détermination de trois époques de la nature par les produits des volcans et sur l'usage qu'on peut faire de ces époques dans l'étude des volcans, *in* Mémoires de l'Institut des Sciences, Lettres, et Arts, Paris, 1806.
v.6, p.219-289. 4 fold. plates. 506 PAD/1806

650 **Desmarest.**
Encyclopédie méthodique. Géographie-physique. Paris, H. Agasse, [1794-95]-1828.
5v. and atlas. 551.03/En19

Library lacks v.1.

651 **Desor, Édouard, 1811-1882.**
Notice sur le dinotherium giganteum, quadrupède mammifère antédiluvien, d'une taille colossale, présentant une anomalie frappante dans la structure du crâne et dans la confi-

guration de la mâchoire inférieure. Paris, Imprimerie de Stahl, 1837.
12p., illus. 568.19/D46n

Cover illustrations.

652 **Desor,** Échinites . . . Des Galérites.
See AGASSIZ, LOUIS, Monographies d'échinodermes . . . 1838-42.

653 **Desor.**
Journal d'une course faite aux glaciers du Mont Rose, et du Mont Cervin, en société de MM. Studer, Agassiz, Lardy, Nicolet et autres. Genève, Tiré de la Bibliothèque Universelle, 1840.
63p., fold. plate. uncat.

654 **Desor.**
Compte rendu des recherches de M. Agassiz pendant ses deux derniers séjours à l'Hôtel des Neuchâtelois sur le glacier inférieur de l'Aar en 1841 et 1842. Genève, Bibliothèque Universelle, 1843.
79p. 551.31/AgleYd

655 **Desor.**
Agassiz geologische Alpenreisen unter Agassiz Mitwirkung verfasst. Deutsch mit einer topogr. Einleitung über die Hochgebirgsgruppen von C. Vogt. Frankfurt am Main: Literarische Anstalt J. Rütten, 1844.
548p., illus., plates. 554.947/Ag16d9Gv

656 **Desor.**
Excursions et séjours dans les glaciers et les hautes régions des Alpes, de M. Agassiz et de ses compagnons de voyage. Neuchâtel, Kissling, 1844.
638p., illus., ports. 551.31/D46e

657 **Desor.**
Nouvelles excursions et séjours dans les glaciers et les hautes régions des Alpes, de M. Agassiz et de ses compagnons de voyage. Accompagnées d'une notice sur les glaciers de l'Allée-Blanche et du Val-Ferret, par M. Agassiz, et d'un Aperçu sur la structure géologique des Alpes, par M. Studer. Avec une carte des glaciers de l'Oberland bernois, une carte géologique de ces mêmes contrées, et une coupe idéale du système alpin. Neuchâtel, J.-J. Kissling; [etc., etc], 1845.
266p., fold. plate, 2 fold. maps. 914.947/D46n

658 **Devèze de Chabriol, J. S.**
Essai géologique et minéralogique sur les environs d'Issoire, département du Puy-de-Dôme, et principalement sur la Montagne de Boulade, avec la description et les figures lithographiées des ossemens fossiles qui y ont été recueillis; par J. S. Devèze de Chabriol . . . et J. B. Bouillet . . . Clermont-Ferrand, Thibaud-Landriot [etc.], 1827.
104p., 30 plates (part. col., incl. col. map). f554.4591/D49e

Jacket is marked "5e et dernière livraison."

659 **Dezallier d'Argenville, Antoine Joseph,** 1680-1765.
L'Histoire naturelle éclaircie dans deux de ses parties principales, la lithologie et la conchyliologie, dont l'une traite des pierres et l'autre des coquillages. Ouvrage dans lequel on trouve une nouvelle méthode & une notice critique des principaux auteurs qui ont écrit sur ces matières. Enrichi de figures dessinées d'après nature. Par M*** de la Société royale des sciences de Montpellier. Paris, De Bure l'Aîné, 1742.
491p., 33 plates. q594/D53h

Title page engraving.
Library owns two copies.

660 **Dezallier d'Argenville.**
L'Histoire naturelle éclaircie dans une de ses parties principales, l'oryctologie, qui traite des terres, des pierres, des métaux, des minéraux, et autres fossiles, ouvrage dans lequel on trouve une nouvelle méthode latine & françoise de les diviser, & une notice critique des principaux ouvrages qui ont paru sur ces matières. Enrichi de figures dessinées d'après nature. Par M*** des Sociétés royales des sciences de Londres & de Montpellier. A Paris, chez De Bure, l'Aîné, 1755.
560p., 26 plates. 560/D53h

Bibliographical notes in margins.
Title page engraving.

661 **Dezallier d'Argenville.**
Enumerationis fossilium, quae in omnibus Galliae Provinciis reperiuntur, tentamina. Parisiis, Apud Joannem de Bure, 1751.
131p. 560.944/D53e

662 **Dezallier d'Argenville.**
Conchyliologie; oder, Abhandlung von den Schnecken, Muscheln und andern Schaalthieren welche in der See, in süssen Wassern und auf dem Lande gefunden werden; nebst der Zoomorphose, oder Abbildung und Beschriebung der Thiere

Försök
till en
Mineralogie,
eller
Mineral Rikets
Upställning.

Andra Uplagan Förbättrad.

A. Snack sc.

STOCKHOLM.

Tryckt hos Johan Arv. Carlbohm 1781.

på Förläggarens bekostnad.

Cat. No. 550

welche die Gehäuse bewohnen, aus dem Französischen übersetzt und mit Anmerkungen vermehret des Herrn Dezallier von Argenville. Wien, Auf Kosten der Kraussischen Buchhandlung, 1772.
[468p. in various pagings], 41 leaves of plates, illus. q594/D53cGk

663 **Dietman, Johann Maximilian.**
Gründliche Untersuchung des Nieder-Oesterreichischen Badner-Bades dessen Gebrauch u. Missbrauch. Anfänglich in Lateinischer Sprache hrsg., nunmehro . . . ins Teutsche übersetzet und mit einer Vorrede Joh. Nic. Weissens, . . . vermehret. Wien, J. A. Schmidt, 1734.
93p. 553.7/D56dGw

Translation of the author's inaugural disputation, Vienna, 1732, entitled, De balneo Badensi in Austria.

664 **Dietman.**
Eigentliche Beschreibung deren berühmten dreyen Gesundheits-Bädern in dem Ertz-Hertzogthum Oesterreich unter der Enns, als Baaden, Teutsch-Altenburg und Pyrenwarth [von J. W. I. Lehr] Ohnlängst in Lateinischer Sprach . . . numnehro . . . in die Teutsche Mutter-Sprach übersetzet, von J. A. C. v.S. Nürnberg, J. P. Krauss, 1735.
187p., front. 553.7/D56dGs

665 **Dixon, Frederick,** 1799-1849.
The geology and fossils of the Tertiary and Cretaceous formations of Sussex. London, Printed for the author by R. and J. E. Taylor, 1850.
422p., illus., 40 [i.e. 41] plates, front. q554.225/D64g

Edited by Sir Richard Owen.
Includes descriptions of fossils by W. Lonsdale, J. de C. Sowerby, P. Egerton, R. Owen, T. Bell.

666 **Dixon.**
The geology of Sussex; or, The geology and fossils of the Tertiary and Cretaceous formations of Sussex. By the late Frederick Dixon . . . New ed., rev. and aug. by T. Rupert Jones . . . Brighton, W. J. Smith, 1878.
469p., front., illus., plates, fold. map. q554.225/D64g/1878

Includes geological papers by R. A. C. Godwin-Austen, A. Bell, J. S. Bowerbank, O. Fisher, J. Prestwich, E. H. Willett, T. R. Jones, W. Topley; and paleontological papers by W. Carruthers, H. B. Brady, T. R. Jones, J. de C. Sowerby, W. Lonsdale, G. P. Deshayes, C. J. A. Meÿer, Sir Richard Owen, E. Forbes, T. Bell, H. Woodward, Sir P. de M. Grey-Egerton, E. T. Newton, W. J. Sollas.

667 **Dolce, Lodovico,** 1508-1568.
Trattato delle gemme che produce la natura; nel quale si discorre della qualità, grandezza, bellezza et virtù loro; compositione . . . in tre libri diviso. Venetia, Appresso G. B. & G. B. Sessa, 1617.
188p. x549/D68t/1618

668 **Dolomieu, Déodat Guy Silvain Tancrède Gratet de,** 1750-1801.
Voyage aux îles de Lipari, fait en 1781; ou, Notices sur les Iles Aeoliennes, pour servir à l'histoire des volcans; suivi d'un mémoire sur une espèce de volcan d'air, & d'un autre sur la température du climat de Malthe, & sur la différence de la chaleur réelle & de la chaleur sensible. Paris, Sous le privilège de l'Académie royale des sciences, 1783.
208p. 551.21/D69v

669 **Dolomieu.**
Mémoires sur les volcans éteints du Val di Noto en Sicile, *in* Observations sur la physique, sur l'histoire naturelle et sur les arts, Paris, 1784.
v.25, p.191-205. 505 JOUP/1784

670 **Dolomieu.**
Memoria del commendatore Deodato de Dolomieu sopra i tremuoti (sic) della Calabria nell'anno 1783. Prima traduzione dal francese. Roma, L. P. Salvioni, stampator Vaticano, 1784.
80p. 551.22/D69ml

671 **Dolomieu.**
Mémoire sur les Isles Ponces, et catalogue raisonné des produits de l'Etna; pour servir à l'histoire des volcans: suivis de la description de l'éruption de l'Etna, du mois de juillet 1787. Ouvrage qui fait suite au Voyage aux Iles de Lipari, 1 vol. in 8°. du même auteur. Paris, Chez Cuchet, 1788.
525[2]p., 4 fold. maps, fold. table. 551.21/D69m

672 **Dolomieu.**
Lettre du Commandant Déodat de Dolomieu, à M. Picot de la Peyrouse . . . sur un genre de pierres calcaires très-peu effervescentes avec les acides, & phosphorescentes par la collision, *in* Observations sur la physique, sur l'histoire naturelle et sur les arts, Paris, 1791.
v.39, p.2-10. 505 JOUP/1791

673 **Dolomieu.**
Mémoire sur les pierres composées et sur les roches, *in*

Observations sur la physique, sur l'histoire naturelle et sur les arts, Paris, 1791.
v.39, p.374-407. 505 JOUP/1791

674 **Dolomieu.**
Distribution méthodique de toutes les matières dont l'accumulation forme les montagnes volcaniques, ou tableau systématique dans lequel peuvent se placer toutes les substances qui ont des relations avec les feux souterrains, *in* Journal de Physique, de chimie et d'histoire naturelle, Paris, 1794.
v.1, p.102-125. 505 JOUP/1794

675 **Dolomieu.**
Mémoire sur les roches composées en général, & particulièrement sur les pétro-silex, les trapps & les roches de corne, pour servir à la distribution méthodique des produits volcaniques, *in* Journal de Physique, de chimie et d'histoire naturelle, Paris, 1794.
v.1, p.175-200, 241-263, 406-428; v.2, 1794, p.82-105.
505 JOUP/1794

676 **Dolomieu.**
See NOSE, KARL WILHELM, Beschreibung . . . vulcanisirten Fossilien . . . 1797.

677 **Dolomieu.**
Sur la philosophie minéralogique, et sur l'espèce minéralogique. Paris, Impr. de Bossange, Masson et Besson, 1801.
128p. 549/D69s

678 **Donati, Vitaliano,** 1713-1763.
Essai sur l'histoire naturelle de la mer Adriatique, par le docteur Vitaliano Donati, avec une lettre du docteur Leonard Sesler, sur une nouvelle espèce de plante terrestre, tr. de l'italien. La Haye, P. de Hondt, 1758.
73p., 11 plates. 593.3/D71sFc

Translation of the author's Della storia naturale marina dell' Adriatico by J. Salvemini de Castillon. cf. Bibliothèque Nationale Cat.

679 **Dondi dall'Orologio, Antonio Carlo,** 1751-1801.
Dello sprofondamento di una costa di monte negli Euganei; lettera del marchese Antonio-Carlo Dondi Orologio . . . diretta al ch. signor Ab. Alberto Fortis . . . Padova, Stamperia Penada, 1787.
40p. 551.43/D71d

CLAUDII DAUSQUII

CAN. TORNACENSIS

TERRA, ET AQUA,

SEU

TERRÆ FLVTANTES.

PARISIIS,

Apud FREDERICUM LEONARD Typographum
Regis, & Cleri Gallicani, viâ Jacobæâ
ſub Signo Scuti Veneti.

M. DC. LXXVII.

Cat. No. 609

680 **Dörckes, Jacob.**
. . . Dissertatio juridica de metallis eorumque fodinis, occasione tituli Codic. metallis & metallariis . . . Jenae, Typis Johannis Wertheri, [1671].
[44p.] uncat.

681 **Douglas, James,** 1753-1819.
A dissertation on the antiquity of the earth, read at the Royal society, 12th May, 1785 . . . By the Rev. James Douglas . . . London, G. Nicol, 1785.
86p., 7 plates. x550.1/D746d

682 **Doulcet.**
Dictionnaire oryctographique, ou Description de toutes les substances connues, qui sont sur la surface ou dans le sein de la terre, avec leur analyse et une table des progressions, des pesanteurs spécifiques des minéraux, celle de l'eau étant supposée 10,000. Paris, Belin [etc.], 1806.
240p. 549/D74d

Caption title: Dictionnaire raisonné universel des minéraux, ou Oryctographie des substances pierreuses, métalliques, salines et inflammables, avec leur analyse.
Manuscript notes inserted.

683 **Drake, Daniel,** 1785-1852.
Natural and statistical view, or picture of Cincinnati and the Miami country, illustrated by maps. With an appendix, containing observations on the late earthquakes, the aurora borealis, and south-west wind. Cincinnati, Printed by Looker and Wallace, 1815.
251p. 917.717/D789n

684 **Dralet, Étienne François,** 1754-1844.
Description des Pyrénées, considérées principalement sous les rapports de la géologie, de l'économie politique, rurale et forestière, de l'industrie et du commerce. Paris, Arthur Bertrand, 1813.
2v., fold. map, tables. 914.4/D78d

Bibliographical footnotes.

685 **Drapiez, Pierre Auguste Joseph,** 1778-1856.
Coup-d'oeil minéralogique et géologique sur la province du Hainaut, royaume des Pays-Bas. Bruxelles, P. J. de Mat, 1823.
164p., 4 plates. 554.93/D79c

686 **Drayton, John,** 1766-1822.
A view of South-Carolina, as respects her natural and civil concerns. Charleston, Printed by W. P. Young, 1802.
252p., plates, fold. maps, tables. x917.57/D77v

687 **Drayton.**
Beschreibung von Süd-Carolina. Aus dem Englischen und mit Anmerkungen und Zusätzen des Uebersetzers begleitet. Weimar, Im Verlag des F. S. Privil Landes-Industrie Comptoirs, 1808.
406p., 1 folded map, tables. x917.57/D77vG

688 **Drian, Aimé.**
Minéralogie et pétrologie des environs de Lyon. Disposées suivant l'ordre alphabétique. Lyon, C. Savy jeune, 1849.
539p. 552/D83m
Bibliography: p.[520]-539.

689 **Drummond, William Hamilton,** 1778-1865.
The Giants' Causeway, a poem: . . . Belfast, Printed by J. Smith for Longman, Hurst, Rees, Orme & Browne, 1811.
204p., 6 plates, 2 maps. 821/D839g

690 **Dube, Paul.**
De mineralium natura in universum ubi praesertim de aqua minerali fontis escarleiaram vulgo des escharlis proprè Montargiũ cuius vires in usum medicum expenduntur . . . Parisiis, Apud Franciscum Piot, 1649.
126p. uncat.

691 **Du Bois de Montpéreux, Frédéric,** 1798-1849.
Conchiologie fossile et aperçu géognostique des formations du plateau Wolhyni-Podolien. Berlin, S. Schropp, 1831.
76p., 8 plates, col. map. 564/D85c

692 **Du Buat, Pierre Louis Georges,** 1734-1809.
Principes d'hydraulique, vérifiés par un grand nombre d'expérience faites par ordre du gouvernement . . . Nouvelle édition, revue & considérablement augmentée, Paris, L'imprimerie du Monsieur, 1786.
2v., 4 fold. plates. 532 D852p/1786

693 **Dubuisson, François René André.**
Essai d'une méthode géologique; ou, Traité abrégé des roches. Nantes, Impr. de Mellinet-Malassis, 1819.
106p. 552/D85e

694 **Du Fouilhoux, Antoine.**
Discours de l'origine des Fontaines. Ensemble quelques histoires de la guarison de plusieurs grandes et difficiles maladies, faicte par l'usaige de l'eau medicinale des fontaines de Pougues en Nivernois . . . Item le traicté de la faculté & manière d'user de ladicte eau de Pougues . . . A Nevers, Par Pierre Roussin, 1592.
103p., front. uncat.

695 **Dufour, Léon,** 1782-1865.
See GRATELOUP, JEAN PIERRE SYLVESTRE DE, Conchyliologie . . . [1836-38].

696 **Dufrénoy, Ours Pierre Armand Petit,** 1792-1857.
Mémoires pour servir à une description géologique de la France, rédigés par ordre de Monsieur Becquey, sous la direction de M. Brochant de Villiers, par MM. Dufrénoy et Élie de Beaumont. Paris, F. G. Levrault, 1830-38.
4v., fold. plates, fold. maps. 554.4/D87m

Titles vary slightly.
Presentation copy signed: Élie de Beaumont.

697 **Dufrénoy.**
Carte géologique de la France . . . *See* FRANCE. SERVICE DE LA CARTE GÉOLOGIQUE DE LA FRANCE, Carte géologique . . . 1841.

698 **Dufrénoy.**
Explication de la carte géologique de la France, *See* FRANCE. SERVICE DE LA CARTE GÉOLOGIQUE DE LA FRANCE, Explication . . . v. 1-3, pt. 1, 1841-73.

699 **Dufrénoy.**
Traité de minéralogie. Paris, Carilian-Goeury et V. Dalmont, 1844-47.
4v. [v.4: atlas of 224 plates]. 549/D87t

V.4 (atlas) lacking.

700 **Dufrénoy.**
Explication des planches de fossiles qui accompagnent le cours lithographié de géologie professé à l'École royale des ponts et chausseés par M. Dufrénoy, par Émile Bayle. [n.p.], 1847 [i.e. 1848].
94p., 12 plates. q550/D87e

Cover dated 1848.

ESPERIENZE
ED
OSSERVAZIONI

INTORNO

ALLA PRESSIONE DELLE TERRE, ED ALLA RESISTENZA DE' MURI, CHE LE MEDESIME TERRE SOSTENGONO, LA DI LORO COMPOSIZIONE NATURALE IMPEDENDO.

FATTE DA

PAOLO DELANGES

CAPITANIO DEGLI INGEGNERI AL SERVIGIO DELLA SERENISSIMA REPUBBLICA DI VENEZIA.

---

in promptu cognita quæ ſunt
. . . . manu ducunt, & credere cogunt.

Lucrez. Lib. II.

IN VERONA

PER GLI EREDI DI MARCO MORONI

Con Licenza de' Superiori.

MDCCLXXIX.

Cat. No. 628

701 **Dugdale, William,** 1605-1686.
The history of imbanking and drayning of divers fenns and marshes, both in foreign parts and in this kingdom, and of the improvements thereby. Extracted from records, manuscripts, and other authentic testimonies. [London, 1662?].
469p., 11 fold. maps. q627.5/D878h

702 **Dugdale.**
The history of imbanking and draining of divers fens and marshes, both in foreign parts and in this kingdom, and of the improvements thereby. Extracted from records, manuscripts, and other authentic testimonies. The 2d ed., rev. and cor., by Charles Nalson Cole . . . London, Printed by W. Bowyer and J. Nichols, at the expence of R. Geast, 1772.
469p., 11 fold. maps. xq627.5/D878h2

Ed. 1 published in 1662.

703 **Du Hamel, Jean Baptiste,** 1624-1706.
De meteoris et fossilibus libri duo. In priore libro mixta imperfecta, quaeque in sublimi aëre vel gignuntur, vel apparent, fusè pertractantur. Posterior liber mixta perfecta complectitur; ubi salium, bituminum, lapidum, gemmarum, & metallorum naturae, causae, & usus inquiruntur. Parisiis, apud Petrum Lamy, 1660.
310p., illus. uncat.

704 **Duhamel, Jean Pierre François Guillot,** 1730-1816.
Géometrie souterraine, élémentaire, théorique et pratique, où l'on traite des filons ou veines minérales, & de leurs dispositions dans le sein de la terre; de la trigonométrie appliquée à la connoissance des filons, à la conduite des travaux de mines & à la confection de leurs plans & profils. Paris, Imprimerie royale, 1787.
294p., 14 plates, tables (183p.). 622/D88g

705 **Dumont, André Hubert,** 1809-1857.
Carte géologique de la Belgique, exécutée par ordre du gouvernement, sous les auspices de l'Académie Royale des Sciences, des Lettres et des Beaux Arts. [Bruxelles], Établissement géographique de P. Vandermaelen, [1849?].
9 [i.e. 8] col. maps on double leaves. uncat.

Signed by author.

706 **Dumont.**
Carte géologique de la Belgique, indiquant les terrains qui se trouvent au-dessous du limon hesbayen et du sable cam-

pinien. [Bruxelles], Établissement géographique de P. Vandermaelen, [1849?].
9 [i.e. 8] col. maps on double leaves. f554.93/D89c

Signed by author.

707 **Dunker, Wilhelm Bernhard Rudolph Hadrian,** 1809-1885.
Monographie der norddeutschen Wealdenbildung. Ein Beitrag zur Geognosie und Naturgeschichte der Vorwelt. Nebst einer Abhandlung über die in dieser Gebirgsbildung bis jetzt gefundenen Reptilien, von Hermann von Meyer. Mit XX nach der Natur gezeichneten Tafeln und einer Tafel mit Gebirgsprofilen. Braunschweig, Oehme und Müller, 1846.
83p., 21 plates. 560.943/D92m

"Literatur": p.[vii]-viii.

708 **Du Noyer, George Victor,** d.1869.
[Proofs of the original plates drawn for, but not published in, Joseph Ellison Portlock's Report on the geology of the county of Londonderry and of parts of Tyrone and Fermanagh.]
[n.p., n.d.], 47 plates. q560.9415/D92p

Additional plate by G. West tipped in at end.

709 **Durocher, Joseph Marie Elizabeth,** 1817-1860.
Voyages de la Commission Scientifique du Nord, en Scandinavie, en Laponie, au Spitzberg et aux Feröe, pendant les années 1838, 1839, et 1840 . . . Géologie, minéralogie, métallurgie et chimie. Paris, A. Bertrand, [1843].
215p. 554.8/D93g

710 **Du Souich, A.**
Essai sur les recherches de houille dans le Nord de la France. Paris, Carilian-Goeury et V. Dalmont; [etc., etc.], 1839.
132p., 3 fold. plates. 553.2/D94e

711 **Du Thoum, Louis.**
Le tremble-terre ou sont contenus ses causes, signes, effets & remedes, Bourdeaus, G. Vernoy, 1616.
250p. x551.22/D95t

Dedicatory letter signed: Louys Thaoum.

712 **Eaton, Amos,** 1776-1842.
An index to the geology of the Northern States, with a transverse section from Catskill Mountain to the Atlantic. Prepared for the geological classes at Williams College, Northampton, Belchertown, Leicester and Worcester, (Mass.). Leicester, printed by Hori Brown, 1818.
52p., col. section. uncat.

713 **Eaton.**
An index to the geology of the Northern States, with transverse sections, extending from Susquehanna River to the Atlantic, crossing Catskill Mountains. To which is prefixed a geological grammar. 2d ed. wholly written over anew, and published under the direction of the Troy Lyceum. Troy, N.Y., W. S. Parker, 1820.
286p., 2 plates. x557/Ea8i/1820

714 **Eaton.**
A geological and agricultural survey of Rensselaer County in the state of New-York. To which is annexed, a geological profile, extending from Onondaga Salt Springs, across said county, to Williams College in Massachusetts. Taken under the direction of the Honourable Stephen Van Rensselaer. Albany, Printed by E. and E. Hosford, 1822.
70p., profile, fold. table. 557/N48ear

Introduction signed: Amos Eaton.

715 **Eaton.**
A geological and agricultural survey of the district adjoining the Erie canal in the state of New York. Taken under the direction of the Hon. Stephen Van Rensselaer. Part I. Containing a description of the rock formations; together with a geological profile, extending from the Atlantic to lake Erie. Albany, Printed by Packard & Van Benthuysen, 1824.
163p., 2 fold. profiles. 557/N48eae

Introductory report signed: Amos Eaton.

716 **Eaton.**
A geological nomenclature for North America; founded upon geological surveys, taken under the direction of the Hon. Stephen Van Rensselaer, prepared for Rensselaerean schools. Albany, printed by Packard and Van Benthuysen, 1828.
31p., 3 col. plates. uncat.

717 **Eaton.**
Geological text-book, prepared for popular lectures on North American geology; with applications to agriculture and the arts. Albany, Printed by Websters and Skinners, 1830.
[9]-63p., illus., fold. map. x550/Ea83g

718 **Eaton.**
Geological text-book for aiding the study of North American geology: being a systematic arrangement of facts, collected by the author and his pupils, under the patronage of the

L'HISTOIRE NATURELLE ECLAIRCIE DANS UNE DE SES PARTIES PRINCIPALES, L'ORYCTOLOGIE, QUI TRAITE DES TERRES, DES PIERRES, DES MÉTAUX, DES MINERAUX, ET AUTRES FOSSILES,

OUVRAGE DANS LEQUEL ON TROUVE une nouvelle méthode Latine & Françoiſe de les diviſer, & une notice critique des principaux Ouvrages qui ont paru ſur ces matières.

Enrichi de Figures deſſinées d'après Nature.

Par M*** des Sociétés Royales des Sciences de Londres & de Montpellier.

A PARIS,
Chez DE BURE l'Aîné, Quai des Auguſtins, du côté du Pont Saint Michel, à Saint Paul.

M. DCC. LV.

AVEC APPROBATIONS ET PRIVILÉGE DU ROY.

Cat. No. 660

Hon. Stephen Van Rensselaer. Second edition, with additional facts, figures and descriptions of essential organized remains, and an alphabetical list of the most important localities. Albany, Websters and Skinners, etc., 1832.
134p., 5 plates. uncat.

719 **Ebel, Johann Gottfried,** 1764-1830.
Ueber den Bau der Erde in dem Alpen-Gebirge zwischen 12 Längen- und 2-4 Breitengraden, nebst einigen Betrachtungen über die Gebirge und den Bau der Erde überhaupt. Mit geognostischen Karten. Zürich, O. Füssli, 1808.
2v., fold. maps. 554.94/Eb3b

Library owns 2 copies. Copy 2 is 2v. in 1, bound with errata and the paper, Ideen über die Organisation und das eigenthümliche Leben des Erdkörpers und über die gewaltsamen Veränderungen seiner Oberfläche von Doctor Johann Gottfried Ebel, Wien, 1811. 38p.

Copy 2 lacks the folded maps.

720 **Eden, Richard,** 1521?-1576, ed.
History of trauayle in the West and East Indies and other countries lying either way... with a discourse of the North-west Passage. Translated by Richard Eden; augmented by Richarde Willis. London, R. Jugge, 1577.
467p. uncat.

Original work titled: The decades of the newe worlde... London, 1555. A miscellany of travels by Martyr, Oviedo, and others published in order to solicit venture capital for English expeditions. Talk of gold and speculation (Oviedo) upon its origin.

721 **Egen, P.N.C.,** 1793-1849.
Ueber das Erdbeben in den Rhein- und Niederlanden vom 23. Februar 1828, *in* Annalen der Physik und Chemie, 1828.
v.89, ser. 2, v. 13, p.153-163. 530.5 AP

722 **Ehrenberg, Christian Gottfried,** 1795-1876.
Neue Untersuchungen über das kleinste Leben als geologisches Moment; mit kurzer Characteristik von 10 neuen Generibus und 66 neuen Arten. Berlin, Zu haben bei L. Voss, Leipzig, 1845.
35p. 563.1/Eh8n

Auszug aus einem Vortrage in der Königl. Akademie der Wissenschaften zu Berlin am 27. Februar 1845.

723 **Ehrenberg.**
Vorlaüfige zweite Mittheilung über die Beziehungen des kleinsten organischen Lebens zu den vulkanischen Massen

der Erde mit kurzer Characteristik von 26 neuen Arten. Berlin; Leipzig, L. Voss, 1845.
25p., fold. table. 552.2/Eh8v

Aus den Monatsberichten der Königl. Akad. d. Wiss. zu Berlin vom April 1845.

724 **Ehrenberg.**
Mikrogeologie. Das Erden und Felsen schaffende Wirken des unsichtbar kleinen selbstständigen Lebens auf der Erde. Leipzig, L. Voss, 1854.
lv. in 2. f593.15/Eh83m

725 **Ehrenberg.**
Zur Mikrogeologie. Einundvierzieg Tafeln . . . gezeichnet vom Verfasser. Leipzig, L. Voss, 1854.
31p., 41 plates. f593.15/Eh83m/atlas

726 **Ehrlich, Franz Carl,** 1808-1886.
Geognostische Wanderungen im Gebiete der nordöstlichen Alpen, besonders in der Umgebung von Spital am Pyhrn und Linz . . . Ein spezieller Beitrag zur Kentniss Oberösterreich's . . . Linz, Joseph Wimmer; in Commission bei H. Hübner, 1852.
144p., illus., 5 plates. 554.36/Eh8g

Title page mutilated with loss of part of imprint.

727 **Eichwald, Carl Eduard von,** 1795-1876.
Ueber das silurische Schichtensystem in Esthland. St. Petersburg, 1840.
210p. 554.74/Ei2u

Aus dem ersten und zweiten Heft der 'Zeitschrift für Natur- und Heilkunde' der Medizinischen Akademie zu St. Petersburg besonders abgedruckt.

728 **Eichwald.**
Die Urwelt Russlands, durch Abbildungen erläutert. Hft. 1-4. St. Petersburg, 1840-48.
4v. 560.947/Ei2u

"Aus dem Russischen uebersetzt."
Library owns v.1 only: reprinted from Schriften der Kaiserlichen St. Petersburgischen Mineralogischen Gesellschaft.

729 **Eichwald.**
Paleontologiîa Rossïi; opisanie Molassoboĭ i namyvnoĭ formatsii Rossïi, po obraztsam khraniîashchimsiîa v muzeîe Imperatorskoĭ mediko-khirurgicheskoĭ akademii. Sochinenie Eduarda Eikhvai'da . . . Sanktpeterburg, V. Tip. E. Pratsa, 1850.

248p. 560.947/Ei2ℓR

Bibliographical footnotes.

730 **Eichwald.**

Lethaea rossica; ou, Paléontologie de la Russie, décrite et figurée par Edouard d'Eichwald . . . Stuttgart, E. Schweizerbart, 1853-68.

3v. in 5. 560.947/Ei2ℓ

(v. 1, 1860)

Contents. — v. 1. Ancienne période. — v. 2. Période moyenne. — v. 3. Dernière période.

731 **Eichwald.**

Lethaea rossica; ou, Paléontologie de la Russie. Atlas. Stuttgart, E. Schweizerbart, 1853-68.

2v. q560.947/Ei2ℓ/atlas

732 **Élie de Beaumont, Jean Baptiste Armand Louis Léonce,** 1798-1894.

Observations géologiques sur les différentes formations qui, dans le système des Vosges, séparent la formation houillère de celle du Lias. Paris, Impr. de Mme Huzard, 1828.

199p., 3 fold. profiles. 554.438/Eℓ40

Extrait des Annales des mines, 1827-1828.

Added t.p., Observations géologiques sur quelques terrains secondaires du système des Vosges.

733 **Élie de Beaumont.**

See DUFRÉNOY, OURS PIERRE ARMAND PETIT, Mémoires pour servir à une description géologique de la France . . . 1830-1838.

734 **Élie de Beaumont.**

Researches on some of the revolutions which have taken place on the surface of the globe; presenting various examples of the coincidence between the elevation of beds in certain systems of mountains, and the sudden changes which have produced the lines of demarcation observable in certain stages of the sedimentary deposits, *in* The Philosophical Magazine, or Annals of Chemistry, Mathematics, Astronomy, Natural History, and General Science, 1831.

v.10, p.241-264. 501 LE

735 **Élie de Beaumont.**

Instructions pour les géologues de l'expédition qui se rend dans le nord de l'Europe. [Paris, Impr. de Bachelier, 1838?].

24p. 550.72/Eℓ4i

TRATTATO

DELLE GEMME

CHE PRODVCE

la natura;

NEL QVALE SI DISCORRE

della qualità, grandezza, bellezza

& virtù loro.

COMPOSITIONE

di M. Lodouico Dolce, da lui

in tre Libri diuiſo.

Hora con diligenza riſtampato.

CON PRIVILEGIO.

IN VENETIA, MD CXVII.

Appreſſo Gio. Batt & Gio. Bernardo Seſſa.

Cat. No. 667

736 **Élie de Beaumont,** Carte géologique de la France . . .
See FRANCE. SERVICE DE LA CARTE GÉOLOGIQUE DE LA FRANCE, Carte géologique . . . 1841.

737 **Élie de Beaumont,** Explication de la carte géologique de la France . . .
See FRANCE. SERVICE DE LA CARTE GÉOLOGIQUE DE LA FRANCE, Explication . . . v.1-3, pt. 1, 1841-73.

738 **Élie de Beaumont.**
Rapport sur un mémoire de M. A. Bravais relatif aux lignes d'ancien niveau de la mer dans le Finmark. Commissaires MM. Biot, Liouville, Élie de Beaumont rapporteur. [Paris], Institut de France, Académie Royale des Sciences, [1842].
33p. 554.81/Eℓ4r

Extrait des Comptes rendus des séances de l'Académie des Sciences, séance du 31 octobre 1842.

739 **Élie de Beaumont.**
Leçons de géologie pratique, professées au Collège de France, pendant l'année scolaire 1843-1844. Paris, P. Bertrand; [etc., etc.], 1845-49.
2v., plates, maps, diagrs. 550/Eℓ4ℓ

1. Geology — Text-books.

740 **Élie de Beaumont.**
Notice sur les systèmes de montagnes. Paris. P. Bertrand, 1852.
3v., fold. maps, diagr. 551.43/Eℓ4n

Extrait du tome XII du Dictionnaire universel d'histoire naturelle, dirigé par M. Ch. d'Orbigny.
Paged continuously.

741 **Élie de Beaumont.**
Études stratigraphiques sur le département de la Haute-Marne, faites par Élie de Beaumont et de Chancourtois, pendant la publication de la Carte géologique de Duhamel. Paris, Mallet-Bachelier, 1862.
84p., diagr., tables. 554.4332/Eℓ4e

742 **Ellicott, Andrew,** 1754-1820.
The journal of Andrew Ellicott, late commissioner on behalf of the United States during part of the year 1796, the years 1797, 1798, 1799, and part of the year 1800; for determining the boundary between the United States and the possessions of His Catholic Majesty in America, containing occasional remarks on the situation, soil, rivers, natural productions, and diseases of the different countries on the Ohio, Mississippi,

and gulf of Mexico, with six maps comprehending the Ohio, the Mississippi from the mouth of the Ohio to the gulf of Mexico, the whole of West Florida and part of East Florida. To which is added an appendix, containing all the astronomical observations made use of for determining the boundary... likewise a great number of thermometrical observations... Philadelphia, Printed by Budd & Bertram, for Thomas Dobson, 1803.
299, 151p., illus., 6 fold. plates, 8 fold. maps. 973.46/E15

743 **Emmons, Ebenezer,** 1799-1863.
Manual of mineralogy and geology: designed for the use of schools; and for persons attending lectures on these subjects, as also a convenient pocket companion for travellers, in the United States of America. Albany, Printed by Websters and Skinners, 1826.
229p. x549/Em6m

Adopted as a text-book in the Rensselaer school.

744 **Emmons.**
Geology of New York, Part 2, comprising the geology of the Second Geological District. Albany, White and Visscher, 1842.
437p., 17 plates (part. col., incl. geol. maps, sections, fossils).
q508.742/N48/v.4, pt.2

Natural History of New York, v.4, pt.2.

745 **Emmons.**
Geological map of the state of New York. By Legislative Authority. New York, 1842.
Col. map, 91 x 100 cm. G3800/.C5/1842/.N4

Scale not given.
4 cross sections.
Accompanied "Geology of New York" by Emmons, Hall, Mather, and Vanuxem. New York, 1842.

746 **Emmons.**
See NORTH CAROLINA. GEOLOGICAL SURVEY, Report of Professor Emmons, on his geological survey of North Carolina... 1852.

747 **Engel, Samuel,** 1702-1784.
Essai sur cette question: quand et comment l'Amérique a-t-elle été peuplée d'hommes et d'animaux? Par E. B. d'E.... Amsterdam, M. M. Rey, 1767.
610p. 572.97/En3e

E. B. d'E. *i.e.* Engel, bailli d'Echalens.
Running title: De la population de l'Amérique.

748 **Engelhardt, Moritz von,** 1779-1842.
See RAUMER, KARL GEORG VON, Geognostiche Fragmente . . . 1811.

749 **Engelspach-Larivière, Auguste,** 1799-1831.
Essai géognostique sur les environs de Saint-Pétersbourg. Bruxelles, P. M. de Vroom, 1825.
34p. 554.74/En3d

750 **Englefield, Henry Charles,** 1752-1822.
A description of the principal picturesque beauties, antiquities, and geological phoenomena, of the Isle of Wight . . . With additional observations on the strata of the island, and their continuation in the adjacent parts of Dorsetshire. By Thomas Webster, esq. Illustrated by maps and numerous engravings by W. and G. Cooke, from original drawings by Sir H. Englefield and T. Webster. London, Printed by W. Bulmer and Co., for Payne and Foss, 1816.
238p., 50 plates (incl. maps). q914.228/En3d

751 **Entzelt, Christoph.**
De re metallica, hoc est, de origine, varietate, & natura corporum, metallicorum, lapidum, gemmarum, atq; aliarum, quae ex fodinis eruuntur, rerum, ad medicinae deservientium, Libri III. Francofurti, apud Chr. Egenolphum, [1551].
271p., tables. uncat.

This copy contains an introductory letter by Philippus Melanchthon to the printer requesting that this work be printed, stating that the author is not competing with Agricola but provides an independent study.

The Latin name of the author is Encelius. This is a rare and important pioneer work in systematic mineralogy.

752 **Ercker, Lazarus,** 1530-1594.
Beschreibung aller fürnemisten mineralischen Ertzt und Bergkwercksarten wie dieselbigen und eine jede in sonderheit irer natur und eigenschaft nach auff alle Metaln Probirt und im kleinen fewer sollen versucht werden mit erklärung etlicher fürnemer nützlicher Schmelzwerk . . . Franckfurt am Mayn, 1580.
135p. uncat.

753 **Ercker.**
Aula subterranea domina dominantium, subdita subditorum. Das ist : untererdische Hofhaltung ohne welche weder die Herren regieren noch die Unterthanen gehorchen können. Oder Gründliche Beschreibung derjenigen Sachen, so in der

E S S A I

S U R

L'HISTOIRE NATURELLE

D E

LA MER ADRIATIQUE

PAR LE DOCTEUR

VITALIANO DONATI,

AVEC UNE LETTRE DU DOCTEUR

LEONARD SESLER,

SUR UNE

NOUVELLE ESPECE

D E

PLANTE TERRESTRE,

TRADUIT DE L'ITALIEN.

A LA HAYE,
CHEZ PIERRE DE HONDT,
M. DCC. LVIII.

Cat. No. 678

Tieffe der Erden wachsen als aller Erzen der Königlichen und gemeinen Metallen, auch fürnehmster Mineralien, durch welche nächst Gott, alle Künste, Übungen und Stände der Welt gehandhabet und erhalten werden da dann fürnehmlich hierinn gelehret wird wie sothanige Erz- und Bergwercksarten jede insonderheit ihrer Natur und Eigenschafft gemäss auff alle metalle probirt und im kleinen Feuer Versucht Werden, nebst Erklärung einiger fürnehmer nützlichen Schmeltzwerck im grossen Feuer, Item Ertz scheider bochen waschen und rösten auch Scheidung Goldes, Silbers und anderer Metallen ingleichem Kupffer seigern Messing brennen Salpeder sieden destillation der Scheidwasser und ihren Brauch auch zu nutzmachung anderer mineralischen Berg- und Saitsarten ... Franckfurt, In Verlegung Johann David Zumners ... Gedruckt bey Paulus Lummen, 1672.
332p., illus. q uncat.

Engraved t.-p. added.
Text in German.

754 **Ercker.**
Aula subterranea, domina dominantium, subdita subditorum. Das ist : Untererdische Hofhaltung, ohne welche weder die Herren regieren, noch die Unterthanen gehorchen können. Oder, Gründliche Beschriebung dererjenigen Sachen, so in der Tieffe der Erden wachsen, als aller Ertzen der Königlichen und gemeinen Metallen, auch fürnehmster Mineralien durche welche, nechst Gott, alle Künste, Übungen und Stände der Welt gehandhabet und erhalten werden, da dann fürnehmlich hierin gelehret wird, wie sothanige Ertz- und Bergwercks-Arten, jede insonderheit ihrer Natur und Eigenschafft gemäsz, auf alle metalla probirt, und in kleinem Feuer versucht werden, nebst Erklärung einiger fürnehmer nützlichen Schmeltzwercke im grossen Feuer, Item Ertz scheiden, puchen, waschen, und rösten, auch Scheidung Goldes, Silbers, und anderer Metallen, ingleichem Kupfer saigem, Messing brennen, distillation der Scheidwasser und ihrem Brauch, auch zu Nutzmachung anderer mineralischen Berg- und Saltz-Arten. Anjetzo aber bey dieser neuesten Auflage ... mit vielen nothwendigen Erläuterungen und Geheimnüssen ... um vieles vermehret ... Zusamt einem neuen angehengten gründlichen Unterricht von dem Salpeter pflantzen ... vermehrt und verbessert durch J. E. C. 5.Aufl. Franckfurt am Mayn, J. D. Jung., 1736.
208p., illus. q669/Er2a/1736

755 **Esmarch, Lauritz.**
Historische Nachricht über den Flugsand in Nord-Jütland, Kopenhagen, In der Schultzischen Officin, 1817.
108p. 551.35/Es52h

756 **Esper, Johann Friedrich,** 1732-1781.
Ausführliche Nachricht von neuentdeckten Zoolithen unbekannter vierfüsiger Thiere, und denen sie enthaltenden, so wie verschiedenen andern denkwürdigen Grüften der Obergebürgischen Lande des Marggrafthums Bayreuth. Nürnberg, G. W. Knorrs Erben, 1774.
148p., 14 col. plates. xq569/Es6a

Bibliographical footnotes.

757 **Estner, Franz Joseph Anton,** 1739-1803.
Versuch einer Mineralogie für Anfänger und Liebhaber. Nach des Herrn Bergcommissionsraths Werner's Methode. Wien, J. G. Oehler, 1794-99.
3v. in 4, illus. (part. col.). 549/Es87v

758 **Etschenreutter, Gallus,** fl. 1565-1571.
Aller heylsamen Bäder, Saurbrunnen, un anderer wasser so in Teutschland bekandt und erfahren, auch ihrer Metallen und Mineralien natur krafft tugent und wirckung. [Strassburg, bey Christian Müllers, Erben, 1580].
192p. uncat.

759 **Evans, Lewis,** 1700?-1756.
Geographical, historical, political, philosophical and mechanical essays. The first, containing an analysis of a general map of the middle British colonies in America, and of the country of the confederate Indians; a description of the face of the country, the boundaries of the confederates, and the maritime and inland navigations of the several rivers and lakes contained therein. Philadelphia, Printed by B. Franklin, and D. Hall, and sold by J. & R. Dodsley in Pall-Mall, London, 1755.
32p., fold. map. xf912.73/Ev1

In a case with the following:

1. Pownall, Thomas, A topographical description of such parts of North America . . . 1776.

2. Stevens, Henry Newton, Lewis Evans, his map of the middle British colonies; a comparative account of ten different editions published between 1755 and 1807 . . . 1905.

3. Ten other editions of Lewis Evans' map of 1755. These maps form the basis for Stevens' 1905 monograph (above). Stevens later raised the number of editions to twenty-six; see Stevens, 2nd edition, 1920 (912.73/Evlys/1920) and Stevens, 3rd edition, 1924 (912.73/

Evlys/1924). For University of Illinois copies, contact Rare Book Room.

760 **Evelyn, John,** 1620-1706.
A philosophical discourse of earth, relating to the culture and improvement of it for vegetation, and the propagation of plants, &c., as it was presented to the Royal Society, April 29, 1675. London, Printed for J. Martyn, 1676.
182p. x631/Ev2p

761 **Evelyn.**
Terra: a philosophical discourse of earth. Relating to the culture and improvement of it for vegetation, and the propagation of plants, as it was presented to the Royal Society. New ed. With notes by A. Hunter. York, Printed by A. Ward, for J. Dodsley, London; 1778.
194p., fold. table. x631/Ev2p/1778

First pub. in 1676 under title: A philosophical discourse of earth. Editor's autograph presentation copy to Sir John Pringle, bart. Error in binding: p. [vii]-viii bound before p. [iii].

762 **Eytelwein, Johann Albert,** 1764-1848.
Praktische Anweisung zur Bauart der Faschinenwerke und der dazu gehörigen Anlagen an Flüssen und Strömen, nebst einer Anleitung zur Veranschlagung dieser Baue. 2.Aufl. Berlin, Realschulbuchhandlung, 1818.
116p., 8 fold. plates. 627.1/Ey8p

First ed. issued in 1800 under title: Praktische Anweisung zur Konstrukzion der Faschinenwerke.

763 **Ezquerra del Bayo, Joaquín,** 1793-1859.
Elementos de laboreo de minas precedidos de algunas nociones sobre geognosia y la descripción de varios criaderos de minerales, tanto de España como de otros reinos de Europa. Madrid, Salvador Albert, 1839.
447p., fold. plates, table. 553/Ez7e

764 **Ezquerra del Bayo.**
Datos y observaciones sobre la industria minera, con una descripción característica de los minerales útiles, cuyo beneficio puede ser objeto de las empresas. Madrid, En la impr. de A. Yenes, 1844.
353p., 3 fold. plates. 338.2/Ez7d

765 **Fabre, Jean Antoine,** 1749-1834.
Essai sur la théorie des torrens et des rivières, contenant les moyens les plus simples d'en empêcher les ravages, d'en retrécir le lit & d'en faciliter la navigation, le hallage & la

DELLO SPROFONDAMENTO
DI UNA
COSTA DI MONTE
NEGLI EUGANEI
LETTERA

Del Marchese

ANTONIO - CARLO DONDI OROLOGIO

Socio dell' Accademia delle Scienze, Arti, e Belle Lettere di Padova, e della R. Accademia di Mantova.

DIRETTA

AL CH. SIGNOR AB.

ALBERTO FORTIS

Delle Accademie di Padova, di Napoli, di Bologna, di Mantova, di Siena, di Berlino, di Bordeaux, di Lunden ec.

IN PADOVA MDCCLXXXVII.

NELLA STAMPERIA PENADA

Con Lic. de' Sup.

Cat. No. 679

flottaison. Accompagné d'une discussion sur la navigation intérieure de la France; et terminé par le projet de rendre Paris, port maritime, en faisant remonter à la voile, par la Seine, les navires qui s'arrêtent à Rouen. Ouvrage mis à la portée de tout le monde. A l'usage des ingénieurs & des élèves des ponts & chaussées. Paris, Bidault, 1797.
284p., 8 fold. plates. uncat.

766 **Fabre, Pierre Jean,** 17th cent.
Hydrographum spagyricum. In quo de mira fontium essentia, origine, & tractatur. Tolosae Tectosagum, Apud P. Bosc, 1639.
260p. x615.853/F114h

767 **Fabricius, Georg,** 1516-1574.
De metallicis rebus ac nominibus observationes variae & eruditae . . . Tiguri, n.p., 1566.
32p. uncat.

Bound with Gesner, Konrad, De omni rerum fossilium genere gemmis, lapidibus, metallis, et huiusmodi, libri aliquot, pleriique nunc primum editi. Tiguri, 1565.

768 **Fabricius, Johann Albert,** 1668-1736.
Hydrotheologie; oder, Versuch, durch aufmerksame Betrachtung der Eigenschaften, reichen Austheilung und Bewegung der Wasser, die Menschen zur Liebe und Bewunderung ihres gütigsten, weisesten, machtigsten Schöpfers zu ermuntern. Nebst einem Verzeichniss von alten und neuen See- und Wasser-Rechten, wie auch Materien und Schriften, die dahin gehören, unter XL.Titul gebracht. Hamburg, König und Richter, 1734.
[26], 436p., front. 231.5/F11h

769 **Fairholme, George,** 1789-1846.
Miscellaneous papers on geology and other branches of science, published in various periodical works, and now collected together. By George Fairholme, Esq., and others, 1837 [i.e. 1832-46].
535p., illus. 550.8/F16m

Title in manuscript.

Binder's title: Papers on geology, &c.

Volume of articles and mounted clippings from British publications, collected by Fairholme; added, in ms., are his copious notes, a table of contents and pagination. Includes reviews and criticism of his "A general view of the geology of Scripture" (London, 1833) and "New and conclusive physical demonstrations, both of the fact and period of the Mosaic deluge" (London, 1837).

"Positions géologiques en vérification directe de la chronologie de la Bible," par George Fairholme. Munich, 1834 (32p.) inserted following p.530.

770 **Falconer, Hugh,** 1808-1865.
Fauna antiqua Sivalensis, being the fossil zoology of the Sewalik Hills, in the north of India [by] Hugh Falconer and Proby T. Cautley. Illustrations, pt. 1-9. London, 1846-49.
1v. of plates. f569/F18f

Title from Brit. Mus. Cat.

771 **Falconer.**
Palaeontological memoirs and notes of the late Hugh Falconer... With a biographical sketch of the author. Comp. and ed. by Charles Murchison... London, R. Hardwicke, 1868.
2v., front. (port.) illus., plates, maps. 560.8/F18p.

Contents. — I. Fauna antiqua sivalensis. — II. Mastodon, elephant, rhinoceros, ossiferous caves, primeval man and his contemporaries.

Portrait signed H. Falconer.

772 **Falloppio, Gabriello,** 1523-1562.
De medicatis aquis, atque de fossilibus. Tractatus pulcherrimus, ac maxime utilis. Ab Andrea Marcolino Fanestri... collectus. Accessit eiusdem Andreae duplex epistola; in quarum altera ad lectorem, & huius libri inter reliqua utilitas, & docendi modus, ac totius rei, quae in hoc ipso opere continetur summa breviter explicatur. Cum indice rerum magis observandarum copiosissimo, ac Capitum omnium, quae in hoc opusculo tractantur, cathalago, quem sequens pagina indicabit. Venetiis, Apud L. Avantium, 1564.
176 leaves. x553.7/F19d

773 **Falloppio.**
De medicatis aquis atque de fossilibus. Tractatus pulcherrimus, ac maxime utilis; ab Andrea Marcolino Fanestri medico ipsius discipulo amantissimo collectus. Accessit eiusdem Andree duplex epistola: in quarum altera ad lectorem, & huius libri inter reliqua utilitas, & docendi modus, ac totius rei, quae in hoc ipso opere continetur, summa breviter explicatur. Cum indice rerum magis observandarum copiosissimo, ac capitum omnium, quae in hoc opusculo tractantur, catalogo. Quem decima octava pagella indicabit. Cum privilegio. Venetiis, Ex officina L. Avantij, 1569.
176 leaves. x553.7/F19d/1569

774 **Fantonetti, Giovanni Battista Bernardo,** 1791-1861.
Considerazioni alle Osservazioni e ricerche mineralogico-chimiche sopra alcune valli dell'Ossola pubblicate in Milano nel 1819 dal chimico Gaetano Rosina, del mineralogista dottor fisico Giovambattista Fantonetti . . . Torino, Vedova Pomba e figli, 1821.
85p. 549/R731oYf

775 **Farey, John,** 1766-1826.
See GREAT BRITAIN. BOARD OF AGRICULTURE, General view of the agriculture and minerals of Derbyshire . . . 1811.

776 **Faujas de Saint-Fond, Barthélemy,** 1741-1819.
See PALISSY, BERNARD, Oeuvres, . . . 1777.

777 **Faujas de Saint-Fond.**
Recherches sur la pouzzolane, sur la théorie de la chaux et sur la cause de la dureté du mortier, avec la composition de différens cimens en pouzzolane, & la manière de les employer, tant pour les bassins, aqueducs, réservoirs, cîternes & autres ouvrages dans l'eau, que pour les terrasses, bétons et autres constructions en plein air. Grenoble, J. Cuchet, 1778.
125p. 552.2/F27r

Extrait des Recherches sur les volcans éteints du Vivarais et du Velay.

778 **Faujas de Saint-Fond.**
Recherches sur la pouzzolane, sur la théorie de la chaux et sur la cause de la dureté du mortier, avec la composition de différens cimens en pouzzolane, & la manière de les employer, tant pour les bassins, aqueducs, réservoirs, cîternes & autres ouvrages dans l'eau, que pour les terrasses, bétons & autres constructions en plein air. Grenoble, J. Cuchet, 1778.
125p. 549/M62m

Bound with Millin, Aubin Louis, Minéralogie homérique; . . . 1790.

779 **Faujas de Saint-Fond.**
Recherches sur les volcans éteints du Vivarais et du Velay; avec un discours sur les volcans brûlans, des mémoires analytiques sur les schorls, la zéolite, le basalte, la pouzzolane, les laves & les différentes substances qui s'y trouvent engagées, &c. Grenoble, J. Cuchet, 1778.
460p., plates. f551.21/F27r

Library owns 2 copies; c.2 contains several plates in facsimile; c.1 is complete.

LE

TREMBLE-TERRE

OV SONT CONTENVS
ſes cauſes, ſignes, effets
& remedes.

*Par Louys du Thoum Docteur
és Droicts, & Aduocat
en la Cour.*

A Monſeigneur l'Illuſtriſſime & Reuerendiſſime LOVYS DE LA VALETE
Archeueſque de Toloſe.

A BOVRDEAVS,

Par GILBERT VERNOY.

M. DC. XVI.

Cat. No. 711

780 **Faujas de Saint-Fond.**
Mémoire sur la manière de reconnoitre les différentes especes de Pouzzolane, et de les employer dans les constructions sous l'eau et hors de l'eau; pour servir de Suite & de Supplement aux Recherches sur la Pouzzolane . . . Amsterdam, Nyon, 1780.
51p., illus. 549/M62m

Bound with Millin, Aubin-Louis, Minéralogie homérique; . . . 1790.

781 **Faujas de Saint-Fond.**
Minéralogie des volcans, ou Description de toutes les substances produites ou rejetées par les feux souterrains . . . Paris, Cuchet, 1784.
511p., 3 plates. 552.2/F27m

782 **Faujas de Saint-Fond.**
Essai sur l'histoire naturelle des roches de trapp, contenant leur analyse, & des recherches sur leurs caractères distinctifs; suivi du tableau systématique de toutes les espèces & variétés de trapps & des roches qui ont pour base cette pierre. Paris, 1788.
159p. 552/F27e

783 **Faujas de Saint-Fond.**
Voyage en Angleterre, en Écosse et aux îles Hébrides; ayant pour objet les sciences, les arts, l'histoire naturelle et les moeurs; avec la description minéralogique du pays de Newcastle, des montagnes du Derbyshire, des environs d'Edinburgh, de Glasgow, de Perth, de S. Andrews, du duché d'Inverary et de la grotte de Fingal. Paris, H. J. Janson, 1797.
2v., fold. plates. 914.2/F27v

784 **Faujas de Saint-Fond.**
Histoire naturelle de la montagne de Saint-Pierre de Maestricht. Paris, H. J. Jansen, [1799].
263p., 54 plates. xq574/F27h

Error in binding: sign. 11 (p.81-88) bound after sign. 12 (p.89-96).

785 **Faujas de Saint-Fond.**
Des émaux, des verres, et des pierres ponces des volcans brûlans et des volcans éteints . . . Paris, A. Belin, imprimeur du jardin du roi, 1816.
36p. 552.1/F27d

From Mémoires du Muséum d'histoire naturelle, t.3, 1817.

786 **Favre, Alphonse,** 1815-1890.
Considérations géologiques sur le Mont Salève et sur les terrains des environs de Genève. Genève, Impr. de J. G. Fick, 1843.
113p., 2 col. plates (incl. map). 554.94/F279c

Extrait du tome X des Mémoires de la Société de physique et d'histoire naturelle de Genève.

787 **Featherstonhaugh, George William,** 1780-1866.
Geological report of an examination made in 1834 of the elevated country between the Missouri and Red rivers . . . Washington, Printed by Gales and Seaton, 1835.
97p., front. (fold. profile). 557/W27f

Issued also as House doc. no. 151, 23d Cong., 2d sess.

788 **Featherstonhaugh.**
Report of a geological reconnaissance made in 1835, from the seat of government, by the way of Green Bay and the Wisconsin Territory to the Coteau de Prairie, an elevated ridge dividing the Missouri from the St. Peter's River. By G. W. Featherstonhaugh, U.S. geologist. Washington, Printed by Gales and Seaton, 1836.
168p., 4 plates, 2 fold. maps. 557/Un3f

Issued also as Senate doc. no. 333, 24th Congress, 1st sess.
Maps lacking.

789 **Feijóo y Montenegro, Benito Jerónimo,** 1676-1764.
Nuevo systhema, sobre la causa physica de los terremotos, explicado por los phenomenos electricos, y adaptado al que padeciò España en primero de Noviembre del año antecedente de 1755. Puerto de Santa Maria, Impr. de la Casa Real de las Cadenas, 1756.
[88], 56p. x551.22/F32n

Dedication signed: Juan Luis Roche.

790 **Feijóo y Montenegro.**
El terremoto, y su uso, dictamen de Benito Feijóo . . . explorado por Juan de Zuñiga. Toledo, Francisco Martin, 1756.
60p. x551.22/F32nY3 and x551.22/F32nYz

791 **Fenoyl.**
Anatomie du monde sublunaire, contenant les demonstrations des dispositions, de la constitution & mouvemens de toutes les parties du globe élémentaire, depuis sa circonference jusqu'à son centre. Lyon, 1707.
198p. 551.1/F36a

792 **Fenwick, Thomas.**
A theoretical and practical treatise on subterraneous surveying and the magnetic variation of the needle. Newcastle upon Tyne, S. Hodgson, 1804.
207p., diagrs., tables. 622.14/F36t

793 **Ferber, Johann Jakob,** 1743-1790.
Briefe aus Wälschland über natürliche Merkwürdigkeiten dieses Landes, an den Herausgeber derselben Ignatz Edler von Born. Prag, W. Gerle, 1773.
407p. 549/F37b

Bound with Born, Ignaz, Briefe über mineralogische Gegenstände auf seiner Reise durch das Temeswarer Bannat . . . 1774.

794 **Ferber.**
See ARDUINO, GIOVANNI, Raccolta de Memorie Chimico-Mineralogiche . . . 1775.

795 **Ferber.**
Lettres sur la minéralogie et sur divers autres objets d'histoire naturelle de l'Italie; écrites par Mr. Ferber à Mr. le Chev. de Born. Ouvrage traduit de l'allemand, enrichi de notes & d'observations faites sur les lieux par Mr. le B. de Dietrich . . . À Strasbourg, Chez Bauer & Treuttel, [Imprimerie de J. H. Heitz, imprimeur de l'Univ.], 1776.
1v. in 2, [507p.], illus. 549/F37bFd

Interleaved with ms. notes.

796 **Ferber.**
Travels through Italy, in the years 1771 and 1772. Described in a series of letters to Baron Born, on the natural history, particularly the mountains and volcanoes of that country. Translated from the German, with explanatory notes, and a preface on the present state and future improvement of mineralogy. By R. E. Raspe. London, for L. Davis, 1776.
377p. x914.5/F373bEr

797 **Ferber.**
. . . Versuch einer Oryktographie von Derbyshire in England . . . Mietau, J. F. Hinz, 1776 . . .
104p., 2 fold. plates, 2 fold. diagr. 554.251/F37v

798 **Ferber.**
Neue Beyträge zur Mineralgeschichte verschiedener Länder. I Bd., der zugleich Nachrichten von einigen chymischen Fabriken enthält . . . Mietau, J. F. Hinz, 1778.
462p., fold. tables. 549.09/F37n

DE RE METALLICA, HOC EST, DE ORIGINE, Varietate, & Natura Corporum Metallicorum, Lapidum, Gemmarum, atq; aliarum, quæ ex fodinis eruuntur, rerum, ad Medicinæ usum deseruientium, LIBRI III.

Autore

CHRISTOPHORO Encelio Salueldensi.

Cum priuilegio Imper.

FRANC. Apud Chr. Egenolphum.

Cat. No. 751

799 **Ferber.**
Physicalisch-Metallurgische Abhandlung über die gebirge und Bergwerke in Ungarn . . . Nebst einer Beschreibung des Steirischen Eisenschmelzens und Stahlmachens von einem Ungenannten. Mit Kupfern . . . Berlin und Stettin, bey F. Nicolai, 1780.
328p., 4 fold. plates. 669/F37p

800 **Ferber.**
Untersuchung der Hypothese von der Verwandlung der mineralischen Körper in einander. Berlin, F. Maurer, 1788.
72p. 549/F37h

801 **Ferber.**
Mineralogische und metallurgische Bemerkungen in Neuchatel, Franche Comté und Bourgogne, im Jahr 1788 angestellt. Berlin, August Mylius, 1789.
77p., 4 fold. diagrs. uncat.

802 **Ferrara, Francesco,** 1767-1850.
Descrizione dell' Etna con la storia delle eruzioni e il catalogo dei prodotti . . . Palermo, Presso L. Dato, 1818.
256p., 5 fold. plates (incl. map). 551.21/F41d

803 **Ferrer Maldonado, Lorenzo,** d. 1625.
Imagen del mundo, sobre la esfera, cosmografia, y geografia, teorica de planetas, y arte de navegar . . . Por el capitan Lorenço Ferrer Maldonado . . . Alcala, J. Garcia y A. Duplastre, 1626.
276p., illus., diagrs., tables. x525/F37i

804 **Férussac, André Étienne Just Pascal Joseph François d'Audebard,** 1786-1836.
Mémoires géologiques sur les terreins formés sous l'eau douce par les débris fossiles des mollusques vivant sur la terre ou dans l'eau non salée . . . Paris, Poulet, 1814.
76p. 564/F41m

805 **Fichtel, Johann Ehrenreich von,** 1732-1795.
Geschichte des Steinsalzes und der Steinsalzgruben im Grossfürstenthum Siebenbürgen, mit einer, das Streichen des unterirrdischen Salzstockes, durch mehrere Länder andeutenden Karte, und andern Kupfern. Nürnberg, Raspischen Buchhandlung, 1780.
134p., 4 fold. col. plates. 554.392/F44b

Also includes 7 fold. plates, one of them colored, separately numbered and not listed with text.

Bound with his Versteinerungen des Grossefürstenthums Siebenbürgen . . . 1780.

806 **Fichtel.**
Nachricht von den Versteinerungen des Grossfürstenthums Siebenbürgen, mit einem Anhange und beygefügter Tabelle über die sämmtlichen Mineralien und Fossilien dieses Landes. Nürnberg, Raspischen Buchhandlung, 1780.
158p. 554.392/F44b

Bound with his Geschichte des Steinsalzes . . . 1780.
Half-title : Beytrag zur Mineralgeschichte von Siebenbürgen.

807 **Fichtel.**
Mineralogische Aufsätze. Wien, M. A. Schmidt, 1794.
374p. 549/F44m

808 **Filson, John,** 1747?-1788.
The discovery and present state of Kentucky. And an introduction to the topography and natural history of that rich and important country; also Colonel Daniel Boon's narrative of the wars of Kentucky; with an account of the Indian Nations within the limits of the United States, their manners, customs, religion, and their origin; and the stages and distances between Philadelphia and the Falls of the Ohio, from Pittsburgh to Pensacola, and several other places. London, for John Stockdale, 1793.
57p., fold. map. x976.9/F48d/1793

809 **Filson.**
Histoire de Kentucke, nouvelle colonie à l'ouest de la Virginie: contenant, 1°. La découverte, l'acquisition, l'établissement, la description topographique, l'histoire naturelle, &c. du territoire: 2°. la relation historique du Colonel Boon, un des premiers Colons, sur les guerres contre les naturels: 3°. l'assemblée des Piankashaws au Poste Saint Vincent: 4°. un exposé succinct des nations indiennes qui habitent dans les limites des treize États-Unis, de leurs moeurs & coutumes, & des reflexions sur leur origine; & autres pièces: avec une carte. Ouvrage pour servir de suite aux Lettres d'un cultivateur américain. Traduit de l'anglois, de John Filson; par M. Parraud. Paris, Buisson, 1785.
234p., fold. map. x976.9/F48dFp/1785

Trans. of The discovery, settlement and present state of Kentucke . . . , Wilmington, Del., 1784. The translation includes several additions (p. 146-231) not found in the original and omits the distance tables given in that edition.

810 **Filson.**
See IMLAY, GILBERT, A topographical description of the western territory of North America . . . Filson supplement . . . 1793 and 1797.

811 **Finch, John,** fl. 1835.
Travels in the United States of America and Canada, containing some account of their scientific institutions, and a few notices of the geology and mineralogy of those countries. To which is added, an essay on the natural boundaries of Empires. London, Longman, Rees, Orme, Brown, Green and Longman, 1833.
455p. 917.3/F49t

"On the natural boundaries of empires; an essay to shew the effects of the geological structure of the earth on the political boundaries of nations, by J. Finch": p.[333]-455.

812 **Fischer von Waldheim, Gotthelf,** 1771-1853.
Oryctographie du gouvernement de Moscou. Ouvrage publié aux frais de la Société impériale des naturalistes de Moscou. Moscou, Impr. d'A. Semen, 1837.
202p., illus., 6 col. maps, plates, port. xf560.94731/F52o/1837

Added t.p., engr., with the dates 1830-1837.

813 **Fischer von Waldheim.**
Chilonopsis; novum genus testarum e familia Helicum, icone et descriptione illustratum [et Notice sur quelques fossiles de Russie. Moscou, 1848].
15p., 3 plates. 564/F52c

Extrait[s] du Bull. de la Soc. Impér. des Natural. de Moscou, tome XXI, 1848.
Bound with "Notice sur quelques fossiles de la Russie," 15p.
Bibliographical footnotes.

814 **Fish, Samuel.**
Granite rock, *See* Scientific tracts, for the diffusion of useful knowledge . . . 1836.

815 **Fish, Samuel.**
Theory of the earth, *See* Scientific tracts, for the diffusion of useful knowledge . . . 1836.

816 **Fitton, William Henry,** 1780-1861.
A geological sketch of the vicinity of Hastings. London, Longman, Rees, Orme, Brown, Green, and Longman, etc. etc., 1833.
94p., illus., col. plate. uncat.

Beschreibung/

Allerfürnemisten Mineralischen Ertzt vnnd

Bergkwercks arten / wie dieselbigen / vnd eine jede in sonderheit / jrer natur vnd eigenschafft nach / auff alle Metaln Probirt / vnd im kleinen fewer sollen versucht werden / mit erklärung etlicher fürnemer nützlicher Schmeltzwerck / im grossen fewer / auch scheidung Goldts / Silbers / vnd anderer Metaln / Sampt einem bericht deß Kupffersaigerns / Messing brennens / vnd Salpeter siedens / auch aller saltzigen Minerischen proben / vnd was denen allen anhengig / in fünff Bücher verfast / Deßgleichen zuuorn niemals in Druck kommen.

Allen Liebhabern der Feuwerkünste / jungen Probirern vnd Bergkleuten zu nutz / mit schönen Figuren vnd abriß der Instrument / trewlich vnd fleissig an tag geben.

Auffs newe an vielen orten mit besserer außführung / vnd mehrern Figurn erklert / Durch

Den Weitberühmten Lazarum Erckern / der Röm. Kay. May. Obersten Bergkmeister vnd Buchhalter / im Königreich Behem / 2c.

Mit Röm. Keys. May. Priuilegien.

Getruckt zu Franckfurt am Mayn / M. D. LXXX.

Cat. No. 752

817 **Fitzroy, Robert,** 1805-1865, ed.
Narrative of the surveying voyages of His Majesty's ships Adventure and Beagle, between the years 1826 and 1836, describing their examination of the southern shores of South America, and the Beagle's circumnavigation of the globe . . . London, H. Colburn, 1839.
3v. and appendix, fronts. (v. 1, 2), plates, maps (part. fold.), diagrs.
x508.3/F58n

Vol. 3 was reissued the same year in separate form.

818 **Flade, Christian Gottlob.**
De re Metallica Midianitarum Edomitarum et Phonenicum. Lipsiae, Officina Klaubarthia, [1791].
549/F59

Dissertatio I, 28p. Dissertatio II, 16p.
Bound with other treatises on mineralogical subjects, by various authors.

819 **Flamichon,** geographical engineer.
Théorie de la terre, déduite de l'organisation des Pyrenées et pays adjacens, rédigée par Jn. Latapie, sur les manuscrits de M. Flamichon. Pau, Tonnet, 1816.
311p. 551.4/F61t

Editor's autograph copy.

820 **Flamsteed, John,** 1646-1719.
A letter concerning earthquakes, written in the year 1693, by the late celebrated astronomer, Mr. John Flamsteed . . . to a gentleman then residing at Turin in Savoy, on occasion of the destruction of Catanea, and many other cities, towns and villages, in Sicily, in the year 1692. London, Printed for A. Millar, 1750.
20p., illus. x551.22/F582ℓ

821 **Fleming, Robert,** 1630-1694.
A discourse on earthquakes, as supernatural and premonitory signs to a nation, especially as to what occurred in the year 1692. With some remarks respecting that assurance of mind which is attainable in the light and power of religion, under the greatest surprizals and terrors of sense. Also an enquiry into the ground of our hopes and fears, as to the public state of the church of Christ in these days . . . London, Printed for G. Terry, [1793].
84p. x252/F629d/1793

Authorities differ as to whether the older or younger Robert

Fleming was the author of this work, the first edition of which was published in 1693.

822 **Flint, Timothy,** 1780-1840.
The history and geography of the Mississippi valley. To which is appended a condensed physical geography of the Atlantic United States, and the whole American continent. 2d ed. Cincinnati, E. H. Flint and L. R. Lincoln, 1832.
2v. in 1. x917.3/F64h2

Vol. 2 has title: The United States and the other divisions of the American continent. By Timothy Flint . . . vol. II.

823 **Flint.**
The history and geography of the Mississippi valley. To which is appended a condensed physical geography of the Atlantic United States, and the whole American continent. 3d ed. Cincinnati, E. H. Flint and Boston, Carter, Hendee & co., 1833.
2v. in 1. x917.3/F64h3

Vol. 2 has title: The United States and the other divisions of the American continent. By Timothy Flint . . . vol. II.

824 **Flint.**
Lectures upon natural history, geology, chemistry, the application of steam, and interesting discoveries in the arts. Boston, Lilly, Wait, Colman and Holden; [etc., etc.], 1833.
408p. x504/F646ℓ

825 **Flourens, Marie Jean Pierre,** 1794-1867.
Cuvier. Histoire de ses travaux . . . 2.éd., rev. et cor. Paris, Paulin, 1845.
524p. 590/F668c

"Liste des ouvrages de M. Cuvier": p.[310]-322.

826 **Flourens.**
See CUVIER, GEORGES, Éloges historiques . . . [18-].

827 **Fontenelle, Bernard le Bovier de,** 1657-1757.
Entretiens sur la pluralité des mondes. Nouvelle édition augmentée. Londres, Paul & Isaak Vaillant, 1710.
125p., illus. x844/F73K/1710

The library owns Fontenelle's *Oeuvres diverses,* La Haye, 3 v., 1728 (x844/F73K/1728) and *Oeuvres,* London, 2 v. (x844/F73/I 1784).

828 **Fontenelle.**
A discovery of new worlds. Trans. A. Behn. London, William Canning, 1688.
158p. x844/F73/OeEb

The following other editions exist with various titles: The theory or system of several new inhabited worlds, London, 1700 (x828/B39h/1700); A plurality of worlds, London, 1702 (x844/F73/OeEg); The theory or system of several new inhabited worlds, London, 1718 (x828/B39a/1718); A conversation on the plurality of worlds, Glasgow, 1749 (x844/F73/OeE); Conversations on the plurality of worlds, Dublin, 1761 (x844/F73/OeE/1761).

829 **Forbes, Edward,** 1815-1854.

A history of British starfishes, and other animals of the class Echinodermata. London, John Van Voorst, 1841.

267p., illus. 593.9/F74

830 **Forbes.**

The relations of natural history to geology and the arts; a lecture introductory to the course to be delivered during the Session 1851-52. London, Eyre and Spottiswoode, for H. M. Stationery Office, 1851.

16p. 504/F74r

At head of title: Museum of Practical Geology. Government School of Mines and of Science Applied to the Arts.

831 **Forbes.**

. . . Tertiary fluvio-marine formation of the Isle of Wight. Pub. by order of the lords commissioners of Her Majesty's Treasury. London, Printed for H. M. Stationery off., Longman, Brown, Green, and Longmans, 1856.

162p., front., fold. table, illus., 10 plates (partly fold., incl. map). 554.2/F74t

Memoirs of the Geological survey of Great Britain, and of the Museum of practical geology.

832 **Forbes, James David,** 1809-1868.

Meteorology: pamphlets. [v.p., 1825-55].

14 pamphlets in 1v., illus., plates (part. fold., 1 col.). London, R. & J. E. Taylor. x551.508/M565

Binder's title.

Contents: [1] Forbes, J. D. Supplementary report on meteorology. 1841. [2] Forbes, J. D. On a remarkable structure observed by the author on the ice of glaciers. [1842]. [3] Forbes, J. D. Historical remarks on the first discovery of the real structure of glacier ice. [1843]. [4] Forbes, J. D. From Professor Forbes to Professor Guyot. 1842. [5] Forbes, J. D. Professor Forbes' account of his recent observations on glaciers. [1842]. [6] Poole, Henry. On the meteorology of the Albion mines, Nova Scotia. [1854]. [7] Richardson, Sir John. Results of thermometrical observations made at Sir Edward Parry's several wintering-places. [183-?]. [8] Richardson, Sir John. Remarks on the climate and vegetable productions of the Hudson's Bay countries. [1825]. [9] Richardson, Sir John. Observations on solar radiation. [1841]. [10] Richardson, Sir John. On the frozen soil of North America. [1841]. [11] Sabine, Sir Edward. Observations made

Geographical, Hiſtorical,
Political, Philoſophical and Mechanical

E S S A Y S.

THE FIRST, CONTAINING

A N

A N A L Y S I S

Of a GENERAL M A P of the

MIDDLE BRITISH COLONIES

I N

A M E R I C A;

And of the COUNTRY of the *Confederate Indians:*

A DESCRIPTION of the Face of the Country;

The BOUNDARIES of the CONFEDERATES;

A N D T H E

Maritime and Inland NAVIGATIONS of the ſeveral R I V E R S and L A K E S contained therein.

By *L E W I S E V A N S.*

P H I L A D E L P H I A:
Printed by B. F R A N K L I N, and D. H A L L. MDCCLV.
And ſold by R. and J. D O D S L E Y, in *Pall-Mall, London.*

Cat. No. 759

at the magnetic observatories of Toronto in Canada . . . 1841. [12] Sabine, Sir Edward. On some of the results obtained at the British colonial magnetic observatories. [1854?]. [13] Greg, R. P. An essay on meteorites. [1855]. [14] Belville, J. H. A manual of the barometer. 1849.

833 **Forbes.**

Professor Forbes' account of his recent observations on glaciers.

26p. x551.508/M565

4 letters, dated; various places, 4 July 1842-5 Oct. 1842.

Communicated to Professor Jameson, Editor of the Edinburgh New Philosophical Journal.

Bound with Meteorology; [pamphlets].

834 **Forbes.**

On a remarkable structure observed by the author in the ice of glaciers. (From the Edinburgh New Philosophical Journal of January 1842).

8p., 1 plate. x551.508/M565

Read to the Royal Society, Edinburgh, Dec. 6, 1841.

Bound with Meteorology; [pamphlets].

835 **Forbes.**

First, Second & Third Letters on Glaciers; Sixth, Seventh & Eighth Letters on Glaciers; Ninth Letter on Glaciers. [Edinburgh, 1842-1845].

uncat.

Extracted from the Edinburgh New Philosophical Journal, Oct., 1842; Oct. 1844; April, 1845.

836 **Forbes.**

Historical remarks on the first discovery of the real structure of glacier ice. [1843].

20p. x551.508/M565

Bound with Meteorology; [pamphlets].

837 **Forbes.**

Travels through the Alps of Savoy and other parts of the Pennine chain, with observations on the phenomena of glaciers. Edinburgh, A. and C. Black; [etc., etc.], 1843.

424p., 13 plates (incl. front.), 2 maps. 914.94/F74tr

838 **Forbes.**

Reisen in den Savoyer Alpen und in anderen Theilen der Penninen-Kette, nebst Beobachtungen über die Gletscher. Bearb. von Gustav Leonhard. Stuttgart, E. Schweizerbart, 1845.

386p., illus., plates, maps. 914.94/F74trG

Translation of: Travels through the Alps of Savoy and other parts of the Pennine chain.

839 **Forbes.**
Norway and its glaciers, visited in 1851; followed by journals of excursions in the high Alps of Dauphiné, Berne and Savoy. Edinburgh, A. and C. Black, 1853.
349p., front., illus., 10 col. plates, 2 maps (1 fold.). 551.31/F74n

840 **Forbes.**
Occasional papers on the theory of glaciers, now first collected and chronologically arranged. With a prefatory note on the recent progress and present aspect of the theory. Edinburgh, Adam and Charles Black, 1859.
278p. 551.31/F74

841 **Forchhammer, Johan Georg,** 1794-1865.
Om de geognostiske forhold i en deel af Sjelland og naboeöerne. [København, 1826].
p.[247]-280, 4 col. plates. uncat.

Disbound from the series, Royal Danish Society of Sciences. Works.

842 **Forchhammer.**
Geognostiske bidrag. [København, 1828].
p.[365]-378. uncat.

Disbound from the series, Royal Danish Society of Sciences. Works.

843 **Forchhammer.**
Danmarks geognostiske forhold, forsaavidt som de ere afhaengige af danneiser, der ere sluttede, fremstillede i et indbydelsesskrift til Reformationsfesten den 14de novbr. 1835. Kjöbenhavn, Trykt hos directeur J. H. Schultz, [1835].
112p., illus., 1 fold. geol. map. uncat.

844 **Fordyce, William.**
See Memoirs concerning Herculaneum . . . illustrated with notes by William Fordyce . . . 1750.

845 **Forsius, Sigfridus Aronus,** 1550-1624.
Minerographia, thet är, mineralers ächskillighe jordeslags, metallers eller malmars och edle steenars beskrifwelse. Aff förnemlige authoribus samman hämptat, och medh . . . Stockholm, hoos Ignattum Meurer, Ahr 1643.
160p. uncat.

On minerals, metals, and precious stones.

846 **Forster, Westgarth,** 1772-1835.
A treatise on a section of the strata, commencing near Newcastle upon Tyne, and concluding on the west side of the mountain of Cross-Fell. With remarks on mineral veins in general, and engraved figures of some of the different species of those productions. To which are added tables of the strata in Yorkshire and Derbyshire. The whole intended to amuse the mineralogist, and assist the miner in his professional researches. Newcastle, Printed for the author, by Preston & Heaton, 1809.
156p., diagrs. x554.2/F77t

847 **Forster.**
A treatise on a section of the strata, from Newcastle-upon-Tyne, to the mountain of Cross Fell, in Cumberland; with remarks on mineral veins in general. Also, tables of the strata, in Yorkshire, Derbyshire, &c. To which is added, a treatise on the discovery, the opening, and the working of lead mines; with the dressing and smelting of lead ores. 2d ed., greatly enlarged. Alston, Cumberland, Printed, for the author, at the Geological Press, 1821.
422p., illus., plates, diagrs. 554.2/F77t/1821

848 **Forster.**
A Treatise on a section of the strata from Newcastle-upon-Tyne to Cross Fell, with remarks on mineral veins; also, Tables of the strata in Northumberland, Durham, Yorkshire, Derbyshire, West Cumberland, and Fifeshire; to which is added, a Treatise on the discovery, the opening, and the working of lead mines, and the dressing and smelting of lead ores. Third edition, revised and corrected to the present time, by the Rev. W. Nall, M.A., with a memoir of the author's life. Newcastle-upon-Tyne, Andres Reid; London, Edward Stanford, 1883.
208p., illus., fold. plates. uncat.

849 **Fortis, Alberto,** 1741-1803.
Della valle vulcanico-marina di Roncà, nel territorio veronese, memoria orittografica del Abate Fortis. Venezia, C. Palese, 1778.
70p., illus. 554.531/F77d

850 **Fortis.**
Mémoires pour servir à l'histoire naturelle et principalement à l'oryctographie de l'Italie, et des pays adjacens. À Paris, Chez J. J. Fuchs, 1802.

GABRIELIS

. FALLOPPII MVTINENSIS, MEDICI, AC PHILOSOPHI PRAESTANTISSIMI,

DE MEDICATIS AQVIS ATQVE DE FOSSILIBVS

TRACTATVS PVLCHERRIMVS, ac maxime utilis; ab Andrea Marcolino Faneſtri medico ipſius diſcipulo amantiſſimo collectus.

Acceſsit eiuſdem Andreę duplex Epiſtola: in quarum altera ad lectorem, & huius libri inter reliqua vtilitas, & docendi modus, ac totius rei, quæ in hoc ipſo opere continetur, ſumma breuiter explicatur.

Cum indice rerum magis obſeruandarum copioſiſsimo, ac Capitum omnium, quæ in hoc opuſculo tractantur, catalogo, quem decima octaua pagella indicabit.

Cum Priuilegio.

VENETIIS,

Ex officina Ludouici Auantij.

MDLXIX.

Cat. No. 773

2v., plates. 554.5/F77m

Includes "Quatre lettres relatives à des expériences hydroscopiques et métalloscopiques," by Charles Amoretti and Prof. Spallanzani, and other letters by Jean Arduino, J. Festari and Comte N. da Rio.

851 **Foster, John Wells,** 1815-1873.
Geological map of Isle Royale, Lake Superior, Michigan. Made pursuant to the Act of Congress, approved March 1st, 1847, by J. W. Foster, G. J. D. Whitney, assisted by S. W. Hill, W. S. Chlatter. New York, Akerman Lithr., [1847?].
Col. map, 42.5 x 61 cm. G4112/.I8/1847/.F7

Scale, ca. 1:1,125,000.

852 **Foster.**
Report on the geology and topography of a portion of the Lake Superior land district, in the state of Michigan: by J. W. Foster and J. D. Whitney, United States geologist. In two parts. Part I. Copper Lands; Pt. II. The Iron Region. Washington, Printed for the House of Representatives, 1850-51.
2v., atlas. 557/M58f
557/M58f/atlas

U.S. 31st Cong., 1st sess. House ex. doc. no. 69.

853 **Fothergill, Anthony,** 1732?-1813.
A new experimental inquiry into the nature and qualities of the Cheltenham water. To which are now added, observations on sundry other waters: shewing how their properties may be ascertained, their virtues preserved, their impurities corrected, etc. With an appendix on the mephitic-alkaline water; a new and approved remedy against the stone and gravel. Bath, Printed by R. Cruttwell, [etc., etc.], 1788.
122p. x553.7/F82n/1788

854 **Fougeroux de Bondaroy, Auguste Denis,** 1732-1789.
Autograph draft of a report to the Académie des sciences by Fougeroux de Bondaroy and Lavoisier on Guettard's Mémoires sur la minéralogie du Dauphiné. [ca. 1778].
[4]p. MS/549/G93mYf

Holograph.
Guettard's Mémoires were published in 1779.
Accompanied by a descriptive note by the bookdealer, Emil Offenbacher.

855 **Fournel, Henri Jérôme Marie,** 1799-1876.
Etude des gîtes houillers et métallifères du Bocage vendéen faite en 1834 et 1835. Paris, Imprimerie royale, 1836.
206p., atlas of maps and diagrams. 553/F827e
553/F827e/atlas

France. Administration des mines. Études de gîtes minéraux.

856 **Fournet, Joseph Jean Baptiste Xavier,** 1801-1869.
Sur le diluvium de la France. Lyon, Impr. de L. Boitel, 1843.
38p. 551.313/F827s

857 **Fournet.**
Histoire de la dolomie. [Lyon, 1847?].
135p., diagr., tables. 549.78/F82h

Detached from Annales de la Société royale d'agriculture, histoire naturelle et arts utiles de Lyon, v.10.

858 **France. Commission Scientifique du Nord,** Voyages . . .
See DUROCHER, JOSEPH MARIE ELIZABETH, Voyages . . . 1843, and ROBERT, EUGÈNE, Voyages . . . 1845.

859 **France. Service de la Carte Géologique de la France.**
Carte géologique de la France, exécutée sous la direction de Mr. Brochant de Villiers par MM.Dufrénoy et Élie de Beaumont . . . terminée en 1840 . . . [Paris, 1841].
554.4/F84e/sup.

Col. map on 6 sheets.

860 **France. Service de la Carte Géologique de la France.**
Explication de la carte géologique de la France. Paris, Imprimerie royale [etc.], 1841-79.
4v. in 5, atlas to v.4. 554.4/F84e

Library owns v.1, v.2, v.3 (pt. 1), v.4 (pt. 2), atlas pt. 1.

861 **François, Jean,** 1582-1668.
L'art des fontaines; c'est à dire, pour trouver, esprouver, assembler, mesurer, distribuer, & conduire les sources dãs les lieux publics & particuliers; d'en rendre la conduite perpetuelle; et de donner par arr des eaux coulantes aux lieux, où elles manquent par nature. Avec l'art de niveler, et par le niveau de connoistre les hauteurs des sources sur le lieu où on veut les conduire; les bornes des estang qu'on veut faire; & les terres qu'on doit inonder; le moyen de dessecher les marais & lieux trop humides; de faire des canaux soit à porter batteaux soit à faire floter le bois, soit à joindre les mers & les

rivieres, & à rendre celles-cy navigables . . . Ed.2. Rennes, P. Hallaudays, 1665.
120, 32p., illus. x627/F84a/1665

862 **Fraser, James,** of Dublin.
Geological copy [of] Fraser's travelling map of Ireland, shewing all the towns, lakes, rivers, roads and railways with distances marked between all the towns, railway stations and other important places. Dublin, James McGlashan, 1853.
G5780/.C5/1853/.11

Scale, 1:633,600.

863 **Frauendiener, Johann Matthias.**
Kurtze Beschreibung des Röthel-Bads bey Geisslingen: worinn von desselben Ursprung und Beschaffenkeit wie auch dessen Gehalt heylsamen Gebrauch und denen schon viele Jahr hero in denen schweresten auch langwürigsten Kranckheiten glücklich verrichteten Würckungen ect. gehandelt wird. Aufgesetzt und auf Verlangen zum Druck befördert von Jo. Matthia Frauendiener. Ulm, Gedruckt bey E. Kühnen, 1729.
66p., fold. table. 553.7/F86k

864 **Fregoso [or Fulgoso], Baptista,** 1453-1504.
De dictis factisque memorabilibus collectanea a Camillo Gilino latina facta, illis exceptis quae luculenter Maximus Valerius edibit. Mediolani, I. Ferrarius, 1509.
336 leaves. xq853/F881/Od.g

865 **Freiesleben, Johann Carl,** 1774-1846.
Geognostiche Arbeiten. Freyberg, Craz und Gerlach, 1807-1815.
6v. in 4, fold. plates, fold. maps, tables. 554.3/F85g

v.1-3 titled: Geognostischer Beitrag zur Kenntnis des Kupferschiefergebirges mit besonderer Hinsicht auf einen Theil der Grafschaft Mansfeld und Thüringen.

v.4 titled: Geognostischer Beitrag zur Kenntnis des Kupferschiefergebirges auch benachbarter Gebirgsformationen, mit besonderer Hinsicht auf Thüringen.

v.5-6 titled: Beiträge zur Mineralogischen Kenntnis von Sachsen.

866 **Freke, John,** 1686-1756.
A plain account of the cause of earthquakes. Being a supplement to a treatise, lately published, on fire. By the same author. London, Printed for W. Innys and J. Richardson, 1756.
37p. x551.22/F884p

RECHERCHES

SUR

LES VOLCANS ÉTEINTS DU VIVARAIS ET DU VELAY;

Avec un Diſcours ſur les Volcans brûlans, des Mémoires analytiques ſur les Schorls, la Zéolite, le Baſalte, la Pouzzolane, les Laves & les différentes Subſtances qui s'y trouvent engagées, &c.

Par M. FAUJAS DE SAINT-FOND.

Dessiné et Gravé par Savart.

A GRENOBLE,

Chez JOSEPH CUCHET, Imprimeur-Libraire de Monſeigneur le DUC D'ORLÉANS.

A PARIS,

Chez { NYON aîné, Libraire, rue Saint-Jean-de-Beauvais.
NÉE ET MASQUELIER, Graveurs, rue des Francs-Bourgeois, Porte St-Michel. }

M. DCC. LXXVIII.

AVEC APPROBATION ET PRIVILÉGE DU ROI.

Cat. No. 779

867 **Frémont, John Charles,** 1813-1890.
Report of the exploring expedition to the Rocky Mountains in the year 1842, and to Oregon and north California in the years 1843-'44. By Brevet Captain J. C. Frémont, of the topographical engineers, under the orders of Col. J. J. Abert, chief of the Topographical bureau. Printed by order of the Senate of the United States. Washington, Gales and Seaton, printers, 1845.
693p., plates, maps. 917.8/F88r

U.S. 28th Cong., 2d sess. Senate. Ex. doc. 174.
Large folded map wanting.

868 **Fridvaldszky, Johann,** 1740?-1784.
Minero-logia magni principatus Transilvaniae, seu metalla, semi-metalla, sulphura, salia, lapides, & aquae conscripta. Claudiopoli, Typis Academicis Societatis Jesu, 1767.
206p. 549/F912m

869 **Friese, Franz.**
Übersicht der österreichischen Bergwerks-Production in den Jahren 1823-1854. Wien, Mauz, 1855.
44p. 549.9/qF91u

870 **Friese.**
Die Bergwerksproduction der österreichischen Monarchie. Wien, Mauz, 1852.
52p. 549.9/qF91b

871 **Frisi, Paolo,** 1723-1784.
Traité des rivières et des torrens. Augmenté du traité des canaux navigables: tr. de l'italien. Paris, Imprimerie royale, 1774.
248p., fold. map, fold. plan. 627.1/F91dFd/1774

Translator's dedication signed: Deserrey.
Translation of Dei fiumi e dei torrenti.

872 **Frizimelica, Francesco,** 1491-1559.
. . . De balneis metallicis artificio parandis, liber postumus novi argumenti e bibliotheca Joannis Rhodii. Norimbergae, Impensis Johannis Ziegeri, 1679.
44p. uncat.

873 **Fromherz, Carl,** 1797-1854.
Geognostische Beobachtungen über die Diluvial-Gebilde des Schwarzwaldes oder über die Geroell-Ablagerungen in die-

sem Gebirge, welche den jüngsten vorgeschichtlichen Zeiträumen angehören. Freiberg, Adolph Emmerling, 1842.
443p., col. map. uncat.

874 **Funk, Franz Ernst Theodor,** 1768-1820.
Beyträge zur allgemeinen Wasser-Baukunst; oder Ausführliche Beschreibung der grossen und zahlreichen hydrometrischen Versuche . . . Lemgo, Im Verlage der Meyerischen Buchhandlung, 1808-09.
2v., illus. 627.12/F96b

Vol. 2 has alternate title: Ausführliche Maschinen-Berechnungen und andere hydraulische Untersuchungen.

875 **Gadbury, John.**
Natura Prodigiorum: or A discourse touching the nature of prodigies together with the kinds, causes, and effects, of comets, eclipses, and earthquakes . . . London, printed by J.C. for Fr. Cossinet, 1660.
200p. uncat.

876 **Gadd, Pehr Adrian,** 1727-1797, praeses.
Specimen geurgiae Fennicae. Aboae, Impressit J. C. Frenckell, 1763.
30p. 554.71/G116s

Imprint date corrected in manuscript to 1768. For verification of date cf. J. H. Liden. Catalogus disputationum, sect. III (1779) p.60.
Diss. — Abo (Jacob Malleen, respondent)
A classification of earths, stones and minerals.

877 **Gadd,** praeses.
Dissertatio academica indicia mineralogiae Fennicae, ab ortu Christianismi ad jacta fundamenta Academiae Aboiensis delineans. Aboae, In Typographiam ad Regiam Academiam, 1767.
12p. 549/G116d

Dissertation — Abo (Daniel Hirn, respondent)

878 **Gaertner, Carl Friedrich von,** 1772-1850, jt. author.
See LEONHARD, KARL CÄSAR VON, Propaedeutik der Mineralogie . . . 1817.

879 **Galeotti, Henri Guillaume,** 1814-1858.
Mémoire sur la constitution géognostique de la province de Brabant, en réponse à la question suivante: Décrire la Constitution géologique de la Province de Brabant, determiner avec soin les espèces minérales et les fossiles que les divers

terrains renferment et indiquer la synonymie des auteurs qui en ont déja traité. Bruxelles, 1837.
192p., 4 plates (1 col.), 2 fold. maps. q554.93/G13m

Académie royale des sciences, des lettres et des beaux-arts de Belgique. Mémoires couronnés et mémoires des savants étrangers, t.12.
Imprint and series statement from Library of Congress. Catalog of printed cards.
Bibliographical footnotes.
Copy imperfect: 2d prelim. leaf removed.

880 **Galvani, Camillo.**
Della pietra fosforica bolognese. [Bologna, Nella stamperia del Longhi, 1780].
91p., 3 fold. tables. 553.8/G13d

Half title.
Title page wanting.

881 **Gans, Johann Ludwig,** fl. 1620-1630.
. . . Corallorum historia, qua mirabilis eorum ortus, locus natalis, varia genera, praeparationes chymicae quàmplurimae, viresque eximae proponuntur. Francofurti, Sumptibus Lucae Iennisi, 1630.
174p. uncat.

882 **Gardner, J.,** publisher.
Geological map of England and Wales reduced[,] by permission[,] from the map in 6 sheets published by the Geological Society. London, 1826.
Scale, ca. 1:1,112,000. G5750/.C5/1826/.G2

Map is a composite of "A geological map of England and Wales, . . ." by George Bellas Greenough.

883 **Garnier, Fr. Abdon J.**
Mémoire géologique sur les terrains du Bas-Boulonnais, et particulièrement sur les calcaires compacts ou grenus qu'il renferme. Boulogne-sur-Mer, P. Hesse, 1823.
39p., 5p., illus., col. map. 553.51/G18m

Catalogue des produits du sol boulonnais: [1] leaf inserted.
Notice sur la Colonne des Bourbons, par la Société d'agriculture, du commerce et des arts de Boulogne-sur-Mer: 5p. at end.

884 **Garnier.**
Traité sur les puits artésiens, ou sur les différentes espèces de terrains dans lesquels on doit rechercher des eaux souterraines. 2.éd., rev. et augm. Paris, Bachelier (Successeur de Mme Ve Courcier), 1826.
264p., 25 plates (incl. fold. map). 628.11/G18t/1826

MINÉRALOGIE DES VOLCANS,

OU

DESCRIPTION

DE TOUTES LES SUBSTANCES PRODUITES OU REJETEES PAR LES FEUX SOUTERRAINS.

PAR M. FAUJAS DE SAINT-FOND.

A PARIS,

Chez CUCHET, rue & Hôtel Serpente.

M. DCC. LXXXIV.

Avec Approbation, & Privilége du Roi.

Cat. No. 781

"Ouvrage contenant la description des procédés qu'il faut employer pour ramener une partie de ces eaux à la surface du sol, à l'aide de la sonde du mineur ou du fontenier."

885 **Gassendi, Pierre,** 1592-1655.
Opera omnia in sex tomos divisa, quorum seriem pagina praefectiones proximè sequens continet. Lugduni, Laurentii Anisson, & Ioan. Bapt. Devenet, 1658.
6v. in 4. q uncat.

886 **Gautieri, Giuseppe,** 1769-1833.
Slancio sulla genealogia della terra e sulla costruzione dinamica della organizzazione, seguito da una ricerca sull'origine dei vermi abitanti le interiora degli animali. Jena, 1805.
136p. 570/G23s

887 **Geikie, Archibald,** 1835-1924.
See MURCHISON, RODERICK IMPEY, First sketch of a new geological map of Scotland . . . 1861 and 1862.

888 **Gellert, Christlieb Ehregott.**
Anfangsgründe zur metallurgischen Chimie, in einem theoretischen und praktischen Theile nach einer in der Natur gegründeten Ordnung. Mit Kupfern. Leipzig, J. Wendler, 1751.
339p., illus. 546.3/G28a

889 **Gellert.**
Metallurgic chymistry. Being a system of mineralogy in general, and of all the arts arising from this science . . . Translated from the original German, by I. S. [i.e. John Seiferth]. London, T. Becket, 1776.
416p., 4 fold. plates, fold. table. 669.9/G27aEs

Translation of Anfangsgründe zur metallurgischen Chimie, in einem theoretischen und praktischen Theile.

890 **Gemmellaro, Carlo,** 1787-1866.
Sopra i vulcani estinti del val di Noto; memoria seconda. Catania, G. Pappalardo, 1835.
35p. 551.21/G28s

891 **Gemmellaro.**
Saggio sulla costituzione fisica dell'Etna. Catania, Dai tipi dell' Accademia Gioenia, presso F. Sciuto, 1847.
65p., fold. map. 551.21/G28sa

Letto nella seduta ordinaria del 18 marzo,1847.
Estratto dal vol. III, ser. II degli Atti dell' Accademia Gioenia.

892 **Gemmellaro.**
Sulla formazione dello scisto di Ali'. Catania, Tipi dell'Accademia Gioenia, 1847.
31p., illus. 552.4/G28s

893 **Gendron, Pedro.**
Tratado da conservaçam da saude dos povos; obra util, e igualmente necessaria aos Magistrados, Capitaens Generaes, Capitaens de Mar, e Guerra, Prelados, Abbadessas, Medicos, e Paysdefamilias. Com hum appendix Consideraçoens sobre os Terremotos, com a noticia dos mais consideraveis, de que faz mençaõ a Historia, e deste ultimo, que se sentio na Europa no I de Novembro de 1755. Lisboa, Officina de J. Filippe, 1757.
568(i.e. 368)p. 614/G282t

894 **Gennetté, Claude Léopold de,** 1706-1782.
Origine des fontaines. [n.p., n.d., 1774].
133p. 551.45/G287o

895 **Genssane, de,** d.1780.
La géométrie souterraine, ou, Traité de géometrie-pratique, appliqué à l'usage des travaux des mines. Montpellier, Rigaud, Pons, 1776.
224p., 5 fold. plates. 513/G28g

896 Geology of New York.
See entries for MATHER, WILLIAM WILLIAMS, Part 1 . . . 1843; EMMONS, EBENEZER, Part 2 . . . 1842; VANUXEM, LARDNER, Part 3 . . . 1842; HALL, JAMES, Part 4 . . . 1843.

897 **Gerhard, Johann Gregorius.**
Kurtze jedoch gründliche Nachricht, von dem zu Kösen an der Saale entdeckten mineralischen Gesund-Brunnen, als: 1) Von dessen Ursprung und Situation, 2) Von dessen wahren Gehalt und Ingredienten, 3) Von dessen daraus herstammenden Kräfften und Eigenschafften, und endlich, 4) Wie dieser Brunnen zur Wiederherstellung der menschlich-verlohrnen Gesundheit nützlich anzuwenden oder zugebrauchen seyn möchte . . . Naumburg, Verlegt von dem Autore, und gedruckt durch B. Bossögeln, 1726.
80p. 613.12/G31k

898 **Germar, Ernst Friedrich,** 1786-1853.
Die versteinerten Insecten Solenhofens. Beschrieben vom Professor Germar in Halle . . . [Breslau, 1839].
[189]-222p., 3 plates (1 fold.) 565.7/G31v

From K. Leopoldinisch-Carolinische Deutsche Akademie der Naturforscher. Nova acta. v.19, pt. 1.

Bound with Goldenberg, Friedrich. Die fossilen Insekten der Kohlenformation von Saarbrücken. [Stuttgart, 1856]. [17-38]p., 4 plates.

899 **Gesner, Abraham,** 1797-1864.
Remarks on the geology and mineralogy of Nova Scotia. Halifax, Nova Scotia, Printed by Gossip and Coade, 1836.
272p., front., 2 plates, fold. map. x557.16/G33r

Title page torn. Bottom ⅓ missing.

900 **Gesner.**
See NEW BRUNSWICK. GEOLOGICAL SURVEY. Report on the Geological survey of the province of New-Brunswick . . . [1838]-1842.

901 **Gesner.**
The industrial resources of Nova Scotia. Comprehending the physical geography, topography, geology, agriculture, fisheries, mines, forests, wild lands, lumbering, manufactories, navigation, commerce, emigration, improvements, industry, contemplated railways, natural history and resources of the province. Halifax, A. & W. MacKinley, 1849.
341p., illus., fold. map, front. x917.16/G33i

902 **Gesner.**
A practical treatise on coal, petroleum, and other distilled oils. New York, Ballière Brothers, etc., 1861.
130p., illus. uncat.

903 **Gesner.**
A practical treatise on coal, petroleum, and other distilled oils. Second edition, revised and enlarged by George Weltden Gesner. New York, Baillière Brothers, 1865.
181p., illus., front. 665.5/G331p/1865

904 **Gesner, George Weltden.**
See GESNER, ABRAHAM, A practical treatise on coal . . . 1865.

905 **Gesner, Konrad,** 1516-1565.
De raris et admirandis herbis, quae sive quod noctu luceant, sive alias ob causas, lunariae nominantur, commentariolus: & obiter de alijs etiam rebus quae in tenebris lucent. Inseruntur & Icones quaedam herbarum nova. Eiusdem descriptio Montis Fracti, sive Montis Pilati, iuxta Lucernam Helvetia. Tiguri, Apud Andream Gesnerum F. & Iacobum Gesnerum, fratres, 1555.
109p. x580/G33d

L'ART DES FONTAINES,

C'EST A DIRE,

Pour trouuer, esprouuer, assembler, mesurer, distribuer, & conduire les Sources dās les lieux publics & particuliers; d'en rendre la conduite perpetuelle; Et de donner par Art des Eaux coulantes aux lieux, où elles manquent par nature.

AVEC L'ART DE NIVELER

Et par le Niveau de connoistre les hauteurs des Sources sur le lieu où on veut les conduire; Les bornes des Estang qu'on veut faire; & les Terres qu'on doit inonder; Le moyen de dessecher les Marais & lieux trop humides; de faire des Canaux soit à porter batteaux soit à faire floter le bois, soit à joindre les Mers & les Rivieres, & à rendre celles-cy navigables.

EDITION SECONDE.

Par le P. Iean François de la Compagnie de IESVS.

A RENNES,

Chez P. HALLAVDAYS, Impr. ruë S. Germain à la Bible d'Or; Ou ce vend le Liure. M. DC. LXV.

Cat. No. 861

906 **Gesner,** ed.
De omni rerum fossilium genere, gemmis, lapidibus, metallis, et huiusmodi, libri aliquot, plerique nunc primum editi. Operâ Conradi Gesneri: quorum catalogum sequens folium continet. Tiguri, excudebat Iacobus Gesnerus, 1565.
8v. in 1, illus. uncat.

Vols. 2-8 have special t.-p.
Contents: Kentmann, Johannes. Catalogus rerum fossilium. — Kentmann, Johannes. Calculorum qui in corpore ac membris hominum innascuntur, genera XII. — Fabricius, Georgius. De metallicis rebus ac nominibus observationes variae . . . ex schedis Georgii Fabricii . . . quae Georgius Agricola praeterijt. — Goebel, Severin. De succino, libri duo. — Cordus, Valerius. De halosantho seu spermate ceti vulgo dicte, liber, nunc primum in lucem aeditus. — Sancti patris Epiphanii episcopi Cypri ad Diodorum Tyri episcopum, De XII. gemmis quae erant in veste Aaronis, liber graecus, & è regione latinus, Iola Hierotarantino interprete; cum corollario Conradi Gesneri. — Franciscus Rueo. De gemmis aliquot, iis praesertim quarum Diuus Ioannes apostolus in sua Apocalypsi meminit. — Gesner, Konrad. De rerum fossilium, lapidum et gemmarum maximè, figuris & similitudinibus liber: non solùm medicis, sed omnibus rerum naturae ac philologiae studiosis, utilis & iucundus futurus.

907 **Gesner.**
De rerum fossilium, lapidum et gemmarum maximè, figuris & similitudinibus liber: non solùm medicis, sed omnibus rerum naturae ac philologiae studiosis, utilis & iucundus futurus.
See GESNER, KONRAD, De omni rerum fossilium genere, gemmis, lapidibus, metallis, et huiusmodi, libri aliquot, plerique nunc primum editi . . . Tiguri, 1565.
uncat.

908 **Ghezzi, Nicolò,** 1683-1766.
Dell'origine delle fontane e dell'addolcimento dell'acqua marina; lettere scritte al Sig. Conte N. N. [i.e. Giulio Cesare Brusati]. Venezia, S. Occhi, 1742.
404p., plates, fold. table. 551.57/G34d

"Lettere . . . [del Conte Giulio Cesare Brusati] con alcune risposte alle obbiezioni fattevi": p.[287]-404.

909 **Gibbes, Robert Wilson,** 1809-1866.
The present earth the remains of a former world. A lecture, delivered before the South Carolina institute, September 6, 1849. Columbia, S.C., Printed by A. S. Johnston, 1849.
31p. x550/G351p

910 **Giebel, Christoph Gottfried Andreas,** 1820-1881.
Paläozoologie. Entwurf, einer systematischen Darstellung der

Fauna der Vorwelt. Merseburg, Nulandt'sche Buchhandlung (L. Garcke), 1846.
359p. 560/G36p

911 **Giebel.**
Fauna der Vorwelt mit steter Berücksichtigung der lebenden Thiere. Monographisch dargestellt von Dr. C. G. Giebel . . . Leipzig, F. A. Brockhaus, 1847-56.
3v. in 5. 560/G36f

No more published.

Contents. — 1. bd. Wirbelthiere. 1. Abth. Säugethiere. 2. Abth. Vögel und Amphibien. 3. Abth. Fische. — 2. Bd. Gliederthiere. 1. Abth. Insecten und Spinnen. — 3. Bd. Mollusken. 1. Abth. Cephalopoden.

Library owns v.1, pt. 2 and v.3, pt. 1.

912 **Giebel.**
Gaea excursoria Germanica. Deutschlands Geologie, Geognosie und Paläontologie. Leipzig, E. Kummer, 1848.
510p., 24 [i.e. 23] plates (incl. maps). 554.3/G36g

Bibliography: p. 449-452.

913 **Giebel.**
Allgemeine Palaeontologie, Entwurf einer systematischen Darstellung der Fauna und Flora der Vorwelt, zum Gebrauche bei Vorlesungen und zum Selbstunterrichte. Leipzig, A. Abel, 1852.
413p. 560/G36a

914 **Gimma, Giacinto,** 1668-1735.
Della storia naturale delle gemme, delle pietre, e di tutti i minerali, ovvero della fisica sotterranea . . . in cui delle gemme, e delle pietre stesse si spiegano la nobiltà, i nomi, i colori, le spezie, i luoghi, la figura, la generazione, la grandezza, la durezza, la madrice, l'uso, le virtù, le favole: se al fuoco resistano: quali sieno nella Sagra Scrittura nominate: quali i simboli . . . Napoli, Stamperia de Gennaro Muzio [etc.], 1730.
2v., front. 549/G42d

915 **Gioeni, Giuseppe,** 1747-1822.
Relazione della eruzione dell'Etna nel mese di luglio MDCCLXXXVII. Catania, Stamperia dell'Accademia degli Etnei, 1787.
40p. 551.21/G43r

Bibliographical footnotes.

916 **Gioeni.**
Saggio di litologia vesuviana dedicato a S. M. la Regina delle Due Sicilie. Napoli [Nella stamperia Simoniana], 1790.
208p. 552/G43s

917 **Gioeni.**
Versuch einer Lithologie des Vesuvs. Aus dem Italienischen übersetzt und mit einigen Anmerkungen begleitet von Leopold von Fichtel. Wien, J. Stahel, 1793.
392p. 552/G43sGf

Translation of Saggio di litologia vesuviana.
Bibliographical footnotes.

918 **Girardin, Jean Pierre Louis,** 1803-1884.
Élémens de minéralogie appliqûee aux sciences chimiques; ouvrage basé sur la méthode de M. Berzélius, contenant l'histoire naturelle et métallurgique des substances minérales, leurs applications à la Pharmacie, à la Médecine et à l'économie domestique; suivi d'un précis elémentaire de géognosie. Par J. Girardin et H. Lecoq. Paris, Thomine, 1826.
2v., fold. plates (part. col.), table. 549/G44e

919 **Girardin.**
Considérations générales sur les volcans: et examen critique des diverses théories qui ont été successivement proposées pour expliquer les phénomènes volcaniques. Rouen, N. Periaux, 1831.
248p. 551.21/G44c

920 **Giraudet, Alexis.**
Nouveau traité de géologie; ou, Exposé de l'état actuel de cette science, considérée dans ses rapports avec la minéralogie, l'agriculture, l'industrie, les arts et la tradition biblique, à l'usage des collèges, des séminaires et des maisons d'éducation. Tours, A. Mame, 1843.
408p., 12 plates. 550/G43n

Bibliothèque de la jeunesse chrétienne.

921 **Gist, Christopher,** d. 1759.
See POWNALL, THOMAS, A topographical description ... middle British Colonies ... 1776.

922 **Giustiniani, Lorenzo.**
I tre rarissimi opuscoli di Simone Porzio, di Girolamo Borgia, e di Marcantonio delli Falconi scritti in occasione della celebre eruzione avvenuta in Pozzuoli nell'anno 1538, colle memorie storiche de'suddetti autori raccolte da Lorenzo Giustiniani. Napoli, dai Torchi di L. Marotta, 1817.

DE OMNI RERVM FOSSILIVM GENERE, GEMMIS, LAPIDIBVS, METALLIS, ET HVIVSMODI, LIBRI ALIQVOT, PLERIQVE NVNC PRIMVM EDITI.

Operâ Conradi Gesneri: Quorum Catalogum sequens folium continet.

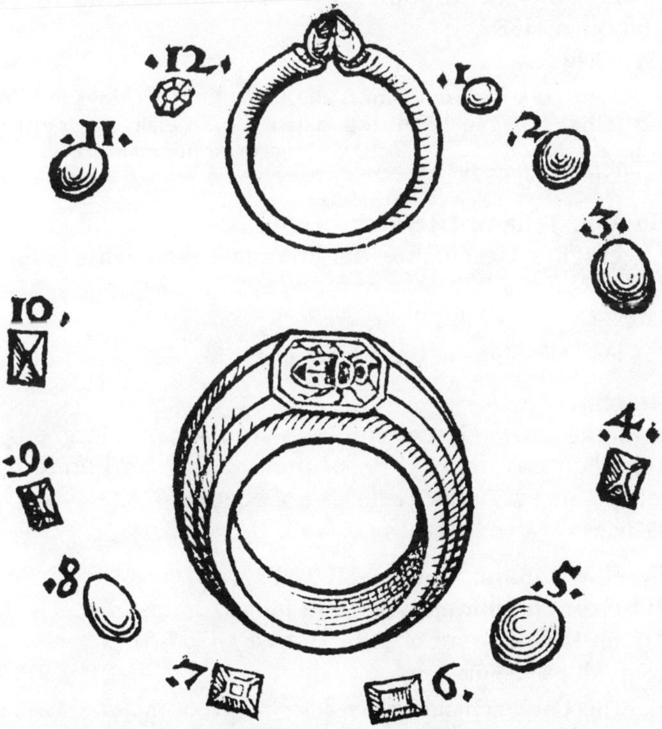

Tiguri, excudebat Iacobus Gesnerus: Anno M. D. LXV.

Cat. No. 906

219p. 551.22/T71

Numerous errors in pagination.

923 **Glauber, Johann Rudolph,** 1604-1670.
Operis mineralis . . . Amstelodami, apud Joannem Janssonium, [1651].
313p. uncat.

This work treats chemistry, alchemy, medicine, and metallurgy. The latter subject is in Part II entitled: De ortu & origine omnium metallorum & mineralium . . . Amstelodami, apud Joannem Janssonium, [1652]. 47p.

924 **Glauber.**
The works of the highly experienced and famous chymist, John Rudolph Glauber: Containing, great variety of choice secrets in medicine and alchymy in the working of metallick mines, and the separation of metals; also, various cheap and easie ways of making salt-peter, and improving of barren-land, and the fruits of the earth . . . Translated into English . . . by Christopher Packe. London, printed by Thomas Milbourn, 1689.
752p., illus. q uncat.

This work includes almost all of Glauber's writings in translation including "Of the birth and nativity of metals," a translation of Operis mineralis . . . De ortu & origine omnium metallorum & mineralium . . . [1652].

925 **Gmelin, Johann Friedrich,** 1748-1804.
Allgemeine Geschichte der mineralischen Gifte. Nürnberg, G. N. Raspe, 1777.
316p. 615.92/G52a

His [Allegemeine Geschichte der Gifte, v.3].

926 **Gmelin.**
Beyträge zur Geschichte des teutschen Bergbaus, vornehmlich aus dem mittlern und spätern Jahrhunderten unserer Zeitrechnung. Halle, J. J. Gebauer, 1783.
452p. x553/G52b

927 **Gmelin, Johann Georg,** 1709-1755.
Reise durch Sibirien, von dem Jahr 1733 bis 1743, Göttingen, Abram Vandenhoecks seel. Wittwe, 1751-1752.
4v., fold. plates, maps. 915.7/G52r

This book contains many descriptions of mines.

928 **Gobet, Nicolas,** 1737-ca. 1781.
See PALISSY, BERNARD, Oeuvres . . . avec des notes par . . . Gobet . . . 1777.

929 **Godeffroy, Carl,** 1787-1848.
Notice sur les glaciers, les moraines et les blocs erratiques des Alpes, avec une table analytique. Paris, A. Cherbuliez, 1840.
112p. 551.31/G54n

930 **Goebel, Severin,** 1530-1612.
De succino libri duo, authore . . . Horum prior liber continet piam commonefactionem, de passione, resurrectione, ac beneficÿs Christi, quae in historia Succini depinguntur. [1558?].
30p. uncat.

Bound with Gesner, Konrad, ed., De omni rerum fossilium genere gemmis, lapidibus, metallis, et huiusmodi, libri aliquot, plerique nunc primum editi. Tiguri, 1565.

The first part is religious, the second part "De generatione succini, petrolei, ac reliquis bituminosis succis" is scientific.

931 **Goldenberg, Carl Friedrich,** 1798-1881.
Die fossilen Insekten der Kohlenformation von Saarbrücken. [Stuttgart, 1856].
[17-38]p., 4 plates. 565.7/G31v

Bound with Germar, Ernst Friedrich, Die versteinerten Insecten Solenhofens . . . 1839.

932 **Goldsmith, Oliver,** 1728-1774.
An history of the earth and animated nature. London, printed for J. Nourse, 1774.
8v., plates. x590/G57h and Nickell/x590/g57h

Library owns 2 copies. Volume 8 is imperfect in the x590/G57h copy; it lacks t.-p. and table of contents; those belonging to vol. 7 have been substituted.

933 **Goldsmith.**
An history of the earth and animated nature. 2nd ed. London, printed for J. Nourse, 1799.
x590/G57h2

Library owns v.1 only.

934 **Goldsmith.**
An history of the earth and animated nature. New York, Thomas Kinnersley, 1825.
880p., 55p., port. 590/G57h/1825

935 **Goldsmith.**
An history of the earth and animated nature. Philadelphia, J.B. Lippincott & Co., 1856.
4v. in 2. 590/G57h/1856

936 **Golitsyn, Dimitriĭ Alexieevich,** 1738-1803.
Traité ou description abrégée et méthodique des minéraux. Présenté à l'Académie impériale & royale des sciences & belles-lettres de Bruxelles par le prince D. de Gallitzin. Maestricht, J. P. Roux, 1792.
244p. 549/G58t

937 **Golitsyn.**
Recueil de noms par ordre alphabétique appropriés en minéralogie aux terres et pierres, aux métaux et demi métaux et aux bitumes. Nouvelle éd., corrigée et augmentée par la nomenclature relative aux mêmes objets, de Mr. Haüy. Brunsvik, impr. de la Maison des Orphelins, 1802.
316p. q549.014/G58r

938 **Grandoni, Stefano,** 1792-1846.
Esperienze fisico-chimiche ed analisi dell'acqua minerale di Bovegno . . . coll'aggiunta di alcune storie de malattie curate colla stessa. Brescia, N. Bettoni e compagni, 1831.
72p. 543.3/G76e

"Descrizione di alcune malattie curate coll'acqua minerale del monte Trovina, dal dottor Giovanni Zantedeschi": p.[51]-72.

939 **Grandoni,** jt. author.
See BALARDINI, LODOVICO, Sulla torba della provincia bresciana . . . 1842.

940 **Grange, Jules,** fl. 19th cent.
Recherches sur les glaciers, les glaces flottantes, les dépôts erratiques, sur l'influence du climat, sur la distribution géographique et la limite inférieure des glaciers, étude du phénomène erratique du nord de l'Europe, suivi de quelques observations sur l'influence des climats, sur la distribution géographique des êtres organisés. Paris, V. Masson; [etc., etc.], 1846.
142p. 551.31/G76r

941 **Grar, Édouard.**
Histoire de la recherche, de la découverte et de l'exploitation de la houille dans le Hainaut français, dans la Flandre française et dans l'Artois, 1716-1791. Valenciennes, Impr. typ. et lithographique de A. Prignet, 1847-50.

DELLA STORIA NATURALE DELLE GEMME, DELLE PIETRE, E DI TUTTI I MINERALI, OVVERO DELLA FISICA SOTTERRANEA

DI

D. GIACINTO GIMMA

Dottore delle Leggi, Avvocato ſtraordin. della Città di Napoli, Promotor-Generale della Scientifica Società Roſſaneſe degl' Incurioſi, *ecc.*

IN CUI

delle Gemme, e delle Pietre steſſe ſi ſpiegano la Nobiltà, i Nomi, i Colori, le Spezie, i Luoghi, la Figura, la Generazione, la Grandezza, la Durezza, la Madrice, l'Uſo, le Virtù, le Favole: ſe al fuoco reſiſtano: quali ſieno nella Sagra Scrittura nominate: quali i Simboli: ed altre notizie, che alle medeſime appartengono. Si dà ancora la cognizione de' Metalli, delle Terre, de' Sali, de' Solfi, de' Bitumi, delle Acque diverſe: di quelche ſi tratta nella Storia de' Foſſili, che dalle Pietre ſi formano: delle Caverne, delle Acque, e de' Fuochi ſotterranei, de' Vulcani del Mondo, e di quanto ſi eſamina nella *Fiſica ſotterranea*; oltre alcuni Trattati valevoli a dilucidare la Storia tutta della Minerale, ed altri, che della Vegetevole, e di quella degli Animali, ſono proprj.

DIVISA IN LIBRI VI. O TOMI II.

colle Tavole de' Capitoli nel primo: de' Nomi delle Pietre, e delle coſe notabili nel ſecondo.

TOMO I.

IN NAPOLI. MDCCXXX.

Nella Stamperia di Gennaro Muzio, erede di Michele-Luigi.

Con Licenza de' Superiori, e Privilegio.

A SPESE DELLO STESSO MUZIO, E DI FELICE MOSCA.

Cat. No. 914

3v., illus., ports., col. maps (part. fold.), geneal. tables. 553.2/G768h

Bibliographical footnotes.

942 **Gras, Scipion,** 1806-187?
Statistique minéralogique du département de la Drôme, ou Description géologique des terrains qui constituent ce département, avec l'indication des mines, des carrières et en général de tous les gîtes de minéraux utiles qui s'y trouvent contenus. Grenoble, Prudhomme, imprimeur-libraire, 1835.
296p., fold. map. 554.4/G76s

943 **Grateloup, Jean Pierre Sylvestre de,** 1782-1861.
Conchyliologie. Fossiles. [Bordeaux, 1836-38].
7 pts. in 1v., 5 plates. 564/G77c

Binder's title.

Five of the 7 parts are extracts from Actes de la Société Linnéenne de Bordeaux. Two have special t.p.

Contents: 1. Conchyliologie fossile du Bassin de l'Adour 2. . . . Mollusques terrestres et fluviatiles . . . Bassin de l'Adour 3. . . . Famille des Mélaniens . . . 4. . . . Famille des Bulléens . . . 5. . . . Géo-zoology sur les Oursins fossiles (Échinides) 6. . . . Excursions au Pic d'Anie et au Pic Amoulat, par M. Léon Dufour 7. Discours sur les sciences et les arts . . .

944 **Grateloup.**
Conchyliologie fossile des terrains tertiaires du bassin de l'Adour, environs de Dax. T. 1er. Univalves. Atlas. Bordeaux, Impr. de Th. Lafargue, 1840.
1v. (chiefly plates). q564.3/G77c

No more published?

Bibliography: p. [xi]-xviii.

945 **Grauel, Johann Philipp,** 1711-1761.
Museum Grauelianum, sive collectionis regni mineralis praecipue historiam naturalem illustrantis. Argentorati, Typis J. H. Heitzii, 1772.
184p. 550/G77m

946 **Gray, John Edward,** 1800-1875.
Spicilegia zoologica; or, Original figures and short systematic descriptions of new and unfigured animals. London, Treüttel, Würtz, [1828-30].
12p., 14 plates. xq590/G793s

No title page.

947 **Great Britain. Board of Agriculture.**
General view of the agriculture and minerals of Derbyshire; with observations on the means of their improvement . . . v.1,

containing a full account of the Surface, Hills, Valleys, Rivers, Rocks, Caverns, Stata, Soils, Minerals, Mines, Collieries, Mining Processes, &c.&c. Together with some Account of the recent Discoveries respecting the Stratification of England; and a Theory of Faults and Denudated Strata, applicable to Mineral Surveying and Mining. Illustrated by five coloured maps, and sections of strata. By John Farey, Sen., Mineral Surveyor... London, Printed by B. McMillan, 1811.
532p., plates, maps, tables. 554.251/G79g

Library owns v.1.

948 **Green, Charles,** rector of Burgh Castle.
A few remarks on the geology of Bacton. Norwich, J. Fletcher, 1842.
39p., illus. 554.261/G82f

949 **Green.**
The history, antiquities, & geology, of Bacton, in Norfolk. Norwich, J. Fletcher; London, Simpkin, Marshall, and co., 1842.
102p., front., 3 plates. 914.261/G82h

950 **Green, Jacob,** 1790-1841.
A monograph of the trilobites of North America, with coloured models of the species. Philadelphia, J. Brano, 1832.
93p., fold. plate. 565.39/G82m

951 **Greenough, George Bellas,** 1778-1855.
A critical examination of the first principles of geology; in a series of essays. London, Longman, Hurst, Rees, Orme, and Brown, 1819.
336p. 550.4/G85c

952 **Greenough.**
A geological map of England and Wales, ... published under the direction of the Geological Society. London, Longman, Hurst, Rees, Orme, and Brown, 1819.
G5750/.C5/1819/.G7

Scale, ca. 1 inch equals 6 British Statute miles.
Map on 6 sheets.
Accompanies his Memoir of a geological map of England ... 1820.

953 **Greenough.**
A Geological map of England and Wales. 2nd edition. [London], The Geological Society, 1839.
G5750/.C5/1839/.G7

Scale, ca. 1:365,000.
Map on 6 sheets.
Explanatory memoir lacking.
Accompanied by index of colors.

954 **Greenough.**
See GARDNER, J., publisher, Geological map of England and Wales . . . 1826.

955 **Greenough.**
Memoir of a geological map of England; to which are added, an alphabetical index to the hills, and a list of the hills arranged according to counties. London, Published under the direction of the Geological Society; by Longman, Hurst, Rees, Orme, and Brown, 1820.
17p. q554.2/G85m

956 **Greg, Robert Phillips.**
Observations on meteorolites or aërolites, considered geographically, statistically, and cosmically; accompanied by a complete catalogue of meteoric falls. Manchester, 1855.
40p., tables. x551.508/M565

Signature of author.
Bound with Meteorology; [pamphlets].

957 **Gregorii, Johann Gottfried,** 1685-1770.
Die curieuse Orographia, oder, accurate Beschreibung derer berühmtesten Berge in Europa, Asia, Africa und America, mit denen auserlesensten theologischen, politischen, physicalischen, moralischen und andern notablen Anmerckungen und Historien, wie auch allen darbey sich ereignenden denckwürdigen Antiquitäten; In alphabethischer Ordnung, samt einem vollständigen Real-Register dargestellet. Franckfurth, Bey H. P. Ritscheln, Buchhändler in Erffurt, 1715.
727p., front. 551.43/G86c

958 **Griffin, John Joseph,** 1802-1877.
A practical treatise on the use of the blowpipe, in chemical and mineral analysis: including a systematic arrangement of simple minerals, adapted to aid the student in his progress in mineralogy, by facilitating the discovery of the names of species. Glasgow, R. Griffin; London, T. Tegg, 1827.
308p., 5 plates. 544.3/G87p

Added t.p., engraving.

DELLA

NATURA DE' FIUMI

TRATTATO FISICO-MATEMATICO

DEL DOTT. DOMENICO GUGLIELMINI

NUOVA EDIZIONE

CON LE ANNOTAZIONI

DI EUSTACHIO MANFREDI

All' Eminentissimo, e Reverendissimo Principe

IL SIG. CARDINALE

NERIO CORSINI

Nepote del regnante sommo Pontefice

CLEMENTE XII.

In Bologna nella Stamperìa di Lelio dalla Volpe. MDCCXXXIX.

Con licenza de' Superiori.

Cat. No. 979

959 **Griffith, Richard John,** 1784-1878.
Geological and mining report on the Leinster coal district. Dublin, Graisberry and Campbell, 1814.
135p. 622.33/G87g

960 **Griffith.**
Geological and mining surveys of the coal districts of the Counties of Tyrone and Antrim in Ireland. Dublin, R. Graisberry, 1829.
77p., illus., fold. sections. 622.33/G87ge

961 **Griffith.**
An address delivered at the ninth annual meeting of the Geological Society of Dublin, on the 12th of February, 1840. To which is subjoined the annual report, etc. Dublin, Hodges & Smith, 1840.
40p. 554.15/G87a

962 **Griffith.**
Geological map of England, to accompany the instructions to valuators appointed under the 15th & 16th Vic., cap 63. Dublin, Forster & Co., 1853.
G5780/.C5/1853/.IO

Scale, 1:1,100,000.
Signed Richard Griffith.

963 **Gruner, Gottlieb Sigmund,** 1717-1778.
Histoire naturelle des glacieres de Suisse, traduction libre de l'allemand de M. Grouner, par M. de Kéralio. Paris, Panckoucke, 1770.
359p., 18 plates (part. fold.), 2 fold. maps, front. x551.31/G92eFk

Translation of Die Eisgebirge des Schweizerlandes.

964 **Gruner.**
Die Mineralgeschichte der Schweiz. Bern, bey der typographischen Gesellschaft, [1774].
183p. 554.9/G92

965 **Gruner.**
Histoire naturelle de la Suisse dans l'ancien monde; tr. de l'allemand. Neuchâtel, J. P. Jeanrenaud & Co., 1776.
159p. uncat.

Avis du traducteur signed: Dulon, ministre.

966 **Gruner.**
Reisen durch die merkwürdigsten Gegenden Helvetiens. (London!) [Bern], 1778.
2v., plates (part. fold.). uncat.

967 **Gualtieri, Niccolò,** 1688-1744.
Riflessioni sopra l'origine delle fontane, descritte in forma di lettera. Lucca, L. Venturini, 1725.
207p. 551.49/G93r

Addressed to Antonio Vallisnieri.

968 **Gualtieri.**
Index testarum conchyliorum quae adservantur in museo Nicolai Gualtieri . . . et methodicae distributae exhibentur tabulis CX . . . Florentiae, ex typographia C. Albizzini, 1742.
5 pt. in 1v., 2 front. (1 port.), illus., 110 (i.e. 120) plates on 110 leaves. f594/G92i

Each part, and each "classis" of pts. 2-4, has special t.-p. Descriptive letterpress on versos facing numbered plates.

969 **Guérin, Joseph Xavier Benerot,** 1775-ca.1850.
Description de la fontaine de Vaucluse. 2. ed. Avignon, Chez F. Seguin, 1813.
312p., illus. 551.49/G936d/1813

970 **Guettard, Jean-Étienne,** 1715-1786.
Mémoire et carte minéralogique sur la nature et la situation des terreins qui traversent la France & l'Angleterre, *in* Mémoires de l'Académie Royale des Sciences, Paris, 1746.
p.363-392, 2 fold. plates. 506 PAH/1746

971 **Guettard.**
Mémoire sur les granits de France, comparés à ceux d'Egypte, *in* Mémoires de l'Académie Royale des Sciences, Paris, 1751.
p.164-210, fold. plate. 506 PAH/1751

972 **Guettard.**
Mémoire sur quelques montagnes de la France qui ont été des volcans, *in* Mémoires de l'Académie Royale des Sciences, Paris, 1752.
p.27-59, 2 fold. plates. 506 PAH/1752

973 **Guettard.**
Mémoire dans lequel on compare le Canada à la Suisse, par rapport à ses minéraux, *in* Mémoires de l'Académie Royale des Sciences, Paris, 1752.
p.189-220, folded plate 7. 506 PAH/1752

The plate is entitled: Carte minéralogique, où l'on voit la nature des terreins du Canada et de la Louisiana . . . dressée par Philippe Buache, 1752.

974 **Guettard.**
Suite du mémoire dans lequel on compare le Canada à la Suisse, par rapport à ses minéraux. Seconde partie. Description des minéraux de la Suisse, *in* Mémoires de l'Académie Royale des Sciences, Paris, 1752.
p.323-360, 5 fold. plates. 506 PAH/1752

Four plates are drawings of fossil fish and fossil fish teeth. The fifth is entitled: Carte minéralogique de la Suisse, par Philippe Buache pour servir aux recherches et un mémoire de Mr. Guettard en 1752.

975 **Guettard.**
Mémoires sur différentes parties des sciences et arts. Paris, L. Prault, 1768-83.
5v. in 6, 262 plates. x508/G938m

In this set, the plates which belong in vols. 4, 5, and 3 are bound, in this order, in a 6th volume, which also contains new title pages: Nouvelle collection de mémoires sur différentes parties intéressantes des sciences et arts. t. 1[-3]. Paris, Lamy, 1786.

976 **Guettard.**
See FOUGEROUX DE BONDAROY, AUGUSTE DENIS, Autograph draft of a report . . . on Guettard's Mémoires sur la minéralogie du Dauphiné . . . [ca. 1778].

977 **Guettard.**
Mémoires sur la minéralogie du Dauphiné. Paris, Impr. de Clousier, 1779.
2v. 19 plates (part. col.). 549/G93m

978 **Guettard.**
Epreuves de cinquante-trois gravures sur cuivre destinées à des travaux de Guettard restés inédits. [19--?].
[10]ℓ., 215 plates. xq508/G938e

In box.

Four sets of engravings, the first of which is described as "premier état avant restauration des planches." The other sets have two versions of plate 48; the 4th set lacks plate 47 and contains 2 copies of plate 25.

In the box with these engravings (Robert sculpsit, 1775) are the 53 original copperplates from which they were made, the plates having been acquired by Lavoisier at the death of Guettard. There is also a typed list of MSS in the Museum national d'histoire naturelle, concerning a projected vol. 6 of Guettard's Memoirs, and a handwritten description of three of these MSS.

The plates include illustrations of marine fossils and living organisms for the projected natural history of the region of Etampes.

979 **Guglielmini, Domenico,** 1655-1710.
Della natura de' fiumi, trattato fisico-matematico del Dott. Domenico Guglielmini; nuova edizione, con le annotazioni

ESSAI
D'UNE THÉORIE
SUR LA STRUCTURE
DES CRYSTAUX,
APPLIQUÉE
A PLUSIEURS GENRES DE SUBSTANCES CRYSTALLISÉES;

Par M. l'Abbé HAÜY, *de l'Académie Royale des Sciences, Professeur d'Humanités dans l'Université de Paris.*

A PARIS,

Chez Gogué & Née de la Rochelle, Libraires, Quai des Augustins, près le Pont Saint-Michel.

M. DCC. LXXXIV.

SOUS LE PRIVILÉGE DE L'ACADÉMIE.

Cat. No. 1020

di Eustachio Manfredi. Bologna, Lelio dalla Volpe, 1739.
427p., fold. plates. uncat.

980 **Guidius, Johannes.**
... De mineralibus tractatus absolutissimus ... Cum duobus indicibus, quorum unus titulorum, alter rerum notabilium. Francofurti, Apud Egenolphum Emmelium, 1627.
202p. uncat.

981 **Guyot, Arnold Henry,** 1807-1884.
Note sur le Bassin erratique du Rhin. Neuchâtel, Henri Wolfrath, 1847.
p.[1-3], 4-12. uncat.

Extrait du Bulletin de la Société des Sciences Naturelles de Neuchâtel, mai et décembre 1845.

The library owns another copy (uncat.).

982 **Hacquet, Balthasar,** 1739-1815.
Oryctographia Carniolica, oder physicalische Erdbeschreibung des Herzogthums Krain, Istrien, und zum Theil der benachbarten Länder. Leipzig, J. G. I. Breitkopf, 1778-89.
4v. in 3. illus., plates, maps. x551.4/H12o

983 **Hagen, Heinrich,** 1709-1772.
Abhandlungen chemischen und physikalischen Inhalts. Königsberg, B. G. L. Hartung, 1778.
203p. 613.12/Ei8a

Bound with Eisfeld, Martin Friedrich Ludwig, Abhandlung von dem Nutzen der Schlackenbäder ... 1776.

984 **Haidinger, Karl,** 1756-1797.
Dispositio rerum naturalium Musei Caesarei Vindobonensis, edita a Carolo Haidinger. Vindobonae, C. F. Wappler, 1782.
61p. 549.074/H12d

Bibliographical footnotes.

985 **Haidinger, Wilhelm Karl von,** 1795-1871.
See MOHS, FRIEDRICH, Treatise on mineralogy ... 1825.

986 **Haidinger.**
Handbuch der bestimmenden Mineralogie, enthaltend die Terminologie, Systematik, Nomenklatur und Charakteristik der Naturgeschichte des Mineralreiches. 2. unveränderte Ausg. Mit 560 Holzschnitten. Wien, W. Braumüller, 1850.
630p., illus. 549.1/H12h/1850

987 **Hales, Stephen,** 1677-1761.
Some considerations on the causes of earthquakes. Which

were read before the Royal Society, April 5, 1750. London, Printed for R. Manby and H. S. Cox, 1750.
23p. x551.22/H13s

988 **Hales.**
Some considerations on the causes of earthquakes. Which were read before the Royal Society, April 5, 1750. The Second Edition, corrected. London, Printed for R. Manby and H. S. Cox, 1750.
23p. x551.22/H13s/1750

989 **Hall, Basil,** 1788-1844.
Account of a voyage of discovery to the west coast of Coren and the great Loo-Choo Island; with an appendix containing charts, and various hydrographical and scientific notices. By Captain Basil Hall . . . And a vocabulary of the Loo-Choo language, by H. J. Clifford . . . London, J. Murray, 1818.
222, 72p., col. front., plates (part. col.), maps (part. fold.), tables. x915.19/H14a

Appendix: Geological Memorandum; being a description of the specimens of rocks collected at Macao and the Ladrone Islands, and on the shores of the Yellow Sea, the west coast of Corba, and the great Loo-Choo Island.

990 **Hall.**
Travels in North America in the years 1827 and 1828. By Captain Basil Hall . . . Edinburgh, Printed for Cadell and Co.; London, Simpkin and Marshall, 1829.
3v., fold. map, fold. tables. Meine/917.3/H136t

991 **Hall, Frederick,** 1780-1843.
Catalogue of minerals, found in the state of Vermont, and in the adjacent states; together with their localities; including a number of the most interesting minerals which have been discovered in other parts of the United States; arranged alphabetically; designed, principally, for the use of persons, who have attended mineralogical lectures, in making collections of specimens. Hartford, P. B. Goodsell, printer, 1824.
44p. x549/H141c

992 **Hall, James,** 1761-1832.
Account of a series of experiments, shewing the effects of compression in modifying the action of heat. [Edinburgh, 1805].
115p., 5 plates. 531.7/H14a

"Read in the Royal Society of Edinburgh, June 3, 1805."

993 **Hall.**
Description d'une suite d'expériences qui montrent comment la compression peut modifier l'action de la chaleur. Traduit de l'anglois par M. A. Pictet. Genève, J. J. Paschoud, 1807.
260p., 5 plates. 531.7/H14aFp

Translation of Account of a series of experiments shewing the effects of compression in modifying the action of heat.

994 **Hall.**
On the vertical position and convolution of certain strata and their relation with granite, *in* Transactions of the Royal Society of Edinburgh, 1815.
v.7, p.79-108, 5 plates. 506 RE

995 **Hall, James,** 1811-1898.
Geology of New York, Part 4, comprising the geology of the Fourth Geological District. Albany, Carroll and Cook, 1843.
683p., 19 plates (part. col., incl. geol. map, sections, fossils).
q508.747/N48/v.4, pt.4

Natural History of New York, v.4, pt.4.
Geol. map is Hall's Geologic Map of the Middle and Western States. Watercolor.

996 **Hall, James.**
See JOHNSON, EDWIN F., Reports on the Bear Mountain Railroad and on . . . Bear Valley Coal Basin . . . 1845.

997 **Halley, Edmond,** 1656?-1743.
An estimate of the quantity of vapour raised out of the sea by the warmth of the sun, derived from an experience shown before the Royal Society, *in* Philosophical Transactions of the Royal Society of London for 1678.
v.16, p.366-370. 506 RO/1685-87

998 **Halley.**
An account of the circulation of watry vapours of the sea, and of the cause of springs, presented to the Royal Society, *in* Philosophical Transactions of the Royal Society of London for 1691.
v.17, p.468-473. 506 RO/1693-94

999 **Halley.**
An account of the evaporation of water, as it was experimented in Gresham Colledge in the year 1693. With some observations thereon, *in* Philosophical Transactions of the Royal Society of London for 1694.
v.18, p.183-190. 506 RO/1685-87

PYRITOLOGIE,

OU

HISTOIRE NATURELLE
DE LA PYRITE,

OUVRAGE DANS LEQUEL ON EXAMINE l'origine, la nature, les propriétés & les usages de ce Minéral important, & de la plûpart des autres Substances du même Regne :

ON Y A JOINT

LE *FLORA SATURNISANS*,

OÙ L'AUTEUR DÉMONTRE L'ALLIANCE QUI SE TROUVE ENTRE LES VÉGÉTAUX ET LES MINÉRAUX;

ET

LES OPUSCULES MINÉRALOGIQUES

Qui comprennent un Traité de L'APPROPRIATION, un Traité de L'ORIGINE DES PIERRES, plusieurs Mémoires sur la CHYMIE & l'HISTOIRE NATURELLE, avec un TRAITÉ des Maladies des Mineurs & des Fondeurs.

Par M. JEAN-FREDERIC HENCKEL, *Docteur en Médecine, Conseiller des Mines du Roi de Pologne, Electeur de Saxe ; de l'Académie Impériale des Curieux de la Nature, & de celle de Berlin.*

OUVRAGES TRADUITS DE L'ALLEMAND.

A PARIS,

Chez JEAN-THOMAS HÉRISSANT, Libraire, rue S. Jacques, à S. Paul & à S. Hilaire.

M. DCC. LX.

Avec Approbation & Privilége du Roi.

Cat. No. 1041

1000 **Hamilton, William,** 1730-1803.
Oeuvres complettes. Commentées par M. l'Abbé Giraud-Soulavie. Paris, Imprimé sous l'Approbation & le Privilege de l'Académie des Sciences, 1781.
506p., fold. map. 551.21/H18oe

1001 **Hamilton.**
Observations on mount Vesuvius, mount Etna, and other volcanos: in a series of letters, addressed to the Royal society, from the Honourable Sir W. Hamilton . . . To which are added, explanatory notes by the author, hitherto unpublished. London, T. Cadell, 1772.
179p., 5 plates, fold. map. x551.21/H18o

1002 **Hamilton.**
Observations on Mount Vesuvius, Mount Etna, and other volcanos: in a series of letters, addressed to the Royal society, from the Honourable Sir W. Hamilton . . . To which are added, explanatory notes by the author, hitherto unpublished. The 2d ed. London, Printed for T. Cadell, 1773.
179p., front. (fold. map), 5 plates. x551.21/H18o/1773

Edited by Thomas Cadell.
Library copy has only 3 plates (fold. map lacking).

1003 **Hamilton.**
Observations on Mount Vesuvius, Mount Etna, and other volcanos: in a series of letters, addressed to the Royal society, from the honourable Sir W. Hamilton . . . To which are added, explanatory notes by the author, hitherto unpublished. A new ed. London, Printed for T. Cadell, 1774.
179p., 5 plates, fold. map. x551.21/H18o/1774

1004 **Hamilton.**
Détails historiques des tremblemens de terre arrivés en Italie depuis le 5 février jusqu'en mai, 1783. Traduits par M. Lefebvre de Villebrune. Paris, T. Barrois le jeune, 1783.
75p. 551.22/H18dFℓ

1005 **Hamilton.**
Neuere Beobachtungen über die Vulkane Italiens und am Rhein, in Briefen . . . nebst merkwürdigen Bemerkungen des Abts Giraud Soulavie . . . Aus dem Französischen zum erstenmal übers. von G. A. R. . . . Mit einer neuen Karte. Frankfurt und Leipzig, bey C. Weigel und A. G. Schneider, 1784.
214p., fold. map. 551.21/H18n

1006 **Hamilton, William,** 1755-1797.
Letters concerning the northern coast of the county of Antrim. Containing a natural history of its basaltes: with an account of such circumstances as are worthy of notice respecting the antiquities, manners and customs of that country. The whole illustrated by an accurate map of the coast, roads, mountains, &c. In these letters is stated a plain and impartial view of the volcanic theory of the basaltes. London, Printed by G. Robinson and Co., 1786.
195p., fold. map. x914.15/H18ℓ/1786

1007 **Hamilton.**
Letters concerning the northern coast of the County of Antrim, containing such circumstances as appear worthy of notice respecting the antiquities, manners and customs of that country. Together with the natural history of the basaltes, and its attendant fossils, in the northern counties of Ireland. The whole illustrated by an accurate map, and engravings of the most interesting objects of the coast. In two parts . . . In these letters is stated a plain and impartial view of the volcanic theory of the basaltes. Dublin, printed by George Bonham, 1790.
Pt. I, 138p., fold. map, fold. plate; Pt. II, 189p., 2 fold. plates. x914.15/H17ℓ/1790

1008 **Hamilton.**
Letters concerning the natural history of the basaltes of the northern coast of the county of Antrim, with an account of its antiquities, manners, and customs. Belfast, printed by Hu. Kirk and Co., 1807.
168p. 552.28/H18L

1009 **Hardy, Robert William Hale,** d. 1871.
Travels in the interior of Mexico, in 1825, 1826, 1827, & 1828. By Lieut. R. W. H. Hardy, R.N. London, H. Colburn and R. Bentley, 1829.
540p., illus., plates, 2 maps (incl. front.). 917.2/H22t

A map of Sonora includes locations of gold and silver.

1010 **Hariot (or Harriot), Thomas,** 1560-1621.
See BRY, THEODOR DE, ed., [America, pt. 1 Latin] Admiranda narratio . . . Virginiae . . . 1590.

1011 **Harris, John,** 1667?-1719.
Remarks on some late papers relating to the universal deluge:

and to the natural history of the earth . . . London, Printed for R. Wilken, 1697.
270p. uncat.

1012 **Hartmann, Philipp Jacob,** 1648-1707.
Succini Prussici physica & civilis historia, cum demonstratione ex autopsia & intimiori rerum experientia deducta. Francofurti, Impensis M. Halleruordi, typis J. Andreae, 1677.
291p., front., 3 fold. plates. x553.29/H25s

Imperfect: front. wanting.

1013 **Hartsoeker, Nicolas,** 1656-1725.
Principes de physique. Paris, J. Anisson, 1696.
236p., illus. x530/H25p

Chapters on formation of the earth, earthquakes, and the origin of fountains and rivers.

1014 **Hasse, Traugott Lebrecht,** 1775-1853.
Denkschrift zur Erinnerung an die Verdienste des in Dresden am 30.Juni 1817 verstorbenen K. S. Bergrath's Werner und an die Fortschritte bei der Berg-academie zu Freiberg, nebst einer übersichtlichen Nebeneinanderstellung der Mineralsysteme Werners und seiner Nachfolger bei dieser Akademie. Mit dem Bildriss von Werner und den Abbildungen zweier Denkmäler. Auch einige Beiträge im Bezug auf mittelbare Folgen der Wernerschen Wirksamkeit. Unter Beihilfe einiger Freunde der Mineralogie. Dresden und Leipzig, In Commission der Arnoldischen Buchhandlung, 1848.
174p., 3 plates (incl. port.). 549.08/D41

1015 **Hasselquist, Frederick,** 1722-1752.
Voyages and travels in the Levant; in the years 1749, 50, 51, 52. Containing observations in natural history, physick, agriculture, and commerce: particularly on the Holy Land, and the natural history of the Scriptures. Written originally in the Swedish language by the late Frederick Hasselquist . . . Published . . . by Charles Linnaeus . . . London, Printed for L. Davis and C. Reymers, 1766.
456p., fold. map. x915.6/H27iE

1016 **Hausmann, Johann Friedrich Ludwig,** 1782-1859.
Handbuch der Mineralogie. Erster-dritter Band . . . Göttingen, Vandenhock und Ruprecht, 1813.
3v. in 1. 549.02/H294h

French and German indexes.

NOUVEAU VOYAGE

d'un Pais plus grand que

L'EUROPE

Avec les reflections des entreprises du Sieur de la Salle, sur les Mines de St. Barbe, &c.

Enrichi de la Carte, de figures expressives, des mœurs & manieres de vivre des Sauvages du Nord, & du Sud, de la prise de Quebec Ville Capitalle de la Nouvelle France, par les Anglois, & des avantages qu on peut retirer du chemin recourci de la Chine & du Japon, par le moien de tant de Vastes Contrées, & de Nouvelles Colonies.

Avec approbation & dedié à sa Majesté

GUILLAUME III.

Roy de la grande

BRETAGNE

PAR LE

R. P. LOUIS HENNEPIN,

Missionaire Recollect & Notaire Apostolique.

A UTRECHT,
Chez ERNESTUS VOSKUYL,
Imprimeur 1698.

Cat. No. 1047

1017 **Hausmann.**
Versuch einer geologischen Begründung des Acker- und Forstwesens. Aus dem Lateinischen übers. von Franz Körte. Berlin, A. Rucker, 1825.
62p. 554.3/H296vGk

1018 **Hausmann.**
Ueber das Gebirgssystem der Sierra Nevada und das Gebirge im südlichen Spanien. Göttingen, Dieterichs, 1842.
46p., plate. 551.43/H29u

1019 **Hausmann.**
Geologische Bemerkungen über die Gegend von Baden bei Rastadt. Göttingen, Dieterichschen, 1844.
42p. 554.3155/B43e

Bound with Bennigsen-Förder, Rudolf von, Erläuterungen zur geognostischen Karte der Umgegend von Berlin, 1843.

1020 **Haüy, René Just,** 1743-1822.
Essai d'une théorie sur la structure des crystaux, appliquée à plusieurs genres de substances crystallisées. Paris, Chez Gogué & Née de la Rochelle, 1784.
236p., 8 fold. plates. x548/H29e/1784

1021 **Haüy.**
Traité de minéralogie . . . Publié par le Conseil des mines. En cinq volumes, dont un contient 86 planches . . . Paris, Chez Louis, 1801.
4v. x549/H29tr

1022 **Haüy.**
Traité de minéralogie . . . Tome cinquième. Caractères minéralogiques. Distribution méthodique des minéraux. Figures géometriques. Paris, Chez Louis, 1801.
10p., 86 plates. xq549/H29tr/atlas

1023 **Haüy.**
See GOLITSYN DMITRII ALEKSIEEVICH, Recueil de noms . . . en minéralogie . . . 1802.

1024 **Haüy.**
See LUCAS, JEAN ANDRÉ HENRI, Tableau méthodique des espèces minérales . . . extrait du Traité de minéralogie de M. Haüy . . . 1806-13.

1025 **Haüy.**
Tableau comparatif des résultats de la cristallographie et

de l'analyse chimique, relativement à la classification des minéraux. Paris, Courcier, 1809.
312p., diagrs. 549/H29t

Bibliography: p. xxxv.

1026 **Haüy.**
Traité des caractères physiques des pierres précieuses, pour servir à leur détermination lorsqu'elles ont été taillées. Paris, Mme Ve Courcier, 1817.
253p., fold. diagrs., tables. 549.2/H29t

Bibliographical footnotes.

1027 **Haüy.**
Traité de cristallographie, suivi d'une application des principes de cette science à la détermination des espèces minérales et d'une nouvelle méthode pour mettre les formes cristallines en projection. Paris, Bachelier et Huzard, 1822.
2v., and atlas (84 plates). 548/H29t
548/H29t/atlas

1028 **Hay, John.**
Calculations introductory to a new theory of the earth, . . . showing, by principles entirely original, that the sacred account of creation is in harmony with natural results; and, in particular, illustrative of the difficulties that occur in accounting for the original formation of the earth, and its constitutional appearances upon principles hitherto known. — Of the earth's original formation. — Of the original formation of strata. — Of the origin of fissures. — Of the formation of the mountains. — Of the formation of the bed of the ocean. — Of the origin of rivers. — Of volcanoes, &c., &c., &c. Edinburgh, Printed for the author, 1824.
10p., 10 plates (part. fold.). 525/H31c

1029 **Hayden, Horace H.,** 1769-1844.
Geological essays; or, An inquiry into some of the geological phenomena to be found in various parts of America and elsewhere. Baltimore, Printed by J. Robinson, for the author, 1820.
412p. x550.4/H324g

1030 **Haywood, John,** 1753-1826.
The natural and aboriginal history of Tennessee, up to the first settlements therein by the white people, in the year 1768. [1st ed.]. Nashville, Printed by G. Wilson, 1823.
390p. x913.768/H33n

Author's presentation copy.

First two chapters give a physical and topographic description of Tennessee and of the formations and fossils.

1031 **Headrick, James,** d. 1841.
View of the mineralogy, agriculture, manufactures and fisheries of the island of Arran. With notices of antiquities, and suggestions for improving the agriculture and fisheries of the Highlands and isles of Scotland, by the Rev. James Headrick. Edinburgh, A. Constable, 1807.
395p., fold. map. 554.139/H34v

1032 **Headrick.**
General view of the agriculture of the county of Angus, or Forfarshire; with observations on the means of its improvement drawn up for the consideration of the Board of Agriculture and internal improvement. Edinburgh, Printed by Neill & Co., 1813.
120p., fold. col. map. 630.941/H34g

1033 **Hebenstreit, Johann Ernst,** 1703-1757.
Museum Richterianum continens fossilia animalia, vegetabilia mar., illustrata iconibus et commentariis. Accedit de gemmis scalptis antiquis liber singularis. Lipsiae, excudi curavit C. Fritsch, 1743.
384p., 34p., front., illus., 17 plates, port. q574.074/H35m

Latin and German in parallel columns.
"Musei Richteriani dactyliotheca gemmas scalptas opere antiquo plerasque complexa interprete Ioh. Frider. Christio": 34 p. at end.

1034 **Heim, Johann Ludwig,** 1741-1819.
Geologischer Versuch über die Bildung der Thäler durch Ströme. Weimar, Hoffman, 1791.
182p. 551.43/H36g

1035 **Hellwig, Theodor Andreas von,** 1694-1721.
Curieuse physicalische Beschreibung, der dreyen beseelten Natur-Reiche, als: Regni animalis, vegetabilis, mineralis . . . Franckfurt und Leipzig, H. P. Ritschel, 1718.
80p. 574.9/H36c

1036 **Helwing, Georg Andrea,** 1666-1748.
Lithographia Angerburgica, sive, lapidum et fossilium, in districtu Angerburgensi & ejus vicinia, ad trium vel quatuor milliarium spatium, in montibus, agris, arenofodinis & in primis circa lacuum littora & fluviorum ripas, collectorum brevis & succincta consideratio. Regiomonti, Literis J. Stelteri, 1717-20.

I. N. I.

HISTORIA BALSAMI MINERALIS ALSATICI SEV PETROLEI VALLIS SANCTI LAMPERT

Germanice

Der Hanauische Erd-Balsam/Lampersloher Oel-oder Bächel-Brunn/

QVAM

AVXILIANTE SVMMO ARCHIATRO

EX CONSENSV GRATIOSISSIMÆ FACVLTATIS MEDICÆ

PRO LICENTIA

SVMMISQVE IN ARTE SALVTARI HONORIBVS
AC PRIVILEGIIS DOCTORALIBVS MORE MAIORVM
RITE OBTINENDIS

SISTIT

A V T O R

IOHANNES THEOPHILVS HOEFFEL

Wœrthensis.

Ad d. [illegible]. Octobr. Anno M DCC XXXIV. Horis locoque consuetis.

A R G E N T O R A T I,

LITERIS Heredum JOHANNIS PASTORII.

Cat. No. 1093

2v. in 1, front., 17 fold. plates. 552/H36ℓ

Vol. 2 has imprint: Lipsiae, Literis Immanuelis Titii, 1720.

1037 **Henderson, Ebenezer,** 1784-1858.
Iceland; or, The journal of a residence in that island, during the years 1814 and 1815. Containing observations on the natural phenomena, history, literature, and antiquities of the island; and the religion, character, manners, and customs of its inhabitants. With an introduction and appendix. Illustrated with a map and engravings. The 2d ed. Edinburgh, Waugh and Innes; [etc., etc.], 1819.
576p., plates, map. 914.91/H381/1819

1038 **Henckel, Johann Friedrich,** 1679-1744.
Introduction à la minéralogie; ou, Connoissance des eaux, des sucs terrestres, des sels, des terres, des pierres, des minéraux, et des métaux: avec une description abrégée des opérations de métallurgie. Ouvrage posthume . . . publié sous le titre de Henckelius in mineralogia redivivus, & traduit de l'Allemand. Paris, G. Cavelier, 1756.
2v. 549/H38hFh

1039 **Henckel.**
Kleine mineralogische und chymische Schriften . . . nebst einer Vorrede von den Bergwercks-Wissenschafften . . . und mit Anmerckungen hrsg. von Carl Friedrich Zimmerman. 2.Aufl. Dresden, F. Hekel, 1756.
619p., fold. plate. 549/H38k/1756

1040 **Henckel.**
Pyritologia: or, A history of the pyrites, the principal body of the mineral kingdom. In which are considered its Names, Species, Beds and Origin; its Iron, Copper, unmetallic Earth, Sulphur, Arsenic, Silver, Gold, original Particles, Vitriol, and Use in smelting. London, A. Millar [etc.], 1757.
326p., front. x549.3/H33pE

1041 **Henckel.**
Pyritologie, ou Histoire naturelle de la pyrite, ouvrage dans lequel on examine l'origine, la nature, les propriétés & les usages de ce minéral important & de la plûpart des autres substances du même regne; on y a joint Le Flora Saturnisans où l'auteur démontre l'alliance qui se trouve entre les végétaux et les minéraux; et les opuscules minéralogiques qui comprennent un Traité de l'Appropriation, un Traité de l'origine des pierres, plusieurs Mémoires sur la chymie &

histoire naturelle, avec un Traité des maladies des mineurs & des fondeurs. Ouvrages traduits de l'Allemand. Paris, Chez Jean-Thomas Herissant, libraire, 1760.
526p., front., 5 fold. plates. 549.3/H38pF

1042 **Hennah, Richard.**
A succinct account of the lime rocks of Plymouth, being the substance of several communications, read before the members of the Geological society, in London, and partly printed in their Transactions. With ten lithographic plates, of some of the most remarkable of the animal remains found in them. By the Rev. Richard Hennah . . . Plymouth, W. Curtis; [etc., etc., 1823].
71p., 11 plates. 553.51/H51s

1043 **Hennepin, Louis,** ca.1640-ca.1705.
Description de la Louisiane, nouvellement decouverte au sud'oüest de la Nouvelle France, par ordre du Roy. Avec la carte du pays: les moeurs et la maniere de vivre des sauvages. Paris, Chez la veuve Sebastien Huré, 1683.
107p., map. uncat.

Early description of Illinois and Mississippi Rivers.

1044 **Hennepin.**
Beschryving van Louisania, nieuwelijks ontdekt ten zuidwesten van Nieuw-Vrankryk . . . Amsterdam, Jan ten Hoorn, 1688.
2v. in 1, 6 plates, fold. map. x917.63/H39dD

First Dutch edition from original French edition of 1683.

1045 **Hennepin.**
Beschreibung der Landschafft Louisiana welche auf Befehl des Königs in Frankreich neulich gegen Sudwesten Neu-Frankreichs in America entdecket worden. Nürnberg, in Verlag Andreä Otto, 1689.
425p., 2 fold. maps. 917.63/H39dG

German translation of Description de la Louisiane . . . 1683.

1046 **Hennepin.**
Nouvelle decouverte d'un tres grand pays situé dans l'Amérique, entre le Nouveau Mexique, et la Mer Glaciale, avec les cartes, & les figures necessaires, & de plus l'histoire naturelle & morale, & les avantages, qu'on en peut tirer par l'établissement des colonies . . . Utrecht, Chez Guillaume Broedelet, 1697.
36p., [296p.], 2 fold. plates. 973.1/H38n

First description of Niagara Falls.

1047 **Hennepin.**
Nouveau voyage d'un pais plus grand que l'Europe, avec les reflections des entreprises du Sieur de la Salle, sur les Mines de St. Barbe, &c. . . . Utrecht, Ernestus Voskuyl, 1698.
389p., 2 plates. uncat.

Title change from previous entry.

1048 **Hennepin.**
Voyage curieux du R.P. Louis Hennepin, missionaire recollect, & notaire apostolique, qui contient une nouvelle decouverte d'un tres-grand pays, situé dans l'Amerique, entre le Nouveau Mexique, & la Mer Glaciale . . . Leide, Chez Pierre vander Aa, 1704.
32p., [296p.], plates, maps. x973.1/H38n/1704

Originally published as Nouvelle decouverte d'une tres grand pays . . . 1697.

A non-geographical work by Sieur De la Borde is appended.

1049 **Hennepin.**
A new discovery of a vast country in America, extending above four thousand miles, between New France and New Mexico. With a description of the great lakes, cataracts, rivers, plants, and animals . . . With a continuation giving an account of the attempts of the Sieur de la Salle upon the mines of St. Barbe . . . To which is added, several new discoveries in North America, not publish'd in the French edition. London, Printed for M. Bentley, etc., 1698.
299p., plates, 2 maps. x973.1/H38nE

Maps lacking.

1050 **Hennepin.**
Nieuwe ontdekkinge van een groot land, gelegen in America, tusschen nieuw Mexico en de Ys-zee . . . Amsterdam, Abraham van Someren, 1699.
220p., fold. plates, fold. maps. x973.1/H38nD

Dutch translation of Nouvelle decouverte d'un tres grand pays . . . 1697.

1051 **Herbinius, Johannes,** ca.1633-1679.
Dissertationes de admirandis mundi cataractis supra & subterraneis, earumque principio, elementorum circulatione, ubi eadem occasione aestus maris reflui vera ac genuina causa asseritur, nec non terrestri ac primigenio Paradiso, locus situsque verus in Palaestina restituitur, in tabula chorographica ostenditur, et contra Utopios, Indianos, Mesopotamios, ali-

MICROGRAPHIA:

OR SOME
Physiological Descriptions
OF
MINUTE BODIES
MADE BY
MAGNIFYING GLASSES.
WITH
OBSERVATIONS and INQUIRIES thereupon.

By *R. HOOKE*, Fellow of the ROYAL SOCIETY.

Non possis oculo quantum contendere Linceus,
Non tamen idcirco contemnas Lippus inungi. Horat. Ep. Lib. 1.

LONDON, Printed for *James Allestry*, Printer to the ROYAL SOCIETY, and are to be sold at his Shop at the *Rose* and *Crown* in *Duck-Lane* MDCLXVII.

Cat. No. 1116

osque asseritur; Amstelodami, Apud Janssonio-Waesbergios, 1678.
267p., illus., fold. maps, fold. plates. uncat.

1052 **Héricart-Farrand, Louis Étienne François,** 1776-1854.
Considérations géologiques et physiques sur la cause du jaillissement des eaux des puits forés ou fontaines artificielles, et recherches sur l'origine ou l'invention de la sonde, l'état de l'art du fontenier-sondeur, et le degré de probabilité du succès des puits forés. Paris, Bachelier, 1829.
339p., illus., fold. col. map. 628.11/H42c

Bibliography: p. [305]-311.

1053 **Hering, Johannes Hermanus,** 1731-1790.
Bespiegeling over Neêrlandsch waternood, tusschen den XXsten en XXIIsten November, MDCCLXXVI. Amsterdam, J. Allart, 1778.
32, 238p., illus. 949.2/H42b

1054 **Hérissant, Louis Antoine Prosper,** 1745-1769.
Bibliothèque physique de la France, ou liste de tous les ouvrages, tant imprimés que manuscrits, qui traitent de l'histoire naturelle de ce royaume: avec des notes critiques & historiques. Paris, Jean-Thomas Hérissant, 1771.
496p. x016.570944/H42b

1055 **Hermann, Benedict Franz Johann von,** 1755-1815.
Versuch einer mineralogischen Beschreibung des uralischen Erzgebürges. Berlin, F. Nicolai, 1789.
2v., fold. tables. 549.94787/H42v

1056 **Hermann, Leonard David,** 1670-1736.
Monumentum gratitudinis marmoreum . . . Das ist: Ein mehr dem Gemühte, als dem besten Marmor eingedrucktes danckbares Andencken wiedmet, mit den beschriebenen Masslichen Muschel-Marmor-Steinen . . . dem Herrn Franz Ernst Brückmann . . . als seinem grossen Freund und Gönner Leonhard David Hermann. Massel, 1729.
31p., illus. 553.51/H42m

1057 **Herrgen, Christiano.**
Descripción geognóstico de las rocas que componen la parte sólida del globo terrestre, extractada de varias obras alemanas de la escuela de Werner, y aumentada con observaciones hechas en la peninsula. Madrid, La Impr. Real, 1802.
224p. 552/H43d

1058 **Herschel, John Frederick William,** 1792-1871, ed.
A manual of scientific enquiry; prepared for the use of Her Majesty's navy; and adapted for travellers in general. Published by authority of the lords commissioners of the Admiralty. London, J. Murray, 1849.
488p., illus., plates, 2 fold. maps, tables, diagrs. 508.3/H43m/1849

1059 **Herttwig, Christoph.**
Neues und vollkommenes Berg-Buch bestehend in sehr vielen und raren Berg-Händeln und Bergwercks-Gebraüchen, absonderlich aber über 200. vorhin noch nicht edirten und ans Licht gegebenen Berg-Urtheln und Abschieden. Dresden, J. C. Zimmermann, 1710.
439p. q622.03/H44n

1060 **Hewitt, William, L.S.A.**
An essay on the encroachments of the German Ocean along the Norfolk coast, with a design to arrest its further depredations . . . Norwich, Printed for the author by Matchett, Stevenson, and Matchett, 1844.
108p., illus. 551.36/H49e

1061 **Heyer, Just Christian Heinrich,** 1746-1821.
Chemische Versuche mit Bernstein [von] Just Christian Heinrich Heyer und Ueber Aquamarin und Topas [von] Johann Carl Wilhelm Voigt. Erfurt, G. A. Keyser, 1787.
24p. 549.8/H51c

Papers presented before the Akademie Nützlicher Wissenschaften zu Erfurt, in 1786.

1062 **Hibbert-Ware, Samuel,** 1782-1848.
A description of the Shetland islands, comprising an account of their geology, scenery, antiquities, and superstitions. With a geological map . . . Edinburgh, A. Constable and Co.; [etc., etc.], 1822.
616p., front., illus., 7 (i.e. 5) plates, fold. map. 914.111/H52d

1063 **Hibbert-Ware.**
History of the extinct volcanoes of the basin of Neuwied, on the lower Rhine. Edinburgh, W. and D. Laing, 1832.
261p., illus., maps (part. col.). 551.21/H52h

1064 **Higgins, William Mullinger.**
The Mosaical and mineral geologies, illustrated and compared. London, Simpkin and Marshall, 1833.
168p. 213/H53m/1833

1065 **Higgins.**
The earth: its physical condition and most remarkable phenomena. New York, Harper & brothers, 1836.
408p., illus. 551.4/H53e

Family library. [vol. 78].

1066 **Higgins.**
The earth: its physical condition and most remarkable phenomena. New York, Harper & brothers, 1839.
408p., illus. 551.4/H53e/1839

Family library. no 78.

1067 **Hill, Ira.**
An abstract of a new theory of the formation of the earth. Baltimore, Printed by John D. Toy for N. G. Maxwell, 1823.
211p. uncat.

1068 **Hill, John,** 1716?-1775.
A general natural history: or, New and accurate descriptions of the animals, vegetables, and minerals, of the different parts of the world; with their virtues and uses, as far as hitherto certainly known, in medicine and mechanics. Illustrated by a general review of the knowledge of the ancients, and the discoveries and improvements of later ages in these studies, including the history of the materia medica, pictoria, and tinctoria, of the present and earlier ages. As also observations on the neglected Properties of many valuable substances known at present; and Attempts to discover the lost Medicines, etc. of former Ages, in a Series of Critical Enquiries into the Materia Medica of the Ancient Greeks. London, Printed for T. Osborne, 1748-52.
3v., col. plates, fold. table. xq570/H55g

Library owns 2 copies; plates not colored in v.2, cop. 2.

1069 **Hill.**
See THEOPHRASTUS, Traité des pierres de Théophraste . . . 1754.

1070 **Hill.**
See THEOPHRASTUS, Theophrastus's history of stones . . . 1774.

1071 **Hill.**
Fossils arranged according to their obvious characters; with their history and description; under the articles of form, hardness, weight, surface, colour, and qualities; the place of their production, their uses, and distinctive English, and

ABSTRACT

OF A

DISSERTATION

READ IN THE

ROYAL SOCIETY OF EDINBURGH,

UPON THE

SEVENTH OF MARCH, AND FOURTH OF APRIL,

M,DCC,LXXXV,

CONCERNING THE

SYSTEM OF THE EARTH, ITS DURATION, AND STABILITY.

Cat. No. 1160

classical Latin names. By J. Hill, M.D. . . . London, Printed for R. Baldwin and P. Elmsly, 1771.
9-51, 56-132, 133*-140*, 133-420p., fold. tables. x560/H55f

1072 **Hinton, John Howard,** 1791-1873, ed.
The history and topography of the United States: ed. by John Howard Hinton, A.M. Assisted by several literary gentlemen in America & England. Illustrated with a series of views, drawn on the spot, and engraved on steel, expressly for this work . . . London, Jennings & Chaplin [etc.]; Philadelphia, T. Wardle; [etc., etc.], 1830-32.
2v., front. (port.), plates, fold. maps, geol. map. 973/H59h/1830

1073 **Hisinger, Wilhelm von,** 1766-1852.
Versuch einer mineralogischen Geographie von Schweden. Umgearbeitete und vermehrte Aufl. Aus der Handschrift übersetzt von F. Wohler. Leipzig: Verlag von J. A. Barth, 1826.
250p. 554.86/H623Gw/1826

1074 **Hisinger.**
Esquisse d'un tableau des pétrifications de la Suède, distribuées en ordre systématique. Stockholm, B. M. Bredberg, 1829.
27p. uncat.

1075 **Hitchcock, Charles Henry,** 1836-1919.
See HITCHCOCK, EDWARD, Elementary geology . . . 1864.

1076 **Hitchcock, Edward,** 1793-1864.
Report on the geology, mineralogy, botany, and zoology of Massachusetts, made and published by order of the government of that State . . . With a descriptive list of the specimens of rocks and minerals collected for the government. Illustrated by numerous wood-cuts and an atlas of plates. Amherst, Press of J. S. and C. Adams, 1833.
700p., illus. and atlas (19 plates). x557/M38h/cop.1

Imperfect: lacking p.693-700.
Atlas missing.
Bound with the author's Report on a re-examination of the economical geology of Massachusetts. Boston, 1838; Maine. Geological Survey (1836-39) First report on the geology of the public lands in the State of Maine. Boston, 1837; and its Second annual report on the geology of the public lands, belonging to the two States of Massachusetts and Maine, 1838.

1077 **Hitchcock.**
Report on the geology, mineralogy, botany, and zoology of Massachusetts, made and published by order of the govern-

ment of that State . . . With a descriptive list of the specimens of rocks and minerals collected for the government. Illustrated by numerous wood-cuts and an atlas of plates. Amherst, Press of J. S. and C. Adams, 1833.
700p., illus. x557/M38h/cop.2

1078 **Hitchcock.**
Report on the geology, mineralogy, botany, and zoology of Massachusetts, made and published by order of the Government of that State. With a descriptive list of the specimens of rocks and minerals collected for the Government. Illustrated by numerous wood cuts and an atlas of plates. 2d ed., corr. and enl. Amherst, J. S. and C. Adams, 1835.
702p., illus. 557/M38h/1835

1079 **Hitchcock.**
Plates illustrating the geology and scenery of Massachusetts. [Boston, Pendleton's Lithography, 1835].
557/M38h/1835/atlas

19 plates including a colored geological map of Massachusetts, 2nd ed., 1834.

Accompanies the Report on the geology . . . of Massachusetts, 2nd ed., 1835 by Edward Hitchcock.

1080 **Hitchcock.**
See DE LA BECHE, HENRY THOMAS, Researches in theoretical geology . . . with a preface and notes by Prof. Edward Hitchcock . . . 1837.

1081 **Hitchcock.**
Report on a re-examination of the economical geology of Massachusetts. Boston, Dutton and Wentworth, 1838.
139p., tables. 557/M38hi

1082 **Hitchcock.**
Report on a re-examination of the economical geology of Massachusetts. Boston, Dutton and Wentworth, 1838.
139p. x557/M38h/c.1

Bound with his Report on the geology, mineralogy, botany, and zoology of Massachusetts . . . 1833.

1083 **Hitchcock.**
Final report on the geology of Massachusetts . . . Northampton, J. H. Butler, 1841.
2v., illus., 55 plates, fold. table. q557/M38

Paged continuously.

Contents. — I. Economical geology. Scenographical geology. — II. Scientific geology. Elementary geology. With an appended catalogue of the specimens of rocks and minerals in the state collection.

1084 **Hitchcock.**

Elementary geology, by Edward Hitchcock and Charles H. Hitchcock. A new ed., remodelled, mostly rewritten, with several new chapters, and brought up to the present state of the science. For use in schools, families, and by individuals. New York, Ivison, Phinney, Blakeman; Chicago, S. C. Griggs, 1864.

430p., illus. 550/H63e/1864

American educational series.
First edition was published in 1840.

1085 **Hitchcock.**

Outline of the geology of the globe, and of the United States in particular: with two geological maps, and sketches of characteristic American fossils. London, S. Low, Son, & Co., 1853.

136p., 5 plates, 2 fold. col. maps. 550/H63o

1086 **Hitchcock.**

Outline of the geology of the globe, and of the United States in particular: with two geological maps, and sketches of characteristic American fossils. 2d ed. Boston, Phillips, Sampson & Company, 1854.

136p., 6 plates, 2 fold. col. maps. 550/H63o/1854

"Prepared chiefly as a sequel to my 'Elementary geology,' especially section XI." — Introductory.

1087 **Hitchcock.**

Illustrations of surface geology. [Washington, Smithsonian Institution, 1857].

155p., illus., 12 plates, incl. maps. q551.4/H63i

Smithsonian contributions to knowledge. [vol. IX, art. 3]
[Smithsonian Institution publication 90]

1088 **Hitchcock.**

See VERMONT, STATE GEOLOGIST, Report . . . 1857 and 1858.

1089 **Hitchcock.**

Ichnology of New England. A report on the sandstone of the Connecticut Valley, especially its fossil footmarks, made to the government of the commonwealth of Massachusetts. Boston, W. White, printer to the state, 1858.

220p., illus., fold. tables, 60 plates, col. map. 557/qM38i

Published under Resolves of the legislature, 1857 and 1858.

HISTORIA NATVRALE

DI FERRANTE IMPERATO NAPOLITANO.

NELLA QVALE ORDINATAMENTE SI TRATTA

Della diuersa condition di Minere, Pietre pretiose, & altre curiosità. Con varie Historie di Piante, & Animali, sin'hora non date in luce.

In questa Seconda Impressione aggiontoui da Gio: Maria Ferro Spetiale alla Sanità, alcune Annotationi alle Piante nel Libro vigesimo ottauo.

DEDICATA ALL' ALTEZZA SER.ma

DI GIOVAN FEDERICO

DVCA DI BRVNSVVICK, ET LVNENBVRG.

VENETIA

Preslo Combi, & La Noù. M. DC. LXXII.

Con licenza de' Superiori, e Priuilegio.

Cat. No. 1172

Bibliography of North American fossil footmarks; or a list of the volumes and papers published on the subject: p. [ix]-xii.

1090 **Hitchcock.**
The religion of geology and its connected sciences. With an additional lecture, giving a summary of the author's present views of the whole subject. London, J. Blackwood, [1860?].
365p., illus. 215/H63r/1860

"Author's copyright edition."

1091 **Hitchcock.**
Report on the geology of Vermont: descriptive, theoretical, economical, and scenographical. By Edward Hitchcock, L.L.D., Edward Hitchcock, Jr., M.D. . . . Albert D. Hager, A.M., Charles H. Hitchcock, A.M. . . . Claremont, N.H., Printed by the Claremont Manufacturing Company, 1861.
2v., illus. q557/V59h

1092 **Hodge, James Thatcher,** 1816-1871.
Report of a tour of exploration through the mineral locations of Lake Superior. New York, 1847.
17p., illus., fold. map. uncat.

1093 **Hoeffel, Johannes Theophilus.**
Historia balsami mineralis Alsatici, seu petrolei Vallis Sancti Lampert Germanice. Der hanauische Erd-Balsam, Lamperslocher Oel- oder Bächel-Brunn. Argentorati, Literis heredum J. Pastorii, 1734.
42p. 553.282/H67h

Bibliographical footnotes.

1094 **Hoffmann, Carl August Siegfried.**
Handbuch der Mineralogie, von C. A. S. Hoffmann [fortgesezt von August Breihaupt]. Freyberg, Craz und Gerland, 1811-18.
4v., illus. 549.02/H67h

Bibliographical footnotes.

1095 **Hoffmann, Friedrich,** 1660-1742, praeses.
De ebore fossili Suevico Halensi. Halae Magdeburgicae, 1734.
29p. 569.6/H67d

Diss. — Halle (J. F. Beyschlag, respondent)

1096 **Hoffmann, Friedrich,** 1797-1836.
Geognostische Charte vom nordwestlichen Deutschland. Berlin, 1828.
G6300/.C5/1828/.H6

Scale not given.

Map essentially taken without revision from his Geognostischen Atlas vom nordwestlichen Deutschland. Simon Schropp et Comp., 1828. 24p.

1097 **Hoffmann.**
Hinterlassene Werke. Berlin, Nicolaischen Buchhandlung, 1837-38.
2v. 550/H67h

v.1 Physikalische Geographie; v.2 Geschichte der Geognosie, und Schilderung der Vulkanischen Erscheinungen. In both — Vorlesungen gehalten an der Universität zu Berlin in den Jahren 1834 und 1835.

1098 **Hoffmann, Karl Friedrich Vollrath,** 1796-1842.
Die Erde und ihre Bewohner, ein Hand- und Lesebuch für alle Stände. 3. berichtigte und verm. Aufl. . . . Stuttgart, Carl Hoffmann; Wien, Carl Gerold'sche Buchhandlung, 1833.
693p., front., 5 plates, fold. maps, tables. 551.4/H677e3

Poor condition.

1099 **Hogard, Henri.**
Description minéralogique et géologique des régions granitique et arénacée du système des Vosges, avec un atlas comprenant une carte géognostique des Vosges, plusieurs vues et coupes. Épinal, Chez Valentin, 1837.
423p., fold. plates, and atlas of 13 plates (part. col., incl. geol. map). uncat.

Text is bound with his Observations sur les moraines et sur les dépôts de transport ou de comblement des Vosges, 1842; and Puton, Ernest, Des métamorphoses et des modifications survenues dans certaines roches des Vosges, Paris, Baillière, 1838.

1100 **Hogard.**
Observations sur les moraines et sur les dépôts de transport ou de comblement des Vosges. Épinal, Imprimerie de Gley, 1842.
81p. uncat.

Bound with his Description minéralogique et géologique des régions granitique et arénacée du système des Vosges . . . 1837.

1101 **Hogard.**
Carte, croquis et coupes pour servir à l'explication de la constitution géologique des Vosges. Planches. [n.p.], [1846].
[8]p., 24 plates (part col.) q554.438/H67c

Plate 1 lacking.
"Avertissement" dated 1848.

1102 **Hogard.**
Principaux glaciers de la Suisse, imprimés en Levis-Aquarelle d'après les originaux dessinés et peints d'après nature par Henri Hogard; aidé des photographies de Mr. Dolfuss-Ausset. Strasbourg, E. Simon; Paris, J. B. Baillière, 1849-1853.
16 art prints, lithograph, col., elephant folio in portfolio.
xF551.312/H67p

1103 **Hogard.**
Coup d'oeil sur le terrain erratique des Vosges. Éd. accompagnée d'un atlas de 32 planches. Épinal, Imprimerie de veuve Gley, 1851.
138p., plus atlas of 32 col. plates in 1 vol. xF551.31309443/H67c

Atlas published in Strasbourg by Dolfuss-Ausset.

Presentation copy from A. Ramsay (with port.) to Sir Archibald Geikie. The text has been inlaid on folio sheets to match the size of the plates. A unique copy.

1104 **Hogard.**
Recherches sur les formations erratiques, 1858. Paris, Librairie de F. Savy, [1860].
233p., and atlas of 19 plates. x551.33/H678r
xf551.33/H678r/plates

1105 **Hogard.**
Recherches sur les glaciers et sur les formations erratiques des Alpes de la Suisse. Épinal, Impr. de veuve Gley, 1858.
319p. 551.31/H76r

Preface dated 1854.

1106 **Hogard.**
Recherches sur les glaciers et sur les formations erratiques des Alpes de la Suisse. Planches. Paris, F. Savy, 1858-62.
35 plates. f551.31/H76r/plates

1107 **Hogg, Thomas,** 1777-1835.
A manual of mineralogy; in which is shown how much Cornwall contributes to the illustration of the science. London, Sold by Longman, Rees, Orme, Brown and Green, 1828.
245p., diagrs. 549/H67m

1108 **Holbach, Paul Henri Thiry d',** 1723-1789.
Systême de la nature. Ou des loix du monde physique & du monde moral. Par M. Mirabaud. Londres, 1770.
2v. x194/H69s/1770

This work was published anonymously under the name of the deceased Jean-Baptiste Mirabaud, perpetual secretary of the Academy at Paris.

DE FOSSILIBVS OPVSCVLVM

AVTHORE

FRANCISCO IMPERATO

IVRE CONSVLTO NEAPOLITANO

FERDINANDI FILIO.

In quo miro ordine continentur Naturalis disciplinæ scitu dignissima, eiusque Professoribus omnino necessaria; ab alijs minimè excogitata. Accesserunt.

Multa, quæ Hieroglyphicè per Fossilia nouiter interpetramur; non nullæque Icones fideliter ad viuum delineatæ.

NEAPOLI, Typis Io. Dominici Roncalioli. M.DC.X.

SVPERIORVM PERMISSV.

Cat. No. 1173

1109 **Holbach.**
Systême de la nature. Ou des loix du monde physique & du monde moral. Par M. Mirabaud. Londres, 1771.
2v. x194/H69s/1771

This work was published anonymously, probably at Neuchâtel, under the name of the deceased Jean-Baptiste Mirabaud, perpetual secretary of the Academy at Paris.

1110 **Holbrook, Josiah.**
Agricultural geology. [Baltimore?, 1851].
36p. 553/H69a

Signed by the author.

1111 **Holland, John,** 1794-1872.
The history and description of fossil fuel, the collieries, and coal trade of Great Britain. By the author of the "Treatise on manufactures in metal," (three volumes), in the Cabinet cyclopaedia . . . 2d ed. London, Whittaker and co.; [etc., etc.], 1841.
485p., illus., tables. 338.2/H719h2

1112 **Holmes, Ezekiel,** 1801-1865.
Report of an exploration and survey of the territory on the Aroostook River, during the spring and autumn of 1838. Augusta, Smith & Robinson, printers to the state, 1839.
78p. 917.41/H73r

1113 **Holmes, J.H.H.**
A treatise on the coal mines of Durham and Northumberland; with information relative to the stratifications of the two counties: and containing accounts of the explosions from fire-damp, which have occurred therein for the last twenty years; their causes, and the means proposed for their remedy, and for the general improvements of the mining system, by new methods of ventilation, &c. London, Printed for Baldwin, Cradock, and Joy, 1816.
259p., 7 plates (incl. map). 622.33/H73t

1114 **Hooke, Robert,** 1635-1703.
The posthumous works of Robert Hooke . . . containing his Cutlerian lectures, and other discourses, read at the meetings of the illustrious Royal society . . . v. Discourses of earthquakes, their causes and effects, and histories of several; to which are annext, physical explications of several of the fables in Ovid's *Metamorphoses,* very different from other mythologick interpreters. Illustrated with sculptures. To these discourses is prefixt the author's life, giving an account of his studies

and employments, with an enumeration of the many experiments, instruments, contrivances and inventions, by him made and produced . . . Publish'd by Richard Waller . . . London, Printed by S. Smith and B. Walford, 1705.
[507] [11]p., 15 plates. x504/H76p

Errors in paging.

1115 **Hooke.**
Micrographia: or some physiological descriptions of minute bodies made by magnifying glasses. With observations and inquiries thereupon . . . London, printed by Jo. Martyn, and Ja. Allestry, 1665.
246p., [10p.], 38 plates part. fold., diagrs. xq578.8/H76m

The library owns a second copy of the 1665 edition.

1116 **Hooke.**
Micrographia: or some physiological descriptions of minute bodies made by magnifying glasses. With observations and inquiries thereupon . . . London, printed for James Allestry, 1667.
246p., [10p.] 38 plates, part. fold., diagrs. xq578.8/H76m/1667

1117 **Hooke.**
Micrographia restaurata: or, the copper-plates of Dr. Hooke's wonderful discoveries by the microscope, reprinted and fully explained: whereby the most valuable particulars in that celebrated author's Micrographia are brought together in a narrow compass; and intermixed, occasionally, with many entertaining and instructive discoveries and observations in natural history . . . London, printed for and sold by J. Bowles, 1745.
65p., 38 plates. xq578.8/H76mi

1118 **Hooker, William Jackson,** 1785-1865.
Journal of a tour in Iceland in the summer of 1809. Yarmouth, Printed by J. Keymer, 1811.
496p., front., illus., 2 plates, plan. 914.91/H76j

1119 **Horrebow, Miels,** 1712-1760.
The natural history of Iceland: containing a particular and accurate account of the different soils, burning mountains, minerals, vegetables, metals, stones, beasts, birds, and fishes; together with the disposition, customs, and manner of living of the inhabitants. Interspersed with an account of the island, by Mr. Anderson . . . To which is added, a meteorological

table . . . Tr. from the Danish original of Mr. N. Horrebow. . . London, Printed for A. Linde [etc.], 1758.
207p., front. (fold. map). xf570.9491/H785tE

Probably a translation of Tilforladelige efterretninger om Island.

1120 **Horsford, Eben Norton,** 1818-1893.
Analyses of grains and vegetables, distinguishing the nitrogenous from the non-nitrogenous ingredients, for the purpose of estimating their separate values for nutrition. Also, on ammonia found in glaciers; and on the action and ingredients of manures. Boston, J. Munroe and company, 1846.
68p. 543.1/H76a

1121 **Horst, Johann,** 1616-1685.
. . . Embser Bades Beschreibung. n.p., 1676.
16p. uncat.

1122 **Horst.**
Kurtzer Bericht vom Embser-Bad an der Lahn in Teutsch und Frantzösisch, so dann Wiss- und Offenauer-Bad . . . D. Georg. Marii . . . Marsilii Weigelii . . . Darmbstadt, Bey Henning Müllern, 1683.
208p. uncat.

1123 **Houghton, Douglass,** 1809-1845.
See MICHIGAN. GEOLOGICAL SURVEY, Annual report . . . 1840-41.

1124 **Houghton, Thomas,** fl. 1680.
Rara avis in terris or the complete miner; in two books. The first containing, the liberties, laws and customs of the lead-mines, within the Wapentake of Wirksworth in Derbyshire; in fifty-nine articles, being all that ever was made. The second teacheth the art of dialling and levelling grooves; a thing greatly desired by all miners; being a subject never written on before by any. London, n.p., 1681.
105p., illus. uncat.

1125 **Houghton.**
Rara avis in terris or the compleat miner; in two books. The first containing, the liberties, laws and customs of the lead-mines, within the Wapentake of Wirksworth in Derbyshire; in fifty-nine articles, being all that ever was made. The second teacheth the art of dialling and levelling grooves; a thing greatly desired by all Miners; being a subject never written on before by any. With an explanation of the miners terms

En Resa Til Norra AMERICA,

På Kongl. Swenska Wetenskaps Academiens befallning, Och Publici kostnad, Förrättad

Af

PEHR KALM,

Oeconomiæ Professor i Åbo, samt Ledamot af Kongl. Swenska Wetenskaps-Academien.

Tom. I.

Med Kongl. Maj:ts Allernådigste *Privilegio.*

STOCKHOLM,

Tryckt på LARS SALVII kostnad 1753.

Cat. No. 1235

of art used in this book. 2d ed., corrected. London, n.p., 1738.

66p., illus. x669/B23aEs/1739

Bound with Barba, Alvaro Alonso, A collection of the scarce and valuable . . . 1739.

1126 **Houghton.**

Rara avis in terris: or the compleat miner; in two books. The first containing, the liberties, laws, and customs of the lead mines, within the Wapentake of Wirksworth in Derbyshire; in fifty-nine articles, being all that ever was made. The second teacheth the art of dialling and levelling grooves; a thing greatly desired by all Miners; being a subject never written on before by any. With an explanation of the miners terms of art used in this book. 2d ed., corrected. n.p., n.d.

p.241-312. x669/B23aEs/1740

Bound with Barba, Alvaro A., A collection . . . 1740.

1127 **Howard, Philip,** d. 1810.

The scriptural history of the earth and of mankind, compared with the cosmogonies, chronologies, and original traditions of ancient nations; an abstract and review of several modern systems; with an attempt to explain philosophically, the Mosaical account of the creation and deluge, and to deduce from this last event the causes of the actual structure of the earth, in a series of letters. With notes and illustrations. London, Printed for R. Faulder, 1797.

602p. x213/H83s

Caption title: Thoughts on the structure of this globe.

"The substance of the following work was published in two letters in the French language . . . 1786. It has since been revised, corrected and considerable enlarged." — Preface.

1128 **Hübner, Johann,** 1668-1731.

Curieuses und reales Natur- Kunst- Berg- Gewerck- und Handlungs-Lexicon . . . Alles dergestalt eingerichtet, dass man dieses als einen andern Theil des Realen Staats-Conversations und Zeitungs-Lexici mit grossem Vortheile gebrauchen kan. 5.Aufl. . . . verb., und . . . vermehret. Leipzig, J. F. Gleditschen Sohn, 1727.

2144 columns, illus. .033/H86c/1727

1129 **Hugi, Franz Joseph,** 1796-1855.

Naturhistorische Alpenreise. Vorgelesen der Naturforschenden Gesellschaft in Solothurn. Mit Titelkupfer und Vignette, 2 Kärtchen, 16 Tafeln Profilansichten und 9 Tabellen be-

rechneter Höhen-Unterschiede. Solothurn, Bei Amiet-Lutiger; [etc., etc.], 1830.
378p., front., 16 plates, 2 fold. maps, 9 fold. tables. x551.31/H87n

1130 **Hugi.**
Ueber das Wesen der Gletscher und Winterreise in das Eismeer. Stuttgart, J. G. Cotta, 1842.
135p. 551.312/H87u

1131 **Hugi.**
Die Gletscher und die erratischen Blöcke. Solothurn, Jent & Gassmann, 1843.
256p., illus. x551.31/H87g

1132 **Hulbert, Charles,** 1778-1857.
Volcanic wonders, and scenes of astonishment: being historic and scientific descriptions of the volcanoes of the Azores, and a general view of burning mountains, in various parts of the globe, selected by C. Hulbert . . . 2d ed. . . . Shrewsbury, C. Hulbert, 1827.
68p., 5 plates (incl. front.), map. 551.21/H872v

1133 **Humboldt, Friedrich Wilhelm Heinrich Alexander von,** 1769-1859.
Gesammelte Werke. Stuttgart, J. G. Cotta'sche Buchhandlung, [pref. 1808-1860].
12v. in 4, front. (port.) 551.H88

Each vol. has also special t.p.

"Anmerkungen" interspersed.

Contents. — 1.-4. Bd. Kosmos [1844]; 5.-8. Bd. Reise in die aequinoktialen Gegenden des neuen Kontinents [1859-60]; 9.-10. Bd. Neuspanien 1.-[11]. Kordilleren [1808]; 11. Bd. Ansichten der Natur [1849], 362p.; 12. Bd. Cuba. Lebensbeschreibung [1849].

1134 **Humboldt.**
Mineralogische Beobachtungen über einige Basalte am Rhein. Mit vorangeschickten, zerstreuten Bemerkungen über den Basalt der ältern und neuern Schriftsteller. Braunschweig, Schulbuchhandlung, 1790.
126p. 552.28/H88m

1135 **Humboldt.**
Political essay on the kingdom of New Spain . . . With physical sections and maps . . . tr. from the original French by John Black. 3d ed. London, Longman, Hurst, Rees, Orme, and Brown [etc.], 1822.
4v., fronts. (v.1, 2, 4), plates, fold. maps, plans, tables, diagrs.
917.2/H88eE/1822

1136 **Humboldt.**
Essai géognostique sur le gisement des roches dans les deux hémisphères. Paris [etc.], F. G. Levrault, 1823.
379p. 552.H88e

1137 **Humboldt.**
Essai géognostique sur le gisement des roches dans les deux hémisphères. 2. éd., conforme à la première. Paris, F. G. Levrault, 1826.
364p. 552/H88e/1826

1138 **Humboldt.**
A geognostical essay on the superposition of rocks in both hemispheres. Tr. from the original French. London, Longmans, Hurst, Rees, Orme, Brown, and Green, 1823.
482p. 552/H88eE

1139 **Humboldt.**
Fragmens de géologie et de climatologie asiatiques. Paris, Gide, 1831.
2v., map, tables. 551.4/H88f

1140 **Humboldt.**
Central-Asien. Untersuchungen über die Gebirgsketten und die vergleichende Klimatologie. Aus dem Französischen übersetzt und durch Zusätze vermehrt, herausgegeben von Dr. Wilhelm Mahlmann . . . Mit einer Karte und vierzehn Tabellen . . . Berlin, C. H. Klemann, 1844.
2v., fold. map., tables. 551.4/H88c

Vol. 1 in two parts, each with t.-p.

1141 **Humboldt.**
Kosmos. Entwurf einer physischen Weltbeschreibung. Stuttgart und Tübingen, J. G. Cotta, 1845-62.
5v., tables (1 fold.). 551/H88k/1845

1142 **Humboldt.**
Kosmos. Entwurf einer physischen Weltbeschreibung . . . (Amerikanische Stereotyp-Ausgabe.) Philadelphia, F. W. Thomas, 1855.
636p. 551/H88k/1855

1143 **Humboldt.**
Kosmos. Entwurf einer physischen Weltbeschreibung. (Amerikanische Jubiläums-Ausgabe.) Philadelphia, F. W. Thomas, 1869.
923p., front. (port.). 551/H88k/1869

ATHANASII KIRCHERI
E SOC. JESU

MUNDUS SUBTERRANEUS,

In XII Libros digestus;

QUO

Divinum Subterrestris Mundi Opificium, mira Ergasteriorum Naturæ in eo distributio, verbo παντάμορφον Protei Regnum,

Universæ denique Naturæ Majestas & divitiæ summa rerum varietate exponuntur. Abditorum effectuum causæ acri indagine inquisitæ demonstrantur; cognitæ per Artis & Naturæ conjugium ad humanæ vitæ necessarium usum vario experimentorum apparatu, necnon novo modo, & ratione applicantur.

TOMUS I.

AD

ALEXANDRUM VII.
PONT. OPT. MAX.

AMSTELODAMI,

Apud JOANNEM JANSSONIUM & ELIZEUM WEYERSTRATEN,
ANNO CIↃ IↃC LXV *Cum Privilegiis.*

Cat. No. 1257

1144 **Humboldt.**
Cosmos, essai d'une description physique du monde; traduit par H. Faye . . . Milan, Chas. Turati, 1846-49.
2v. 551/H88kFf

1145 **Humboldt.**
Cosmos; essai d'une description physique du monde. Paris, Gide, 1846-
4v. 551/H88kFf/1846

Library owns v.1 only.

1146 **Humboldt.**
Cosmos; essai d'une description physique du monde. Traduction de H. Faye et de Ch. Galuski. 4.éd., mise dans un meilleur ordre que les précédentes et augm. d'une notice biographique sur Humboldt, avec des fragments inédits de la correspondance de l'auteur. [Paris], L. Guerin, 1866-67.
4v. 551/H88kFf/1866

Translation of Kosmos.

1147 **Humboldt.**
Cosmos: sketch of a physical description of the universe. Translated under the superintendence of Edward Sabine [by Mrs. Sabine]. 7th ed. London, Printed for Longman, Brown, Green and Longmans, 1849-52.
3v. 551/H88kEs/1849

1148 **Humboldt.**
Ansichten der Natur, mit wissenschaftlichen Erläuterungen. 3. verb. und verm. Ausg. Stuttgart, und Augsburg, Cotta, 1849-1859.
2v. 551.4/H88a

Library lacks v.2.

1149 **Humboldt.**
Ansichten der Natur, mit wissenschaftlichen Erläuterungen. 3. verb. und verm. Ausg. Stuttgart, Cotta, 1871.
2v. in 1. 551.4/H88a/1871

1150 **Humboldt.**
Aspects of nature, in different lands and different climates; with scientific elucidations. Tr. by Mrs. Sabine . . . London, Printed for Longman, Brown, Green, and Longmans [etc.], 1850.
2v. in 1. 551.4/488aEs

1151 **Humboldt.**
Views of nature: or, Contemplations on the sublime phenomena of creation; with scientific illustrations. Tr. from the German by E. C. Otté and Henry G. Bohn . . . London, H. G. Bohn, 1850.
452p., col. front. x551.4/H88aEo

[Bohn's scientific library]

1152 **Humboldt.**
Tableaux de la nature. Edition nouvelle avec changements et additions importantes et accompagnée de cartes. Traduite par Ch. Galusky. Paris, Gide et J. Baudry, 1851.
2v., illus. (part col.), maps. 551.4/H88aFg/1851

1153 **Humboldt.**
Tableaux de la nature. Tr. par Ferdinand Hoefer. Milan, Charles Turatti, 1851.
434p. 551.4/H88aFh/1851

1154 **Humboldt.**
Kleinere Schriften. Erster Band. Geognostische und physikalische Erinnerungen. Mit einem Atlas . . . Stuttgart und Tübingen, J. G. Cotta'scher Verlag, 1853.
474p., fold. tables and atlas of 12 plates (part col., incl. maps).
551.21/H89k
551.21/H89k/atlas

Atlas has title: Umrisse von Vulkanen aus den Cordilleren von Quito und Mexico. Ein Beitrag zur Physiognomik der Natur.
No more published. *cf.* Catalogue of the library of the British museum (Natural history).

1155 **Humboldt.**
Reisen in Amerika und Asien; eine Darstellung seiner wichtigsten Forschungen, von H. Kletke. Berlin, Hasselberg, 1855-56.
4v., front. (port.). 918/H88vGk

Each volume has also special t.p.
Includes volumes of various editions. [v.1, 1856, v.3, 1855].
Translation of Voyage aux régions équinoxiales du nouveau continent.

1156 **Huot, Jean Jacques Nicolas,** 1790-1845.
Nouveau cours élémentaire de géologie. Paris, Librairie encyclopédique de Roret, 1837-39.
2v., illus., atlas of 24 plates. 550/H92n
550/H92n/atlas

1157 **Husson, Camille,** pharmacist.
Esquisse géologique de l'arrondissement de Toul suivie d'un aperçu botanique des environs de cette ville. Toul, Impr. de Ve Bastien, 1848.
105p., tables. 554.4382/H96e

Imperfect: t.p. wanting. Title supplied from Cat. gén. de la Bibliothèque nat.

1158 **Hutchins, Thomas,** 1730-1789.
A topographical description of Virginia, Pennsylvania, Maryland, and North Carolina, comprehending the rivers Ohio, Kenhawa, Sioto, Cherokee, Wabash, Illinois, Mississippi, &c., the climate, soil and produce, whether animal, vegetable, or mineral; the mountains, creeks, roads, distances, latitudes, &c., and of every part laid down in the annexed map. Published by Thomas Hutchins, captain in the 60th regiment of foot. With a plan of the rapids of the Ohio, a plan of the several villages in the Illinois country, a table of the distances between Fort Pitt and the mouth of the Ohio, all engraved upon copper. . . . London, Printed for the author, 1778.
67p., 2 fold. plans, fold. table. x917.5/H97t/1778

First impression.
A col. map in 4 sheets, issued separately, entitled "A new map of the western parts of Virginia, Pennsylvania, Maryland and North Carolina; comprehending the river Ohio, and all the rivers, which fall into it; part of the river Mississippi, the whole of the Illinois River, Lake Erie, part of the lakes Huron, Michigan, &c., and all the country bordering on these lakes and rivers. London, T. Hutchins, 1778.

1159 **Hutchins.**
See IMLAY, GILBERT, A topographical description of the western territory of North America . . . [with papers by T. Hutchins] . . . 1797.

1160 **Hutton, James,** 1726-1797.
Abstract of a dissertation read in the Royal Society of Edinburgh, upon the seventh of March, and fourth of April, M,DCC, LXXXV, concerning the system of the earth, its duration and stability. [Edinburgh, 1785?].
30p. x551.7/H979a

Signed by the author.
Abstract of his Theory of the earth; or An investigation of the laws observable in the composition, dissolution, and restoration of land upon the globe (Royal Society of Edinburgh. Transactions, v.1, pt. 2, 1788, p. [209]-304) on which he afterwards based his The theory of the earth, with proofs and illustrations (Edinburgh, 1795).

1165 **Hutton.**
A dissertation upon the philosophy of light, heat, and fire. Edinburgh, Printed for Messrs. Cadell, Jr. and Davies, London, 1794.
326p. x500/H979d

1166 **Hutton.**
Observations on granite, *in* Transactions of the Royal Society of Edinburgh, 1794.
v.3, p.77-85. 506 RE/1794

1167 **Hutton.**
Of the flexibility of the Brazilian stone, *in* Transactions of the Royal Society of Edinburgh, 1794.
v.3, p.86-94. 506 RE/1794

1168 **Hutton, William,** 1798-1860, jt. author.
See LINDLEY, JOHN, The fossil flora of Great Britain . . . 1831-37.

1169 **Imlay, Gilbert,** 1754?-1828?
A topographical description of the western territory of North America; containing a succinct account of its climate, natural history, population, agriculture, manners and customs, with an ample description of the several divisions into which that country is partitioned, and an accurate statement of the various tribes of Indians that inhabit the frontier country. To which is annexed, a delineation of the laws and government of the state of Kentucky. Tending to shew the probable rise and grandeur of the American empire. In a series of letters to a friend in England. London, Printed for J. Debrett, 1792.
247p. x917.6/Im5t

First edition.

1170 **Imlay.**
A topographical description of the western territory of North America. New-York, Printed by S. Campbell, 1793.
2v., 3 fold. maps (incl. front., v.2). x917.6/Im5

Vol. 2 has title: The discovery, settlement, and present state of Kentucky . . . Being a supplement to Imlay's description of the western territory.

v.1. Imlay, G. Letters I-XI. — Notes.

v.2. Filson, John. The discovery, settlement, and present state of Kentucky. — Imlay, G. Postscript. — U.S. Dept. of state. Report of the secretary of state, to the President . . . of the quantity and situation of the lands not claimed by the Indians, nor granted to nor claimed by any citizens, within the territory of the United States . . . 1791. — Thoughts on emigration. To which are added, miscellaneous observations, relating to the United States of America; and a short account of . . . Kentucky.

Imlay's work appeared first, without that of Filson, London, 1792. The first edition of Filson's "Discovery" was published at Wilmington, Del., in 1784.

1171 **Imlay.**

A topographical description of the western territory of North America . . . Illustrated with correct maps of the western territory of North America; of the state of Kentucky, as divided into counties, from actual surveys by Elihu Barker; a map of the Tenasee government; and a plan of the rapids of the Ohio. The 3d ed., with great additions. London, Printed for J. Debrett, 1797.

598, [28]p., 4 fold. maps. x971.6/Im5t/1797

Contents. — A topographical description of the western territory . . . [by G. Imlay] to which are added, I. The discovery, settlement, and present state of Kentucky . . . by J. Filson. — II. An account of the Indian nations inhabiting within the limits of the xiii states. — III. The culture of Indian corn . . . tobacco . . . — IV. Observations on the ancient works, the native inhabitants of the western country, &c., by J. Heart. — V. Historical narrative . . . of Louisiana and West-Florida . . . by T. Hutchins. — VI. Account of the soil . . . timber and other productions . . . of the Genesee tract . . . — VII. Remarks for the information of . . . settlers in America, by Dr. Franklin. — VIII. Topographical description of Virginia, Pennsylvania, Maryland, and North-Carolina, by . . . T. Hutchins. — IX. P. Kennedy's Journal up the Illinois river, &c. — X. Description of the state of Tenasee. — XI. An act for establishing Knoxville. — XII. Treaty . . . between the United States . . . and . . . Spain, for the free navigation of the Mississippi. — XIII. Plan of association of the North American land company, &c.

The above is the order of contents as given on the t.-p., but differs from the numbering and order of articles as given in the book.

1172 **Imperato, Ferrante,** 1550-1625.

Historia naturale di Ferrante Imperato Napolitano. Nella quale ordinatamente si tratta della diversa condition di minere, pietre pretiose, & altre curiosità . . . Venetia, Presso Combi & La Noù, 1672.

696p., illus. (new acquisition)

1173 **Imperato, Francisco.**

De fossilibus opusculum . . . in quo miro ordine continentur naturalis disciplinae scitu dignissima, eiusque professoribus omnino necessaria: ab alijs minimè excogitata. Accesserunt. Neapoli, Typis Io. Dominici Roncalioli, 1610.

98p. uncat.

RECVEIL
DE
MONVMENS
DES
CATASTROPHES
QVE LE GLOBE DE LA TERRE

A ÉSSVIÉES,

CONTENANT

DES

PÉTRIFICATIONS

ET

D'AVTRES PIERRES CVRIEVSES,

DESSINÉES, GRAVÉES ET ENLVMINÉES,

D'APRÈS LES ORIGINAVX,

AVEC LEVR DESCRIPTION,

PAR

GEORGE WOLFGANG KNORR

à

NVREMBERG.

Cat. No. 1284

On minerals, gems, fossils, rocks, and metals with a table of classification.

1174 **Imrie, Ninian.**
A description and section of the strata of the Grampian Mountains, from the plain of Kincardineshire to the summit of Mount Battock. [Edinburgh], 1811.
39p., illus., fold. plate. x554.1/Im8d

1175 **Indiana. State Geologist.**
Report of a geological reconnoissance of the State of Indiana, made in the year 1837, in conformity to an order of the legislature. By D. D. Owen. Part first. Indianapolis, John C. Walker, 1859.
68p.

Part second. Continuation of report of a geological reconnoissance of the State of Indiana, made in the year 1838, in conformity to an order of the legislature. Indianapolis, John C. Walker, 1859.
68p. 557/In24r

1176 **Indiana. State Geologist.**
Second report of a geological survey of the state of Indiana, made in the year 1838, in conformity to an order of the Legislature, by David Dale Owen, M.D., geologist of the state. Indianapolis, Osborn and Willets, printers, 1839.
54p. x557/In24s

Another edition published in 1859 has title: Continuation of report of a geological reconnoissance of . . . Indiana, made in . . . 1838.

1177 **Insulin, Stephan.**
Verlds-Beskrifningens III:dje des om folkslagens seder ach Lynne. Pä cosmographiske sällskapets ästundan. Strengnäs, L. A. Collin, 1772.
438p. x910/M288w/[v.3]

1178 **Ippolito, Francesco,** jt. author.
See HAMILTON, WILLIAM, Détails historiques des tremblemens de terre . . . 1783.

1179 **Italy. Direzione Generale Di Pubblica Istruzione.**
Catalogo di una collezione di minerali disposta secondo il sistema del celebre Werner, ed acquistata per uso de'licei del regno d'Italia, a Freyberg dalla Direzione generale di pubblica istruzione. Milano, Stamperia reale, 1813.
316p. 549/It1c

Half title: Catalogo mineralogico Werneriano.

1180 **Jachmann, John Benjamin.**
Tentamen chemicum inaugurale, de natura crystallisationis. Quod . . . ex auctoritate . . . D. Gulielmi Robertson . . . Academiae Edinburghae praefecti . . . pro gradu doctoris . . . eruditorum examini subjicit Joannes Benjamin Jachmann, Regiomonto-Borussus . . . Ad diem 12. septembris [1789] . . . Edinburgi, apud Balfour et Smellie, academiae typographos, 1789.
33p. uncat.

1181 **Jackson, Charles Thomas,** 1805-1880, Natural history of water. See Scientific tracts, for the diffusion of useful knowledge . . . 1836.

1182 **Jackson.**
See MAINE. GEOLOGICAL SURVEY, Second [— third annual] report on the geology of the State of Maine . . . 1838-39.

1183 **Jackson.**
Report on the geological and agricultural survey of the state of Rhode Island, made under a resolve of legislature in the year 1839. Providence, B. Cranston & co., 1840.
312p., illus., plates, fold. maps, tables. x557/R34j

Includes hand-colored map of New England.

1184 **Jackson.**
See NEW HAMPSHIRE, GEOLOGICAL SURVEY, 1840-1844, First annual report . . . 1841.

1185 **Jackson.**
See NEW HAMPSHIRE. GEOLOGICAL SURVEY, 1840-1844, Final report . . . 1844.

1186 **Jackson.**
Report on the geological and mineralogical survey of the mineral lands of the United States in the state of Michigan. Washington, Printed for the Senate, 1849.
26 plates, 6 maps. 557/M58u

Issued in Senate Executive Document 1, v.3, 31st Cong., 1st Sess., p. 371-801.

1187 **Jackson, John Richard.**
See JACKSON, JULIAN, 1790-1853.

1188 **Jackson, Julian,** 1790-1853.
Observations on lakes, being an attempt to explain the laws of nature regarding them; the cause of their formation and gradual diminution; the different phenomena they exhibit,

&c., with a view to the advancement of useful science. By Colonel J. R. Jackson. London, Bossange, Barthès, and Lowell; [etc., etc.], 1833.
88p., diagrs. q551/482/J13o

The author has been wrongly called "John Richard" Jackson. Dict. nat. biog.

1189 **Jackson, Lowis D'Aguilar.**
See AQUARIUS.

1190 **James, Edwin,** 1797-1861, compiler.
Account of an expedition from Pittsburgh to the Rocky Mountains, performed in the years 1819, 1820. By order of the Hon. J. C. Calhoun, secretary of war, under the command of Maj. S. H. Long, of the U.S. top. engineers. Comp. from the notes of Major Long, Mr. T. Say, and other gentlemen of the party, by Edwin James . . . London, Printed for Longman, Hurst, Rees, Orme, and Brown, 1823.
3v., fronts., illus., plates, maps, profile. 917.8/J232a/1823

"Observations on the mineralogy and geology of a part of the United States west of the Mississippi; extracted from a report to Major Long" v.3, p. [269]-330.

1191 **Jameson, Robert,** 1774-1854.
An outline of the mineralogy of the Shetland Islands, and of the island of Arran. Illustrated with copper-plates. With an appendix; containing observations on peat, kelp, and coal. Edinburgh, Printed for W. Creech; [etc., etc.], 1798.
202p., plate, 2 maps. 549.941/J23o

1192 **Jameson.**
Mineralogy of the Scottish Isles; with mineralogical observations made in a tour through different parts of the mainland of Scotland, and dissertations upon peat and kelp. Edinburgh, Printed by C. Stewart for B. White, London; and W. Creech, Edinburgh, 1800.
2v. in 1, plates, maps (part fold.). 549.941/J23m

1193 **Jameson.**
Mineralogische Reisen durch Schottland und die Schottischen Inseln. Aus dem Englischen übersetzt und von einem Auszuge aus Bergrath Werner's Geognosie, die Lehre von den Gebirgsarten betreffend, als Einleitung begleitet, von Heinrich Wilhelm Meuder. Leipzig, S. L. Crusius, 1802.
256p., illus., fold. map. 549.941/J23mGm

Translation of Mineralogy of the Scottish Isles.

CAROLI NICOLAI LANGII

Lucernens. Helvet. Phil. & Medici

Acad. Cæs. LEOPOLD. Nat. Curios. German. & Physio-Crit. Senens.

HISTORIA
LAPIDUM FIGURATORUM
HELVETIÆ,
EJUSQUE VICINIÆ,

In quâ non solùm enarrantur omnia eorum GENERA, SPECIES ET VIRES ÆNEISQUE TABULIS REPRÆSENTANTUR,.

Sed insuper adducuntur eorum

LOCA NATIVA,

IN QUIBUS REPERIRI SOLENT, UT CUILIBET FACILE SIT EOS COLLIGERE, MODO ADDUCTA LOCA ADIRE LIBEAT.

VENETIIS MDCCVIII.

Sumptibus Authoris, Typis Jacobi Tomasini.

SUPERIORUM PERMISSU.

Lucernæ, apud Hæredes Gottofredi Hautt, & Joannem Jodocum Halter.

Cat. No. 1321

1194 **Jameson.**
System of mineralogy . . . Edinburgh, 1804-08.
3v., tables (part. fold.), diagrs. (part. fold.). 549/J23s

1195 **Jameson.**
A system of mineralogy . . . 2d ed. Edinburgh, Printed by Neill for A. Constable, 1816.
3v., 13 fold. plates. 549/J23s/1816

Bibliography: v.1, p. ix-xiv.

1196 **Jameson.**
A system of mineralogy, in which minerals are arranged according to the natural history method. 3d ed., enl. and improved . . . Edinburgh, A. Constable & co.; [etc., etc.], 1820.
3v., 11 fold. plates. 549/J23s/1820

"List of works quoted": p. xi-xvi.

1197 **Jameson.**
A mineralogical description of the county of Dumfries. Edinburgh, Bell & Bradfute [etc.]; London, Longman, Hurst, Rees, and Orme, 1805.
185p., 4 plates. 549.941/J23mi

Bound with Stephens, Walter. Notes on the mineralogy of part of the vicinity of Dublin. London, 1812.

1198 **Jameson.**
A treatise on the external characters of minerals. Edinburgh, Printed at the University press for Bell & Bradfute, Guthrie & Tait, and W. Blackwood; [etc., etc., 1805].
84, 32p., 2 fold. diagr. 549/323t

"Tabular view of the different generic and subordinate special external characters of minerals"; 32p. at end.

1199 **Jameson.**
A treatise on the external, chemical, and physical characters of minerals. 2d ed. Edinburgh, Printed by Neill for A. Constable, 1816.
304p., 7 fold. plates. 549/J23t/1816

Bibliography: p.vi-vii.

1200 **Jameson.**
A treatise on the external, chemical, and physical characters of minerals. 3d ed. Edinburgh, printed by Neill for A. Constable and company; [etc., etc.], 1817.
314p., 8 plates. 549/J23t/1817

Bibliography: p. vi-ix.

1201 **Jameson.**
Mineralogical travels through the Hebrides, Orkney and Shetland Islands, and mainland of Scotland, with dissertations upon peat and kelp . . . Edinburgh, Printed for A. Constable and co.; [etc., etc.], 1813.
2v., illus., 9 plates, 4 maps (incl. fold. fronts.). 554.1/J23m

1202 **Jameson, Robert.**
See CUVIER, GEORGES, Essay on the theory of the earth . . . 1813, 1817, and 1818.

1203 **Jameson.**
Manual of mineralogy; containing an account of simple minerals, and also a description and arrangement of mountain rocks. Edinburgh, Printed for A. Constable, 1821.
501p. 549/J23m

1204 **Jefferson, Thomas,** 1743-1826.
Notes on the State of Virginia. London, printed for John Stockdale, 1787.
382p., illus., fold. map. 975.5/J35n

1205 **Jefferson.**
Notes on the state of Virginia. Philadelphia, Printed and sold by Prichard and Hall, 1788.
244p., illus. x975.5/J35n/1788

1206 **Jefferson.**
Notes on the State of Virginia. 2d American ed. Philadelphia, printed for Mathew Carey, 1794.
336p., illus. x975.5/J35n2

1207 **Jefferson.**
Notes on the State of Virginia with an appendix. 3d American ed. New York, M.L.C.W.A. Davis, 1801.
392p., illus. x975.5/J35n/1801

1208 **Jefferson.**
Notes on the State of Virginia with an appendix. 8th American ed. Boston, printed by David Carlisle, 1801.
364p., illus., fold. map, front. x975.5/J35n8

1209 **Jefferson.**
Observations sur la Virginie, par M. J***. Traduites de l'Anglais. Paris, Chez Barrois, 1786.
291p., fold. map. x975.5/J35nFm

1210 **Jeffries, David.**
A treatise on diamonds and pearls. In which their importance is considered: and plain rules are exhibited for ascertaining the value of both: and the true method of manufacturing diamonds. The 2d ed., with large improvements. London, Printed by C. and J. Ackers, 1751.
155p., 30 leaves, 10 plates, tables. x553.8/J382t/1751

1211 **Jobert, Antoine Claude Gabriel.**
See CROIZET, JEAN BAPTISTE, Recherches sur les ossemens fossiles du département du Puy-de-Dôme . . . 1828.

1212 **Jobert.**
The philosophy of geology. Second edition. London, Simpkin, Marshall, and Co.; [etc., etc.], 1847.
[15]-184p. 550/J57p/1847a

1213 **Johnson, Edwin F.,** civil engineer.
Reports on the Bear Mountain Railroad, by Edwin F. Johnson and William R. Casey, civil engineers, and on the coal and iron ores of the Bear Valley Coal Basin, by James Hall, New York State Geological Engineer, with map, profile, and geological cross section: also, an appendix containing the charter of the company, and statistics of the coal trade of the United States and Great Britain. New York, published for private circulation, Jared W. Bell, printer, 1845.
88p., fold. maps. uncat.

1214 **Johnson, Walter Rogers,** 1794-1852.
Report of a survey and exploration of the coal and ore lands, belonging to the Allegheny coal company, in Somerset County, Pennsylvania, accompanied by maps, profiles, and sections. Philadelphia, J. & W. Kite, 1841.
64p., 2 fold. maps, tables. uncat.

1215 **Jolyclerc, Nicolas,** d. 1817.
Cours de minéralogie, rapporté au tableau . . . méthodique des minéraux, donné par Daubenton; ou, Démonstrations élémentaires et naturelles de minéralogie. Paris, La Ve Panckoucke, 1802.
368p. 549/J59c

1216 **Jones, Thomas Rupert,** b. 1819.
See MANTELL, GIDEON ALGERNON, The wonders of geology . . . 8th ed. . . . 1864-66.

CAROLI NICOLAI LANGII

Lucernens. Helvet. Phil. & Med.

Acad.. Cæs. LEOPOLD. Societ. Reg. Prussicæ & Physio-Crit. Senens.

TRACTATUS

DE

ORIGINE

LAPIDUM FIGURATORUM

in quo diffusè disseritur, utrum nimirum sint

CORPORA MARINA A DILUVIO AD MONTES TRANSLATA,

& tractu temporis petrificata

vel an

A SEMINIO QUODAM E MATERIA LAPIDESCENTE

intra terram generentur,

Quibus accedit accurata

DILUVII,

EJUSQUE IN TERRA EFFECTUUM

Descriptio cum Dissertatione

DE GENERATIONE VIVENTIUM,

Testaceorum præcipuè, plurimorumq; corporum,

à vi plasticâ auræ seminalis hinc inde delatæ

extra consuetam matricem

productorum.

LUCERNÆ, MDCCIX.

Sumptibus Authoris, Typis Annæ Felicitatis Hautt.

SUPERIORUM PERMISSU.

Cat. No. 1322

1217 **Jonstonus, Johannes,** 1603-1675.
Thaumatographia naturalis, in decem classes distincta, in quibus admiranda. Amsterdami, apud G. Blaeu, 1632.
501p. uncat.

1218 **Jonstonus.**
Thaumatographia naturalis, in classes decem divisa: in quibus admiranda. Editio secunde priore auctior. Amstelodami, apud J. Janssonium, 1633.
578p. uncat.

1219 **Jonstonus.**
Thaumatographia naturalis, in decem classes distincta, in quibus admiranda. I. Coeli. II. Elementorum. III. Meteororum. IV. Fossilium. V. Plantarum. VI. Avium. VII. Quadrupedum. VIII. Exanguium. IX. Piscium. X. Hominis. Amstelodami, Apud J. Janssonium, 1661.
498p. x500/J73t/1661

1220 **Jonstonus.**
Thaumatographia naturalis, in decem classes distincta, in quibus admiranda I. Coeli. II. Elementorum. III. Meteororum. IV. Fossilium. V. Plantarum. VI. Avium. VII. Quadrupedum. VIII. Exanguium. IX. Piscium. X. Hominis. Amstelodami, Apud J. Janssonium, 1665.
495 (i.e. 497)p. x500/J73t/1665

Added illustrated t.p., engraved.
Error in paging: p.479-89 repeated.

1221 **Jonstonus.**
An history of the wonderful things of nature: set forth in ten severall classes . . . London, John Streater, 1657.
354p. uncat.

Translation of Thaumatographia naturalis . . .

1222 **Jonstonus.**
. . . Notitia regni mineralis, seu, Subterraneorum catalogus, cum praecipuis differentiis. Lipsiae, Typis Haered Colerianorum, 1661.
101p. 549/J73no

1223 **Jorden, Edward.**
A discourse of naturall bathes, and mineral waters. Wherein first the originall of fountains in generall, is declared. Then the nature and differences of minerals, with examples of particular bathes from most of them. Next the generation of minerals in the earth, from whence both the actuall heat of

bathes, and their vertues are proud to proceed. Also by what meanes mineral waters are to bee examined and discovered. And lastly, of the nature and uses of bathes, but especially of our bathes at Bath in Sommerset-shire. The second edition in many points enlarged. London, Printed by Thomas Harper, 1632.
142p. uncat.

1224 **Jorden.**
A discourse of natural bathes, and mineral waters. Wherein, the original of fountains in general is declared. The nature and difference of minerals, with examples of particular bathes. The generation of minerals, in the earth, from whence both the actual heat of bathes, and their virtues proceed. By what means mineral waters are to be discover'd. And lastly, of the nature, and uses of bathes, but especially, of our bathes at Bathe in Somerset-shire . . . And in this Fourth Edition, a quaere concerning drinking bath-water at Bathe, is resolved. To which is added, an appendix concerning Bathe, wherein the antiquity, both of the bathes and city, is more fully discours'd; with a brief account of the nature, and the virtues of the hot waters there. By Thomas Guidot. London, for George Sawbridge, 1673.
167p. uncat.

Includes A quaere . . . by Eugenius Philander. London, for Geo. Sawbridge, 1673. 26p.

1225 **Jugel, Johann Gottfried,** 1707-1786.
Gründliche Nachricht von dem wahren metallischen Saamen; oder, Prima materia metallorum, wie aus derselbe das gantze mineralische Reich seinen Ursprung hat. Nach eigener Erfahrung geprüfet, durch ordentliche Würkungen der Natur bestätiget, und auf eine vorher noch unbekannte Art der Welt mitgetheilet. Leipzig, J. J. Schöps, 1754.
184p. 549.2/J94g

1226 **Jugel.**
Natürliche Berg-Schmelz-und Figier-Kunst in drey Theilen abgefasset . . . Leipzig, J.P. Krauss, 1766.
3v. in 1, port. x549.2/J94n

Contents: I von der Natur aller wesentlichen Kräfte des mineralischen Reichs; II. von dem Bergwesen und der Beschaffenheit der Metalle und Mineralien; III. wie dieselben untersuchet, ihrer Natur gemäss geröstet und geschmelzet werden sollen.

1227 **Jusseraud, Eugène.**
Note sur quelques gisements de minerais de fer qui se trouvent aux environs du bassin houiller de Brassac (Puy-de-Dôme et Haute-Loire). Paris, Garilian-Goeury et V. Dalmont, 1848.
58p., fold. plate. 553.3/J98n

1228 **Jussieu, Antoine de,** 1686-1758.
Examen des causes des impressions des plantes marquées sur certaines pierres des environs de Saint-Chaumont dans le Lionnois, *in* Mémoires de l'Académie Royale des Sciences, Amsterdam, 1718.
p.363-367, illus. 506 PAH/1718

Jussieu recognized that some fossil ferns in coal basins had no counterparts in the present flora of France and deduced that these plants must have floated here from some exotic place by ocean currents.

1229 **Jussieu.**
Recherches physiques sur les pétrifications qui se trouvent en France de diverses plantes & d'animaux étrangers, *in* Mémoires de l'Académie Royale des Sciences, Amsterdam, 1721.
p.89-98, illus. 506 PAH/1721

1230 **Jussieu.**
De l'origine et de la formation d'une sorte de pierre figurée que l'on nomme Corne d'Ammon, *in* Mémoires de l'Académie Royale des Sciences, Amsterdam, 1722.
p.319-331, 380, illus. 506 PAH/1722

1231 **Justi, Johann Heinrich Gottlob von,** 1720-1771.
Grundriss des gesamten Mineralreiches worinnen alle Fossilien in einem, ihren wesentlichen Beschaffenheiten gemässen, Zusammenhange vorgestellet und beschrieben werden. Göttingen, Wwe. Vandenhöck, 1757.
232p. 549/J98g

1232 **Justi.**
Grundriss des gesamten Mineralreiches worinnen alle Fossilien in einem, ihren wesentlichen Beschaffenheiten gemässen, Zusammenhange vorgestellet und beschrieben werden. 2.Aufl. Göttingen, Verlag der Witwe Vandenhöck, 1765.
232p. 549/J98g/1765

1233 **Justi.**
Geschichte des Erd-Cörpers aus seinen äusserlichen und

Versuch
einer
Geschichte
von
Flötz-Gebürgen,
betreffend
deren Entstehung, Lage,
darinne befindliche
Metallen, Mineralien und Foßilien,
größtentheils
aus eigenen Wahrnehmungen,
chymischen und physicalischen Versuchen,
und aus denen Grundsätzen der Natur-Lehre
hergeleitet,
und
mit nöthigen Kupfern versehen,
von

D. Johann Gottlob Lehmann,
Königl. Preuß. Berg-Rath, Mitglied der Königl. Acad. der Wißenschaften, wie auch der Churmäyntzischen Acad. nützlicher Wißenschaften.

Mit Königl. Pohl. Churf. Sächs. allergn. Privilegio.

Berlin,
zu finden bey Gottlieb August Lange.
1756.

Cat. No. 1356

unterirdischen Beschaffenheiten hergeleitet und erwiesen. Berlin, C. F. Himburg, 1771.
386p. 551/J98g

1234 **Kalm, Pehr,** 1715-1779.
See BARTRAM, JOHN, Observations . . . To which is annex'd, a curious account of the cataracts at Niagara [by] Mr. Peter Kalm . . . 1751.

1235 **Kalm.**
En resa til Norra America, på Kongl. Swenska wetenskaps academiens befallning, och publici kostnad, förrättad. Stockholm, Tryckt på L. Salvii kostnad, 1753-61.
3v., illus. x917.3/K12r

1236 **Kalm.**
Travels into North America; containing its natural history, and a circumstantial account of its plantations and agriculture in general, with the civil, ecclesiastical and commercial state of the country, the manners of the inhabitants, and several curious and important remarks on various subjects. Translated into English by John Reinhold Forster . . . London, The editor, 1770-71.
3v., plates, fold. map. 917.3/K12rEf and uncat.

Vol. 1 published at Warrington, printed by W. Eyres.
"This translation omits a great number of details, and everything relating to England. It contains drawings of American birds and animals not found in the original." — Nat'l. Union Cat.

1237 **Kalm.**
Reis door Noord Amerika. Vercierd met kopern platen . . . Te Utrecht, J. van Schoonhoven en comp. [etc.], 1772.
2v. in 1, plates, fold. map. 917.3/K12rD

Added t.-p., engraved.
Translation from the Swedish, The portion relating to England omitted.

1238 **Kalm,** praeses.
Kort afhandling om gipsen. Med wederbörandes minne under . . . hr. Pehr Kalms inseende til offentelig granskning . . . utgifwen af Johan Fridric Müller . . . Åbo, Tryckt hos J. Merckell, 1757.
14p. uncat.

Akademisk avhandling — Åbo.

1239 **Kant, Immanuel,** 1724-1804.
Immanuel Kant's Physische Geographie; auf Verlangen des

Verfassers, aus seiner Handschrift hrsg. und zum Theil bearb. von Friedrich Theodor Rink. Königsberg, 1802.
2v. in 1. 551.4/K12p/1802

Caption and running title: Physische Erdbeschreibung.

1240 **Kant.**
Physische Geographie. Zweite durchaus umgarbeitete Auflage. Von Joh. Jak. Wilh. Vollmer. Mainz, G. Vollmer, [1803]-1805.
7v. in 4. 551.4/K12p

Vol. 1, pts. 1-2 have added title pages: Joh. Jak. Wilh. Vollmer's Physische Geographie nach Kantischen Ideen.
Vol. 4 also has special title page: Neue Anmerkungen zur Erlaüterung der Theorie der Winde, bisher in keiner der grössern oder kleinern Sammlungen seiner Werke gedrückt.

1241 **Karsten, Dietrich Ludwig Gustav,** 1768-1810.
Des Herrn Nathanael Gottfried Leske . . . hinterlassenes Mineralienkabinet; systematisch geordnet und beschrieben, auch mit vielen wissenschaftlichen Anmerkungen und mehreren äussern Beschreibungen der Fossilien begleitet. Leipzig, Im Verlage der I. G. Müllerschen Buchhandlung, 1789.
2v., illus. 549/K149h

Issued also as v.2 of his Museum Leskeanum.

1242 **Karsten.**
Tabellarische Übersicht der mineralogisch-einfachen Fossilien, zum Behuf seiner Vorlesungen. Berlin, gedruckt in der König. Hofbuchdruckerei, 1791.
27p. q549.1/K14t

1243 **Karsten.**
Mineralogische Tabellen mit Rücksicht auf die neuesten Entdekkungen ausgearbeitet und mit erläuternden Anmerkungen versehen. 2. verb. und verm. Aufl. Berlin, H. A. Rottmann, 1808.
103p. q549.1/K14m/1808

1244 **Karsten, Jacob Christian Gustav.**
Des Marockaners Abdulvaheb Temimi Fragmente über Spanien. Rostock, Adlerschen Officin, 1801.
27p. 549/F59

Bound with Flade, Christian Gottlob, De re Metallica . . . [1791].

1245 **Keating, William Hypolitus,** 1799-1840.
Narrative of an expedition to the source of St. Peter's River, Lake Winnepeek, Lake of the Woods, &c., performed in the Year 1823, by order of the Hon. J. C. Calhoun, Secretary of

War, under the command of Stephen H. Long, U.S.T.E. London, G. B. Whittaker, 1825.
2v., illus. 917.7/K22n

1246 **Keckermann, Bartholomaus,** 1571/73-1609.
Meditatio de insolito et stupendo ille terrae-motu, quo anno praeterito, VIII. Septembris intra secundum & primam noctishoram tota penè Europa & Asiae (quantum adhuc compertum est) pars non exigua, uno propè momento contremuit . . . Respondente Samuele Nebelthavio. Haidelbergae, typis Voegelinianis, 1602.
[48p.] uncat.

1247 **Keferstein, Christian,** 1784-1866.
Tabellen über die vergleichende Geognosie. Ein Versuch von Christian Keferstein . . . Halle, Bey Hemmerde und Schwetschke, 1825.
60p. 551/K25t

1248 **Keilhau, Balthasar Matthias,** 1797-1858, ed.
Gaea Norvegica von mehreren Verfassern. Christiania, J. Dahl, 1838-1850.
2v., illus., fold. col. maps. xq554.81/K26g

The other authors are: C. Boeck, K.F. Böbert, Th. Scheerer, and A. Vibe.

1249 **Keill, John,** 1671-1721.
An examination of Dr. Burnet's theory of the earth: with some remarks on Mr. Whiston's New theory of the earth. Also an examination of the reflections on the theory of the earth; and a defence of the remarks on Mr. Whiston's New theory. 2d ed. cor. and all the figures (25 in number) engraved on copper plates. To the whole is annexed a dissertation on the different figures of the celestial bodies, & with a summary exposition of the Cartesian and Newtonian systems. By Mons. de Maupertuis . . . Oxford, Printed for H. Clements; [etc., etc.], 1734.
347, 67p., fold. plates. x551/K262e/1734

1250 **Kendall, Francis.**
A descriptive catalogue of the minerals, and fossil organic remains of Scarborough and the vicinity, including the line of coast from Hornsea to Mulgrave, and extending into the interior as far as Malton. Printed by and for T. Coultas, 1816.
316p., plates. 549.942/K33d

SVMMI POLYHISTORIS

GODEFRIDI GVILIELMI LEIBNITII

PROTOGAEA

SIVE DE PRIMA FACIE TELLVRIS ET ANTIQVISSIMAE HISTORIAE VESTIGIIS IN IPSIS NATVRAE MONVMENTIS DISSERTATIO
EX SCHEDIS MANVSCRIPTIS

VIRI ILLVSTRIS

IN LVCEM EDITA

A

CHRISTIANO LVDOVICO SCHEIDIO.

GOETTINGAE
SVMPTIBVS IOH. GVIL. SCHMIDII, BIBLIOPOLAE VNIVERSIT.
A. S. H. CIƆIƆCCXXXXVIIII.

Cat. No. 1358

1251 **Kentmann, Johannes,** 1518-1574.
Calculorum qui in corpore ac membris hominum innascuntur, genera XII. depicta descriptaque, cum historijs singulorum admirandis.
22p. uncat.

Bound with Gesner, Konrad, De omni rerum fossilium genere . . . Tiguri, 1565.

1252 **Kentmann.**
Catalogus rerum fossilium . . .
95p., front. uncat.

Bound with Gesner, Konrad, De omni rerum fossilium genere . . . Tiguri, 1565.

1253 **Keyserling, Alexander Friedrich Michael Leberecht Nikolaus Arthur von,** 1815-1891, jt. author.
See MURCHISON, RODERICK IMPEY, The geology of Russia . . . 1845.

1254 **Kidd, John,** 1775-1851.
Outlines of mineralogy. Oxford, Printed by N. Bliss, for J. Parker and R. Bliss, 1809.
2v. in 1. 549/K53o

Bibliography: v.1, p. x.
Library owns 2 copies: c.1 — 2v. in 1; c.2 — v.1, v.2.

1255 **King, Charles,** fl. 1705, supposed author.
An account of the origin and formation of fossil-shells, &c. Wherein is proposed a way to reconcile the two different opinions, of those who affirm them to be exuviae of real animals, and those who fancy them to be lusus naturae. London, Printed by W. Botham for James Knapton, 1705.
88p., front. uncat.

1256 **Kircher, Athanasius,** 1602-1680.
Iter extaticum II. qui & mundi subterranei prodromus dicitur. Quo geocosmi opificium, sive terrestris globi structura, unà cum abditis in ea constitutis arcanioris naturae reconditorijs, per ficti raptus integumentum exponitur ad veritatem. In III. dialogos distinctum ad serenissimum Leopoldum Ignatium. Romae, Typis Mascardi, 1657.
237p. uncat.

1257 **Kircher.**
Mundus subterraneus, in XII libros digestus; quo divinum subterrestris mundi opificium, mira ergasteriorum naturae in eo distributio, verbo *παντάμορφον* protei regnum, universae

denique naturae majestas & divitiae summa rerum varietate exponuntur. Abditorum effectuum causae acri indagine inquisitae demonstrantur; cognitae per artis et naturae conjugium ad humanae vitae necessarium usum vario experimentorum apparatu, necnon novo modo, & ratione applicantur. Amstelodami, Apud Joannem Janssonium & Elizeum Weyerstraten, 1665.

2v. in 1, port., illus. q uncat.

Engr. t.-p. added.

1258 **Kircher.**

Mundus subterraneus, in XII libros digestus; quo divinum subterrestris mundi opificium, mira ergasteriorum naturae in eo distributio, verbo . . . protei regnum, universae denique naturae majestas & divitae summa rerum varietate exponuntur. Abditorum effectuum causae acri indagine inquisitae demonstrantur; cognitae per artis & naturae conjugium ad humanae vitae necessarium usum vario experimentorum apparatu, necnon novo modo & ratione applicantur. Editio tertia. Amstelodami, J. Janssonium a Waesberge & filios, 1678.

2v. in 1. q uncat.

1259 **Kircher.**

Iter exstaticum coeleste, quo mundi opificium, id est, coelestis expansi, siderumque tam errantium, quam fixorum natura, vires, proprietates, singulorumque compositio & structura, ab infimo telluris globo, usque ad ultima mundi confinia, per ficti raptus integumentum explorata, novâ hypothesi exponitur ad veritatem . . . hâc secundâ editione praelusionibus & scholiis illustratum; ac schematismis necessariis, qui deerant, exornatum; nec non à mendis quae in primam Romanam editionem irrepserant, expurgatum, ipso auctore annuente, P Gaspara Schotto. Accessit ejusdem auctoris Iter exstaticum terrestre, & synopsis mundi subterranei. Herbipoli, Sumptibus J. A. Endteri, 1671.

[23], 698, [15]p., diagrs. x523/K632i/1671

1260 **Kircher.**

The vulcano's: or, Burning and fire-vomiting mountains, famous in the World: with their remarkables. Collected for the most part out of Kircher's subterraneous World; and expos'd to more general view in English, upon the relation of the late wonderful and prodigious eruptions of Aetna. Thereby to occasion greater admirations of the wonders of nature

(and of the God of nature) in the mighty element of fire . . . London, Printed by J. Darby, for John Allen, 1669.
68p., front., illus. uncat.

1261 **Kirwan, Richard,** 1733-1812.
Elements of mineralogy. London, Printed for P. Elmsley, 1784.
412p. 549/K63e

1262 **Kirwan.**
Elements of mineralogy. 2d ed. with considerable improvements and additions. London, Printed by J. Nichols, 1794-96.
2v. x549/K63e/1794

Vol. 2 has imprint: Printed for P. Elmsley.

1263 **Kirwan.**
Elements of mineralogy. The 3d ed. . . . London, Printed for J. Mackinlay, 1810.
2v., illus. 549/K63e/1810

1264 **Kirwan.**
Élémens de mineralogie; traduits de l'anglois par M. Gibelin. Paris, Cuchet, 1785.
432p. 549/K63eFg

1265 **Kirwan.**
Elementos de mineralogia . . . Traducidos al frances . . . par Mr. Gibelin, . . . y de este idioma al español por Francisco Campuzano. Madrid, P. B. Lopez, 1789.
384p., illus. 549/K63eSPc

1266 **Kirwan.**
Anfangsgrunde der Mineralogie. 2.Ausg. mit beträchtlichen Verbesserungen und Zusätzen. Aus dem Englischen übers., und mit Anmerkungen und einer Vorrede versehen von Lorenz von Crell. Berlin, F. Nicolai, 1796-99.
3v. 549/K63eGc

Translation of Elements of mineralogy.

1267 **Kirwan.**
An essay on the analysis of mineral waters. London, printed by J. W. Myers, 1799.
279p. x553.7/K63e

1268 **Kirwan.**
Geological essays. London, Printed by T. Bensley for Dr. Bremner, 1799.
502p. x550.4/K639g

Speculum Lapidum Clariſſimi Artium Et Medicine Doctoris Camilli Leonardi Piſaurenſis.

¶ Valerii Superchii Piſaurenſis Phyſici Epigramma.

¶ Quicquid in humanos gemmarum parturit uſus
Terra parens: uaſti quicquid & unda maris.
Quālibet exiguo claudis Leonarde libello
Mirandum/& ſeræ poſteritatis opus.
Quod poſitis Cæſar interdum perlegat armis:
Seruariq; ſuas imperet inter opes
Et tibi pro meritis æquos decernar honores:
Conſulat & famæ tempus in omne tuæ.

Cat. No. 1367

1269 **Klaproth, Martin Heinrich,** 1743-1817.
Beitrage zur chemischen Kenntnis der Mineralkörper. Posen, bei Decker; Berlin, bei Heinrich August Rottmann, 1795-1815.
6v. x549.13/K66b

Vol.5 includes index to v.1-5. Vol.6 has added t.p.: Sammlung chemischen Abhandlung gemischten Inhalts.

1270 **Klaproth.**
Analytical essays towards promoting the chemical knowledge of mineral substances. Translated from the German. London, Printed for T. Cadell, Jun. and W. Davies, 1801-04.
2v. 549.13/K66bE

Translation of Beiträge zur chemischen Kenntniss der Mineralkörper.
Bibliographical footnotes.

1271 **Klein, Jacob Theodor,** 1685-1759.
Descriptiones tubulorum marinorum in quorum censum relati lapides caudae cancri Gesneri, & his similes; Belemnitae, eorumque alveoli secundum dispositionem Musei Kleiniani. Addita est dissertatio epistolaris de pilis marinis. Gedani, Apud Knochium, 1731.
26p., 10 plates. 564/K67d/1731

"Conspectus dispositionis echinorum marinorum Musei Kleiniani": [4]p. at end.

1272 **Klein.**
. . . Descriptiones tubulorum marinorum in quorum censum relati lapides caudae cancri Gesneri, et his similes Belemnitae, eorumque alveoli secundum dispositionem Musei Kleiniani. Addita est dissertatio epistolaris de pilis marinis . . . Gedani et Lipsiae, in officina Libraria Gleditschiana, 1773.
40p., 10 plates. 564/K67d

"Conspectus dispositionis echinorum marinorum Musei Kleiniani.": [4]p. at end.

1273 **Klein.**
Naturalis dispositio Echinodermatum. Accessit Lucubratiuncula de aculeis echinorum marinorum, cum spicilegio de belemnitis. Gedani, Typis T. J. Schreiberi, 1734.
78p., 36 plates. 593.9/K67n and x593.9/K67n/1778

Supplemented with Leske, Nathanael Gottfried, Additamenta. Lipsiae, 1778. [200p., 18 plates]. Page numbering continues Klein.
Library owns 2 copies. The copy x593.9/K67n/1778 has colored plates. The original ones by Klein (1734) are printed in sepia.

1274 **Klein.**
Ordre naturel des oursins de mer et fossiles, avec des observations sur les piquans des oursins de mer, et quelques remarques sur les bélemnites. Ouvrage traduit du latin, avec le texte de l'auteur. Paris, C. J. B. Bauche, 1754.
233p., port., 28 plates. 593.95/K67nF

Latin and French on opposite pages.

1275 **Klein.**
Lucubratiuncula suterranea prior de lapidibus macrocosmi proprietalibus. Petropoli, Typis Academiae Scientiarum, 1758.
38p., diagrs. 552/K67ℓ

Bibliographical footnotes.

1276 **Klein.**
Specimen descriptionis petrefactorum Gedanensium cum syllabo tabularum. Probe einer Beschriebung und Abbildung der in der Danziger und umliegenden Gegend befindlichen Versteinerungen. Nürnberg, Erben, 1790.
1v. (unpaged), 24 col. plates. q560.94312/K67s

Text in Latin and German in parallel columns.
Each plate accompanied by leaf of text.

1277 **Klipstein, August von,** 1801-1894.
Beiträge zur geologischen Kenntniss der östlichen Alpen. Giessen, G. F. Heyer, 1843.
311p., 21 leaves of plates. q554.364/K689b

1278 **Klöden, Karl Friedrich von,** 1786-1856.
Die Versteinerungen der Mark Brandenburg: insonderheit diejenigen welche sich in Rollsteinen und Blöcken der südbaltischen Ebene finden. Berlin, C. G. Lüderitz, 1834.
378p., 10 leaves of plates. 554.315/K69v

1279 **Knight, William,** 1786-1844.
Facts and observations towards forming a new theory of the earth. Edinburgh, A. Constable, 1819.
335p. 550.1/K74f

1280 **Knipe, James A.**
Geological and mineralogical map of England and Wales, with parts of Scotland, Ireland, and France, showing also the inland navigation. London, 1841.
G5740/.C5/1841/.K5

Scale, ca. 1:780,000.

1281 **Knoll, Heinrich Christoph Friedrich.**
Unterhaltende Naturwunder. Erfurt, G. A. Keyser, 1786-88.
2v. in 1. 551/K75u

Vol. 2 issued after the author's death with notes and a biography of the author by Johann Christian Wiegleb.

1282 **Knorr, Georg Wolfgang,** 1705-1761.
Sammlung von Merckwürdigkeiten der Natur und Alterthümern des Erdbodens, welche petrificirte Cörper enthält aufgeweisen und beschrieben. Nürnberg, Zu finden bey dem Author, gedruckt bey A. Bieling, 1755-1773.
5v. in 4, port. xq560/K75s

Half title: Die Naturgeschichte der Versteinerungen. Zur Erläuterung der Knorrischen Sammlung von Merkwürdigkeiten der Natur, herausgegeben von J.E.I. Walch, Nürnberg, Felsecken, 1769.
Walch continued and improved Knorr's work.
Imperfect: port. and Latin-German t.p. to v.1 wanting.

1283 **Knorr.**
Sammlung von Merckwürdigkeiten der Natur... Nürnberg, [1755-1771].
4v. in 3. of plates (part. fold., part. col.). xq560/K75s/plates

1284 **Knorr.**
Recueil de monumens des catastrophes que le globe de la terre a éssuiées, contenant des pétrifications et d'autres pierres curieuses, dessinées, gravées et enluminées, d'après les originaux, avec leur description, par George Wolfgang Knorr. Nuremberg, 1768-1778 [v.1, 1777].
4v. in 5, plates (part. fold., part. col.), ports. xf560/K75sF

Translation of Sammlung von Merkwürdigkeiten der Natur...
Second title-page reads as follows: Recueil de monumens des catastrophes que le globe terrestre a éssuiées contenant des pétrifications dessinées, gravées et enluminées d'après les originaux, commencé par feu George Wolfgang Knorr, et continué par ses héritiers avec l'histoire naturelle de ces corps par Jean Ernest Emanuel Walch. Traduit de l'allemand.

1285 **Knorr.**
De natuurlyke historie der versteeningen, of Uitvoerige afbeelding en beschryving van de versteende zaaken, die tot heden op den aardbodem zyn ontdekt: aangevangen door Georg Wolfgang Knorr. Vervolgd en omstandig beredeneerd door Johann Ernst Immanuel Walch. Uit het hoog- in't nederduitsch vertaald door M. Houttuyn, met verscheide aanmerkingen. Amsterdam, J. C. Sepp, 1773.
3v. in 4, plates (part. col.). f560/K75sDh

EDUARDI LUIDII
APUD
Oxonienſes Cimeliarchæ Ashmoleani

LITHOPHYLACII BRITANNICI ICHNOGRAPHIA

SIVE

Lapidum aliorumq; Foſſilium Britannicorum ſingulari figura inſigni-um, quotquot hactenus vel ipſe invenit vel ab amicis accepit,

DISTRIBUTIO CLASSICA,

Scrinii ſui lapidarii Repertorium cum locis ſingulorum natalibus exhibens.

Additis rariorum aliquot figuris ære inciſis cum Epiſtolis ad Clariſſimos Viros de quibusdam circa marina foſſilia & ſtirpes minerales præſertim notandis.

Nusquam magis erramus quam in falſis inductionibus: ſæpe enim ex aliquot exemplis Univerſale quiddam colligimus; idq; perperam cum ad ea, quæ excipi poſſunt, animum non attendimus.

Du Hamel.

LONDINI

Ex Officina M. C. CIƆ IƆC XCIX

LIPSIÆ.

Sumt. Joh. Ludw. Gleditſch & Weidmann.

Cat. No. 1382

Each volume has also special engraved t.p. in German.
Translation of Sammlung von Merkwürdigkeiten der Natur...

1286 **Kobell, Wolfgang Xaver Franz von,** 1803-1882.
Tafeln zur Bestimmung der Mineralien mittelst einfacher chemischer Versuche auf trockenem und nassem Wege. 7.verm. Aufl. München, J. Lindauer, 1861.
102p. 549.1/K79t/1861

The first edition of this work was published in 1833.

1287 **Kobell.**
Tafeln zur Bestimmung der Mineralien mittelst einfacher chemischer Versuche auf trockenem und nassem Wege. 12.neu bearb. und verm. Aufl. von K. Oebbeke. München, J. Lindauersche Buchhandlung, 1884.
108p. 549.1/K79t

"Literatur": p. xx-xxiii.

1288 **Kobell.**
Instructions for the discrimination of minerals by simple chemical experiments. Translated from the German by Robert Corbet Campbell. Glasgow, R. Griffin, 1841.
51p., illus. 549.1/K79tEc

Translation of: Tafeln zur Bestimmung der Mineralien.

1289 **Kobell.**
Skizzen aus dem Steinreiche; geschrieben für des gebildete Gesellschaft. München, C. Kaiser, 1850.
256p. 549/K79s

1290 **Kobell.**
Die Mineral-Namen und die mineralogische Nomenklatur. München, J. G. Cotta, 1853.
160p. 549.03/K79m

1291 **Koch, Albrecht Karl.**
Description of the *Missourium theristocaulodon* (Koch), or Missouri leviathan, *(Leviathan missouriensis)*, together with its supposed habits, and Indian traditions, also comparisons on the whale, crocodile, and *Missourium,* with the leviathan, as described in the 41st chapter of the Book of Job. 4th ed. London, E. Fisher, 1842.
23p. 569.6/K81d

1292 **Köhler, Friedrich Wilhelm,** 1805-1871.
De nonnullis diallagi varietatibus. Marburgi, 1828.
32p. 549.6/K82d

Diss. inaug. — Marburg.

1293 **Koksharov, Nikolaĭ Ivanovich,** 1818-1893.
Materialien zur Mineralogie Russlands. St. Petersburg, Gedruckt bei A. Jacobson, 1853-91 [1853-93].
2v. in 1, illus., fold. plates, diagrs., and atlas of 87 (i.e. 93) plates.
549.947/K82mG

"Dem andenken N. I. von Kokscharow": v. 11: p. [129]-137.
Translation of Materialy dlia mineralogii Rossii.
Imperfect: t.p. of vol.1 wanting.

1294 **König, Emanuel,** 1658-1731.
Regnum minerale, generale et speciale. Quorum illud naturalem et artificialem mineralium productionem cum parallelismo alchymico verorum philosophorum, tractatibus hucusque ineditis, commentario super jntroitum philalethae, &c. candidè sistit, olim sub nomine regni quarti sulphurum fixorum metallicorum promissum. Hoc verò metalla, lapides, salia, sulphura, terras, quin & acidulas, thermas, physicè, chymicè, practicè recludit. Basileae, E. König, Senioris, 1703.
428p., port. (tinted). 549/K83r

1295 **Koninck, Laurent Guillaume de,** 1809-1887.
Description des coquilles fossiles de l'argile, de Basele, Boom, Schelle, etc. [Bruxelles?, Académie royale des sciences, des lettres, et des beaux-arts de Belgique?], 1837.
37p., 4 plates. 564/K836d

Detached from Académie royale des sciences, des lettres et des beaux arts de Belgique. Nouveaux mémoires, t.11.

1296 **Koninck.**
Description des animaux fossiles qui se trouvent dans le terrain carbonifère de Belgique. Liège, Impr. et lit. de H. Dessain; [etc., etc.], 1842-1844.
650p., and atlas of 69 plates. q560.9493/K83d
q560.9493/K83d/plates

1297 **Koninck.**
Description des animaux fossiles qui se trouvent dans le terrain carbonifère de Belgique. Supplément. Liège, 1851.
716p. q560.9493/K83d/sup.

1298 **Koninck.**
Recherches sur les animaux fossiles. Liège, H. Dessain, 1847 and Bruxelles, 1873.
2v. in 1, plates. q560/K83r

Première partie: Monographie des genres Productus et Chonetes, 1847.

Deuxième partie: Monographie des fossiles carbonifères de Bleiberg en Carinthie, 1873.

1299 **Kopp, Johann Heinrich,** 1777-1858, jt. author.
See LEONHARD, KARL CÄESAR VON, Systematisch-tabellarische . . . 1806, and Propaedeutik der Mineralogie . . . 1817.

1300 **Krüger, Johann Gottlob,** 1715-1759.
Geschichte der Erde in allerältesten Zeiten. Halle, Lüderwald, 1746.
186p., 3 plates. 551/K93g

1301 **Krüger.**
Histoire des anciennes revolutions du globe terrestre. Avec une Relation chronologique et historique des tremblemens de terre, arrivés sur notre globe depuis le commencement de l'ère chrétienne jusqu'à present. Amsterdam, Se vend à Paris, Damonneville, 1752.
328p., 3 fold. plates. 551/K93gFd

Translation by F. A. Deslandes of the author's Geschichte der Erde in allerältesten Zeiten.

1302 **Kühn, Heinrich,** 1690-1769.
Méditations sur l'origine des fontaines, l'eau des puits: et autres problèmes qui ont du rapport à ce sujet. A Bordeaux, Chez Pierre Brun, imprimeur-aggrégé de l'Académie Royale, rue Saint James, 1741.
248p., fold. plates. 551.498/K955m

1303 **Kurr, Johann Gottlob von,** 1798-1870.
Grundzüge der ökonomisch-technischen Mineralogie. Ein Lehr- und Handbuch für Oekonomen und Gewerbsmänner sowie für Polytechnische, Real, Gewerbs, Land und Forstwirthschaftliche Lehranstalten. 2.Aufl. Leipzig, Baumgärtners, 1844.
619p., illus., plates. 549/K96g2

1304 **Kutorga, Stepan Semenovich,** 1805?-1861.
Beitrag zur Geognosie und Palaeontologie Dorpate's und seiner naechsten Umgebungen. St. Petersburg, 1835.
45p., 7 plates. 554.741/K96b

1305 **Lacoste, Pierre François,** 1754-1826.
Observations sur les volcans de l'Auvergne, suivies de notes sur divers objets, recueillies dans une course minéralogique, faite l'année dernière, an 10 (1802). Clermont-Ferrand, Vve. Delcros et fils [etc.], 1803.
196p. 551.21/L119o

OBSERVATIONS

ON THE

GEOLOGY

OF THE

UNITED STATES OF AMERICA;

WITH

SOME REMARKS

ON THE

EFFECT PRODUCED ON THE NATURE AND FERTILITY OF

SOILS,

BY THE DECOMPOSITION OF THE DIFFERENT

CLASSES OF ROCKS;

AND AN APPLICATION TO THE FERTILITY OF EVERY STATE IN THE UNION, IN REFERENCE TO THE ACCOMPANYING GEOLOGICAL MAP.

WITH TWO PLATES.

BY WILLIAM MACLURE.

Read as a Memoir before the American Philosophical Society, and inserted in the 1st vol. of their Transactions, New Series.

PHILADELPHIA:

PRINTED FOR THE AUTHOR,

BY ABRAHAM SMALL, No. 112, CHESNUT STREET.

AND SOLD BY HIM AND J. MELISH.

1817.

Cat. No. 1449

1306 **Lacoste.**
Lettres minéralogiques et géologiques sur les volcans de l'Auvergne, écrites dans un voyage fait en 1804. Clermont, L'impr. de Landriot, 1805.
459p., 2 tables. 554.4/L11ℓ

1307 **Laet, Joannes de,** 1593-1649.
De gemmis et lapidibus libri duo quibus praemittitur Theophrasti liber De lapidibus, graece & latine cum brevibus annotationibus. Lugduni Batavorum, ex officina I. Maire, 1647.
210p., illus. x553.8/L12d

1308 **Lamarck, Jean Baptiste Pierre Antoine de Monet de,** 1744-1829.
Recherches sur les causes des principaux faits physiques, et particulièrement sur celles de la combustion, de l'élévation de l'eau dans l'état de vapeurs . . . de l'origine des composés et de tous les minéraux; enfin de l'entretien de la vie des êtres organiques, de leur accroissement, de leur état de vigueur, de leur dépérissement et de leur mort. Avec une planche. Paris, Maradan, [1794].
2v., plates, fold. tab. 530/L16r

1309 **Lamarck.**
Mémoires de physique et d'histoire naturelle, établis sur des bâses de raisonnement indépendantes de toute théorie; avec l'exposition de nouvelles considérations sur la cause générale des dissolutions; sur la matière du feu; sur la couleur des corps; sur la formation des composés; sur l'origine des minéraux; et sur l'organisation des corps vivans. Lus à la première classe de l'Institut National, dans ses séances ordinaires. Paris, L'Auteur, 1797.
410p. x530/L16m

Author's autographed presentation copy to Monsieur Delaport.

1310 **Lamarck.**
Mémoires présentant les bases d'une nouvelle théorie, physique et chymique, fondée sur la considération des molécules essentielles des composés, et sur celle des trois états principaux du feu dans la nature; servant en outre de développement à l'ouvrage intitulé: Réfutation de la théorie pneumatique. Lus à la première classe de l'Institut National, dans ses séances ordinaires. Paris, Chez l'auteur, 1797.
410p., fold. tables. x530.1/L16m

1311 **Lamarck.**
Système des animaux sans vertèbres, ou tableau général des classes, des ordres et des genres de ces animaux. Présentant leurs caractères essentiels et leur distribution, d'après la considération de leurs rapports naturels et de leur organisation, et suivant l'arrangement établi dans les galeries du Muséum d'Histoire Naturelle, parmi leurs dépouilles conservées; précédé du discours d'ouverture du cours de zoologie, donné dans le Muséum national d'histoire naturelle l'an 8 de la République. Paris, Deterville, 1801.
432p., fold. tables. 592/L16s

1312 **Lamarck.**
Hydrogéologie; ou, Recherches sur l'influence qu'ont les eaux sur la surface du globe terrestre; sur les causes de l'existence du bassin des mers, de son déplacement et de son transport successif sur les différens points de la surface de ce globe; enfin sur les changemens que les corps vivans exercent sur la nature et l'état de cette surface. Paris, L'auteur, [1802].
268p. x551.35/L16h

1313 **Lamarck.**
Recherches sur l'organisation des corps vivans, et particulièrement sur son origine, sur la cause de ses développemens et des progrès de sa composition, et sur celle qui, tendant continuellement à la détruire dans chaque individu, — amène nécessairement sa mort; — Précédé du Discours d'ouverture du cours de zoologie donné dans le Muséum National d'Histoire Naturelle. Paris, L'Auteur, 1802.
216p., 1 fold. table. 570.1/L16r

1314 **Lamarck.**
Mémoire sur les fossiles des environs de Paris, comprenant la détermination des espèces qui appartiennent aux animaux marins sans vertèbres, et dont la plupart sont figurés dans la collection des velins du Muséum. [Paris, 1802-09].
284p., 28 plates. 560.944/L16m

Caption title.

Imperfect: last 3(?) pages of text wanting. Plates numbered 13, 19 and 25 in original wanting.

Reprint; text originally published in Annales du Muséum d'hist. nat. [Paris] t.1-8, 1802-06; plates, in t.6-9, 12 and 14, 1805-09. *Cf.* Brit. Mus. (Nat. Hist.) Library. Cat. v.3 (1910), p.1048.

Plates have descriptive legends pencilled in ms.; scattered ms. notes on 96ℓ. [not Lamarck's] numbered in ms. and inserted at end. Stamped on fly leaf: L(?) Faujas S'Fond.

1315 **Lamarck.**
Extrait du cours de zoologie du Muséum d'histoire naturelle sur les animaux sans vertèbres, présentant la distribution et la classification de ces animaux, les caractères des principales divisions, et une simple liste des genres . . . Paris, D'Hautel, 1812.
127p. 592/L16e

1316 **Lamarck.**
Histoire naturelle des animaux sans vertèbres, présentant les caractères généraux et particuliers de ces animaux, leur distribution, leurs classes, leurs familles, leurs genres, et la citation des principales espèces qui s'y rapportent; précédée d'une introduction offrant la détermination des caractères essentiels de l'animal, sa distinction du végétal et des autres corps naturels, enfin, l'exposition des principes fondamentaux de la zoologie. Paris, Verdière, 1815-22.
7v. 596/L16h

Vol. 6 in 2 parts.

1317 **Lamarck.**
Histoire naturelle des animaux sans vertèbres. 2.éd. rév. et augm. des notes présentant les faits nouveaux dont la science s'est enrichie jusqu'à ce jour; par MM. G. P. Deshayes et H. Milne Edwards . . . Paris, Londres, J. B. Bouillière, 1835-45.
11v. 596/L16h2

1318 **La Métherie, Jean Claude de,** 1743-1817.
Théorie de la terre. Paris, Maradan, 1795.
3v., illus. 549/L18t

1319 **La Métherie.**
Théorie de la terre. 2.éd., cor., et augm. d'une minéralogie. Paris, Maradan, 1797.
5v., front. (port.), fold. plates, fold. map, fold. diagrs.
549/L18t/1797

1320 **La Métherie.**
See BERTRAND, PHILIPPE, Nouveaux principes de géologie . . . 1797 and 1803.

1321 **Lang, Carl Nicolaus,** 1670-1741.
Historia lapidum figuratorum Helvetiae, ejusque viciniae, in quâ non solùm enarrantur omnia eorum genera, species et vires aeneisque tabulis repraesentantur, sed insuper adducuntur eorum loca nativa, in quibus reperiri solent, ut cuilibet facile sit eos colligere, modo adducta loca adire libeat. Venetiis, sumptibus authoris, Typis J. Tomasini; [etc., etc.], 1708.

MARBO=
DEI GALLI POETAE VE-
tustissimi de lapidibus pretiosis Encheri-
dion, cum scholijs Pictorij Vil-
lingensis.

EIVSDEM PICTORII DE
lapide molari carmen.

Lectori.

Qui cupis emunctim gemmarum scire medullas,
Huc uenias, totum continet iste liber:
Qui decies senis capitellis nomina dicit,
Et species, patrias, quid ualeantq; simul.

ANNO M. D.
XXXI.

Cat. No. 1495

80p., 53 plates. 560.9494/L25h

Bound with his Tractatus de origine lapidum figuratorum . . . 1709.

1322 **Lang.**
Tractatus de origine lapidum figuratorum, in quo diffusè disseritur, utrum nimirum sint corpora marina a diluvio ad montes translata & tractu temporis petrificata vel an a seminio quodam e materia lapidescente intra terram generentur, quibus accedit accurata diluvii . . . Lucernae, Sumptibus authoris, typis A. F. Hautt, 1709.
80p. 560/L25t

Library owns a second copy which is bound with his Historia lapidum figuratorum Helvetiae . . . 1708 (560.9494/L25h).

1323 **Lange, Johann Joachim,** 1698-1765, praeses.
Lithographia Halensis, exhibens lapides circa Halam Saxonum reperiendos, systematice digestos. Secundum: Classes et ordines, genera et species, cum synonymis selectis et descriptionibus speciarum. Praefactus est Joachim Langius. Halae, Joh. Jac. Curt, 1759.
80p., 2 plates in 1. uncat.

Respondent: Johann Christian Daniel Schreber.

1324 **Langsdorf, Karl Christian von,** 1757-1834.
Vollständige auf Theorie und Erfahrung gegründete Anleitung zur Salzwerkskunde. Altenburg, Richter, 1784.
482p., 18 plates. 622.363/L26v/1784

Plate no. 18 bound in incorrect order.

1325 **Lapérouse, Jean François de Galaup,** 1741-1788.
Voyage de La Pérouse autour du monde, publié conformément au décret du 22 avril 1791, et rédigé par M. L. A. Millet-Mureau. Paris, Plassan, 1798.
4v. 910.41/L31v/1798

1326 **Lapham, Increase Allen,** 1811-1875.
A geographical and topographical description of Wisconsin; with brief sketches of its history, geology, mineralogy, natural history, population. Milwaukee, Wis., P. C. Hale, 1844.
255p., front. (fold. map). x917.75/L31g

2d edition published 1846 under title: Wisconsin: its geography and topography . . .

1327 **Lapham.**
Wisconsin: its geography and topography, history, geology and mineralogy; together with brief sketches of its antiquities,

natural history, soil, productions, population, and government. 2d ed., greatly improved. Milwaukee, I. A. Hopkins; New York, Paine & Burgess, etc., 1846.
208p. 917.75/L31w

1328 **Laplace, Pierre Simon,** 1749-1827.
Exposition du Système du Monde. Paris, Impr. du Cercle-social, 1796.
2v. in 1. 520/L31e

1329 **Laplace.**
The system of the world. Tr. from the French, and elucidated with explanatory notes. By the Rev. Henry H. Harte . . . Dublin, Printed at the University Press for Longmans, Rees, Orme, Brown, and Green, London, 1830.
2v. 523/L31

Contents. — v.1. book I. Of the apparent motion of the heavenly bodies. book II. Of the real motions of the heavenly bodies. book III. Of the laws of motion. — v.2 book IV. Of the theory of universal gravitation. book V. Summary of the history of astronomy.

1330 **Laplace.**
Traité de mécanique céleste. Paris, J. B. M. Duprat, [1798]-1825.
5v. x521.1/L31t/1798

1331 **Larkin, Nathaniel John,** 1781-1855.
An introduction to solid geometry and to the study of crystallography. Containing an investigation of some of the properties belonging to the platonic bodies independent of the sphere. Illustrated by four plates from original drawings by the author. London, Printed for the author: Longman, Hurst, Rees, Orme, and Brown, 1820.
140p., 4 fold. plates. 548.7/L32i

1332 **La Rue, François,** ca.1520-1585.
De gemmis aliquot, iis praesertim quarum Diuus Ionnes Apostolus in sua Apocalypsi meminit: De alys quoque, quarum usus hoc qui apud omnes percrebruit, Libri duo: Theologis non minùs utiles quàm philosophis, omnino felicioribus ingenÿs periucundi, è non vulgaribus utriusque philosophiae adytis deprompti . . . Tiguri, n.p., 1565.
85p. uncat.

Bound with Gesner, Konrad, ed., De omni rerum fossilium genere gemmis, lapidibus, metallis, et huiusmodi, libri aliquot, plerique nunc primum editi . . . 1565.

1333 **La Tourrette, Marc Antoine Louis Claret de Fleurieu de,** 1729-1793.
Voyage au Mont Pilat dans la Province du Lyonnois, contenant des observations sur l'histoire naturelle de cette montagne, & des lieux circonvoisins; suivies du catalogue raisonné des plantes qui y croissent. Avignon, Reynault, 1770.
223p. 570.94458/L35v

1334 **Launay, Louis de,** 1740?-1805?
Minéralogie des anciens. Bruxelles, Weissenbruch, [1803].
2v., fold. table. 549/L37m

Bibliographical footnotes.

1335 **Laurance, John.**
Geology in 1835; a popular sketch of the progress, leading features, and latest discoveries of this rising science. London, Simkin, Marshall, 1835.
139p., illus. 550/L375g

1336 **Lavizzari, Luigi,** 1814-1875.
Istruzione popolare sulle principali rocce; ossia, sulle pietre e terre più comuni del cantone Ticino e loro use nelle arti. Lugano, Tipografia di G. Bianchi, 1849.
101p., 2 fold. tables. 552/L39i

1337 **Lavoisier, Antoine-Laurent,** 1743-1794, jt. author.
See FOUGEROUX, DE BONDAROY, AUGUSTE DENIS, Autograph draft . . . on Guettard . . . ca. 1778.

1338 **Lavoisier.**
Observations générales sur les couches horizontales, qui ont été déposées par la mer, et sur les conséquences qu'on peut tirer de leurs dispositions, relativement à l'ancienneté du globe terrestre, *in* Mémoires de l'Académie Royale des Sciences, Paris, 1789.
p.351-371. 506 PAH/1789

1339 **Lawrence, Byrem.**
A geological map of the western states. J. C. Sharp, printer, 1843.
G4045/.C5/1843/.L3

Col. map.
Scale not given.
Includes structure sections.

1340 **Lawson, John.**
See LESLIE, WILLIAM, Minister of St. Andrews and Longbride, A manual of the antiquities . . . Moray . . . 1823.

HISTOIRE PHYSIQUE DE LA MER.

Ouvrage enrichi de figures dessinées d'après le Naturel.

PAR

LOUIS FERDINAND
COMTE DE MARSILLI,

MEMBRE DE L'ACADEMIE ROYALE DES SCIENCES DE PARIS.

A AMSTERDAM,
Aux DE'PENS DE LA COMPAGNIE.

M. DCC. XXV.

Cat. No. 1504

1341 **Lea, Isaac,** 1792-1886.
Contributions to geology. Philadelphia, Carey, Lea and Blanchard, 1833.
227p., 6 plates. 564/L46c

Contents. — Tertiary formation of Alabama. — New Tertiary fossil shells from Maryland and New Jersey. — New genus of fossil shell from New Jersey. — Tufaceous lacustrine formation of Syracuse, Onondaga County, New Jersey.

1342 **Lea.**
Fossil foot-marks in the red sandstone of Pottsville, Pa. Philadelphia, Printed by T. K. and P. G. Collins, 1855.
16p., double col. plates. f557/P38ℓ

1343 **Le Brun, L. S. D.**
Observations critiques sur un ouvrage intitulé: Examen de la houille, considerée comme engrais des terres, par M. Raulin. Instruction sur l'usage des houilles d'engrais & de leurs cendres. Expériences et observations sur la maladie du seigle nommée ergot . . . Amsterdam, Charle, 1777.
150p. 631.81/L49o

1344 **Le Brun, Pierre.**
Lettres qui decouvrent l'illusion des philosophes sur la baguette, et qui detruisent leurs systémes. Paris, Jean Boudot, 1693.
309p. uncat.

1345 **Lecoq, Henri,** 1802-1871.
Vues et coupes des principales formations géologiques du département du Puy-de-Dôme, accompagnées de la description et des échantillons des roches qui les composent, par H. Lecoq et J. B. Bouillet. Paris, M. F. G. Levrault, 1830.
266p., 31 col. plates. 554.4591/L49v

1346 **Lecoq.**
Description du volcan de Pariou. Clermont-Ferrand, Pelisson, 1833.
91p., 2 plates. 551.21/L49d

1347 **Lecoq.**
Élémens de géologie et d'hydrographie, ou; Résumé des notions acquises sur les grandes lois de la nature; faisant suite et servant de complément aux élémens de géographie physique et de météorologie. Paris, J. B. Baillière, 1838.
2v., 8 plates, front. 550/L492e

1348 **Lecoq.**
Des glaciers et des climats; ou, Des causes atmosphériques en géologie. Recherches sur les forces diluviennes, indépendantes de la chaleur centrale, sur les phénomènes glaciaire et erratique. Paris, P. Bertrand, 1847.
566p. 551.31/L49g

1349 **Lecoq.**
Les eaux minérales, considérées dans leurs rapports avec la chimie et la géologie. Paris, J. Rothschild, éditeur, 1864.
463p. uncat.

Signed by the author.

1350 **Lecoq.**
Les époques géologiques de l'Auvergne. Avec 170 planches ou figures, dont plusieurs coloriées et des autographes de Dolomieu, D'Haüy & de De Saussure et un dessin facsimile de Madame Necker De Saussure. Paris, New York, Baillière, 1867.
5v., plates: part. col., part. fold. (incl. facsims.). 554.459/L49e

1351 **Ledermüller, Martin Frobenius,** 1719-1769.
Traité phisique et microscopique de l'asbeste, l'amiante, le lin de pierre ou de terre, et de quelques autres fossiles qui y ont du rapport. Traduit de l'allemand. Avec VI. planches enluminées. Nuremberg, A. W. Winterschmidt, 1775.
16p., 6 col. plates. 549.6/L49pFh

Translated by Martinus Houttuyn.
Title of original: Physicalisch-mikroskopische Abhandlung vom Asbest, Amiant, Stein- oder Erdfläche.

1352 **Lee, Sarah (Wallis) Bowdich,** 1791-1856.
Mémoires du Baron George Cuvier, publiés en anglais par Mistress Lee, et en français par Théodore Lacordaire, sur les documens fournis par sa famille. Paris, H. Fournier, 1833.
369p., port. B/C992ℓEℓ

Bound with Brunel, Adolphe. Biographie d'Aimé Bonpland. Paris, 1871.

1353 **Lehmann, Johann Christian von.**
Grundsätze der Mineralogie. Frankfurt am Main, Barrentrapp und Wenner, 1791.
569p. 549/L528g

1354 **Lehmann, Johann Gottlob,** 1719-1767.
Kurtze Einleitung in einige Theile der Bergwercks-Wissen-

schaft; Anfängern zum Besten abgefasset. Berlin, bey Christoph Gottlieb Nicolai, 1751.

8, 192, 5p., 2 plates, illus. x553.4/L52k

Bound with: Abhandlung von den Metall-Müttern und der Erzeugung der Metalle . . .

1355 **Lehmann.**

Abhandlung von den Metall-Müttern und der Erzeugung der Metalle aus der Naturlehre und Bergwerckswissenschaft hergeleitet und mit chymischen Versuchen erwiesen. Berlin, verlegt Christoph Gottlieb Nicolai, 1753.

[16], 268, [8]p., 2 folded plates, illus. x553.4/L52k

Bound with: Kurtze Einleitung in einige Theile der Bergwercks-Wissenschaft . . .

Library owns another copy with the call number 553.4/L52a.

1356 **Lehmann.**

Versuch einer Geschichte von Flötz-Gebürgen; betreffend deren Entstehung, Lage, darinne befindliche Metallen, Mineralien und Fossilien gröstentheils aus eigenen Wahrnehmungen, chymischen und physicalischen Versuchen, und aus denen Grundsätzen der Natur-Lehre hergeleitet, und mit nöthigen Kupfern versehen. Berlin, Gottlieb August Lange, 1756.

[80], 240, [8]p., 8 plates, illus. x551.43/L52v

Imperfect: one side of folded plate trimmed off, with loss of key numbers.

1357 **Lehmann.**

Traités de physique, d'histoire naturelle, de minéralogie et de métallurgie; ouvrages traduits de l'Allemand. Paris: Chez Jean-Thomas Hérissant, 1759.

3v., illus. x553.4/L52t:Fh

Each vol. has separate t.p.; general title from half title page.

Translated by P. H. T. Baron d'Holbach.

Vol. 1 includes: Traité des moufettes, ou des exhalaisons pernicieuses qui se font sentir dans les souterreins des mines, traduit du Latin de Zacharie Théobald, enrichi de remarques par M. Lehmann.

Contents. — T.1. L'Art des mines, ou, Introduction aux connaissances nécessaires pour l'exploitation des mines métalliques. — T.2. Traité de la formation des métaux, et de leur matrices ou minieres. — T.3. Essai d'une histoire naturelle des couches de la terre . . .; avec des considérations physiques sur les causes des tremblemens de terre . . .

1358 **Leibniz, Gottfried Wilhelm von,** 1646-1716.

Summi polyhistoris Godefridi Guilielmi Leibnitii Protogaea,

LA FIGURE DE LA TERRE,

DETERMINÉE

PAR LES OBSERVATIONS

De Meſſieurs DE MAUPERTUIS, CLAIRAUT, CAMUS, LE MONNIER, de l'Académie Royale des Sciences, & de M. l'Abbé OUTHIER, Correſpondant de la même Académie,

Accompagnés de M. CELSIUS, Profeſſeur d'Aſtronomie à Upſal:

FAITES PAR ORDRE DU ROI

AU CERCLE POLAIRE.

Par M. DE MAUPERTUIS.

A AMSTERDAM,

Chez *JEAN CATUFFE.*

MDCCXXXVIII.

Cat. No. 1521

sive de prima facie telluris et antiquissimae historiae vestigiis in ipsis naturae monumentis dissertatio, ex schedis manuscriptis viri illustris in lucem edita a Christiano Ludovico Scheidio. Goettingae, Sumptibus I. G. Schmidii, 1749.
86p., 12 plates (part. fold.). 550/L53p

1359 **Leigh, Charles,** 1663-1701.
The Natural History of Lancashire, Cheshire, and the Peak, in Derbyshire; with an account of the British, Phoenician, Armenian, Gr. and Rom. antiquities in those parts. Oxford, printed for the author, 1700.
3 parts paged separately. q570.942/L533n

Some pages are out of sequence.

1360 **Leithart, John.**
Practical observations on the mechanical structure, mode of formation, the repletion or filling up, and the intersection and relative age of mineral veins; with the application of several new theoretical principles to the art of mining. By John Leithart, Mine Agent. London, J. Weale; [etc., etc.], 1838.
83p., 11 plates, diagr. 553.1/L53p

1361 **Lelarge de Lignac, Joseph Adrien,** 1710-1762.
Lettres à un Amériquain sur l'Histoire naturelle, générale et particulière de monsieur de Buffon. Hambourg, 1751.
3v. 500/B86hYℓ

Errata slip inserted in v.1.

1362 **Le Mascrier, Jean Baptiste,** 1697-1760.
Description de l'Egypte contenant plusieurs remarques curieuses sur la geographie ancienne et moderne de ce païs, sur ses monumens anciens, sur les moeurs, les coutumes, & la religion des habitans, sur le gouvernement & le commerce, sur les animaux, les arbres, les plantes, &c. Composée sur les mémoires de M. de Maillet. Paris, chez Louis Genneau et Jacques Rollin, 1735.
570p. illus. fold. map, with portrait of de Maillet. 916.2/L54d

1363 **Lemnius, Levinus,** 1505-1568.
De miraculis occultis naturae, libri IIII. De vita cum animi et corporis incolumitate recte instituenda. Ienae, Tobias Steinman, 1588.
868p. x613 L54d/1588

A small part of this medieval work treats the medicinal use of stones.

1364 **Lemnius.**
Occulta naturae miracula. Wunderbarliche Geheimnisse der Natur in des Menschen Leibe und Seel, auch in vielen andern natürlichen Dingen, als Steinen, Ertzt, Gewechs und Thieren. Allen frommen Hauswirthen, verstendigen Hausfrauen, fleissigen Naturkündigern, guten Hausärtzten, Liebhabern der Gesundheit, und gemeinem Vaterland zum besten, nicht allein aus dem Latein in Deutsche Sprache gebracht, sondern auch zum dritten Mal vermehret und eines grossen Theils von newes selbs geschrieben, durch Iacobum Horstium. Leipzig, 1588.
629p. x613/L54dGh3

The library owns another copy of the same title published in 1593 (x613/L54dGh/1593). Both works are translations of De miraculis occultis naturae . . . 1588.

1365 **Lenz, Johann Georg L.,** 1748-1832.
Mustertafeln der bis jetzt bekannten einfachen Mineralien, worauf dieselben nach ihren Gestalten und natürlichen Farben abgebildet, und ihre übrigen Verhältnisse gegen einander bestimmt werden. Jena, Auf Kosten des Verfassers, 1794.
111p., col. illus. 549.1/L54m

1366 **Lenz.**
System der äussern Kennzeichen der Mineralien, in deutscher, lateinischer, italienischer, französischer, dänischer und ungarischer Sprache. Bamberg, T. Göbhard, 1800.
32p. 549.03/L549s

1367 **Leonardus, Camillus,** fl. 1502.
Speculum lapidum clarissimi artium et medicine . . . Venetiis, Melchior Sessam & Petrum de Rauanis sociis, 1516.
66p. uncat.

1368 **Leonardus.**
The mirror of stones: in which the nature, generation, properties, virtues and various species of more than 200 different jewels, precious and rare stones, are distinctly described. Also certain and infaillible rules to know the good from the bad, how to prove their genuineness, and to distinguish the real from the counterfeits. Extracted from the works of Aristotle, Pliny, Isidorus, Dionysius Alexandrinus, Albertus Magnus, &c . . . dedicated to Caesar Borgia. First translated into English. London, J. Freeman, 1750.
240p. 553.8/L55sE

1369 **Leonhard, Gustav von,** 1816-1878.
Ueber die älteren oder Paläozoischen Gebilde im Norden von Deutschland und Belgien, verglichen mit Formationen desselben Alters in Grossbritannien von Sedgwick und Murchison, nebst einer Uebersicht der Fauna der Paläozoischen Gebilde in den Rheinlanden und einer Tabelle der organischen Reste des Devonischen Systemes in Europa von Archiac und Verneuil. Stuttgart, E. Schweizerbart, 1844.
248p., 3 plates, fold. col. map, col. profiles. 554.3/L55u

1370 **Leonhard.**
Geognostische Skizze des Grossherzogthums Baden. Ein Leitfaden für Vorträge in Höheren und Mittelschulen jeder Art, . . . Stuttgart, E. Schweizerbart'sche Verlagshandlung, 1861.
2.verm. u. verb. Aufl. 168p., fold. map. 554.346/L55g2

Literatur: p. [104]-108.

1371 **Leonhard, Karl Cäsar von,** 1779-1862.
Systematisch-tabellarische Uebersicht und Charakteristik der Mineralkörper. In oryktognostischer und orologischer Hinsicht aufgestellt von C. C. Leonard, K. F. Merz und Dr. J. H. Kopp. Frankfurt am Main, J. D. Hermann, 1806.
125 (i.e. 208)p., 16 plates. f549/L55e

Pages 1-83 numbered in duplicate.

1372 **Leonhard.**
Propaedeutik der Mineralogie. Von Dr. C. C. Leonhard, Dr. J. H. Kopp und C. L. Gaertner. Mit 10 schwartzen und ausgemalten Kupfertafeln. Frankfurt am Main, In der J. C. Hermannschen Buchhandlung, 1817.
315p., 10 plates. f549/L55p

"Literatur der Mineralogie": p. 229-277.

"Alphabetische Uebersicht der Autoren, Uebersetzer und Herausgeber in der Literatur der Mineralogie, sowie der andern mit den angeführten Schriften in Beziehung stehenden Gelehrten": p. 278-309.

Added t.p.: Einleitung und Vorbereitung zur Mineralogie. Als erster Theil den Systematisch-tabellarischen Uebersicht und Charakteristik der Mineralkoerper.

Imperfect: 1st plate wanting.

1373 **Leonhard.**
Handbuch der Oryktognosie. Für akademische Vorlesungen und zum Selbststudium. Mit sieben Steindruck-Tafeln. Heidelberg, Bei Mohr und Winter, 1821.
720p., 7 fold. plates. 549/L55h

MICHAELIS MERCATI
SAMMINIATENSIS
METALLOTHECA.
Opus Posthumum,
Auctoritate, & Munificentiâ
CLEMENTIS
UNDECIMI
PONTIFICIS MAXIMI
E' tenebris in lucem eductum;
Opera autem, & studio
JOANNIS MARIÆ LANCISII
ARCHIATRI PONTIFICII
ILLUSTRATUM.
CUI ACCESSIT APPENDIX
CUM XIX. RECENS INVENTIS ICONIBUS.

ROMÆ MDCCXIX.
Apud Jo: Mariam Salvioni Typographum Vaticanum
In Archigymnasio Sapientiæ.

SUPERIORUM FACULTATE.

Cat. No. 1541

1374 **Leonhard.**
Grundzüge der Geologie und Geognosie; Lehrbuch für öffentliche Vorträge besonders auch in Gymnasien und Realschulen so wie zum Selbstudium. 2., verm. und verb. Aufl. Heidelberg, J. Engelmann, 1831.
308p., 6 plates. 550/L55gr/1831

His Naturgeschichte des Mineralreich, 2. Abth.

1375 **Leonhard.**
Lehrbuch der Geognosie und Geologie. Stuttgart, E. Schweizerbart's Verlagshandlung, 1835.
[5]-869p., illus., diagrs. 552/L55ℓ

Naturgeschichte der drei Reiche zur allgemeinen Belehrung . . . Bd. 3.

1376 **Leonhard.**
Geologie; oder, Naturgeschichte der Erde auf allgemein fassliche Weise abgehandelt. Mit Stahlstichen. Stuttgart, E. Schweizerbart, 1836-44.
5v., illus., 97 plates. 550/L55g

Populäre Vorlesungen über Geologie.

1377 **Leonhard,** ed.
Taschenbuch für Freunde der Geologie in allgemein fasslicher Weise, bearbeitet von Karl Cäsar v. Leonhard. v.1-3, 1845-47. Stuttgart, E. Schweizerbart'sche Verlagsbuchhandlung.
3v., illus., plates (part. col.). 550/L554

Annual.
Ceased publication with v.3, 1847.
Library owns v.1-2.

1378 **Leske, Nathanael Gottfried,** 1751-1786.
See KLEIN, JACOB THEODOR, Naturalis dispositio Echinodermatun . . . Additamenta [by] Leske [1778] . . . 1734.

1379 **Leske.**
See KARSTEN, DIETRICH LUDWIG GUSTAV, Des Herrn Nathanael Gottfried Leske . . . hinterlassenes Mineralienkabinet . . . 1789.

1380 **Leslie, William,** Minister of St. Andrews and Longbride.
A manual of the antiquities, distinguished buildings and natural curiosities of Moray in its ancient prelatic extent; with an outline of the geology and mineralogy of the county and an itinerary of the province for the information of inquisitive travellers and others [by John Lawson]. 2d ed. adjusted to the

passing time. Elgin, Printed by R. Johnston for I. Forsyth, 1823.
144p., front. x914.123/L56m/1823

1381 **Lesser, Friedrich Christian,** 1692-1754.
Lithotheologie, das ist; natürliche Historie und geistliche Betrachtung derer Steine, also abgefasst, dass daraus die Allmacht, Weissheit, Güte und Gerechtigkeit des grossen Schöpffers gezeuget wird, anbey viel Sprüche der Heiligen Schrifft erklähret, und die Menschen allesamt zur Bewunderung, Lobe und Dienste des grossen Gottes ermuntert werden. Neu-verb. Aufl. Hamburg, C. W. Brandt, 1751.
2v., 9 plates. 550/L552ℓ/1751

1382 **Lhwyd, Edward,** 1660-1709.
Lithophylacii Britannici Ichnographia, sive, lapidum aliorumque fossilium Britannicorum singulari figura insignium; quotquot hactenus vel ipse invenit vel ab amicis accepit, distributio classica, scrinii sui lapidarii repertorium cum locis singulorum natalibus exhibens. Additis rariorum aliquot figuris aere incisis; cum epistolis ad clarissimos viros de quibusdam circa marina fossilia & stirpes minerales praesertim notandis. Nusquam magis erramus quam in falsis inductionibus: saepe enim ex aliquot exemplis Universale quiddam colligimus; idque perperam cum ad ea, quae excipi possunt animum non attendimus. Londini, Ex officina M. C., Lipsiae, Sumt. Joh. Ludw. Gleditsch & Weidmann, 1699.
139p., 23 leaves of plates (one folded), illus. (engravings).
x560.942/L61i/1699

This is the 1st ed., published in 120 copies.

1383 **Lhwyd.**
Lithophylacii Britannici Ichnographia sive lapidum aliorumq; fossilium Britannicorum singulari figura insignium, quotquot hactenus vel ipse invenit vel ab amicis accepit, distributio classica, scrinii sui lapidarii repertorium cum locis singulorum natalibus exhibens. Additis rariorum aliquot figuris aere incisis cum epistolis ad clarissimos viros de quibusdam circa marina fossilia & stirpes minerales praesertim notandis. Nusquam magis erramus quam in falsis inductionibus; sape enim ex aliquot exemplis Universale quiddam colligimus; idque perperam cum ad ea, quae excipi possunt animum non attendimus. Lipsiae, Sumt. J. L. Gleditsch & Weidmann, 1699.
145p., illus., 14 fold. plates. uncat.

1384 **Lhwyd.**
Lithophylacii Britannici Ichnographia. Sive, Lapidum aliorumque fossilium Britannicorum singulari figura insignium,

quotquot hactenus vel ipse invenit vel ab amicis accepit, distributio classica; scrinii sui lapidarii repertorium cum locis singulorum natalibus exhibens. Additis rariorum aliquot figuris aere incisis; cum epistolis ad clarissimos viros de quibusdam circa marina fossilia & stirpes minerales praesertim notandis. Ed. altera: novis quorundam speciminum iconibus aucta. Oxonii, E Typographeo Clarendoniano, 1760.
156p., 25 fold. plates. x560.942/L61i/1760

1385 **Liebenroth, Friedrich Ernst Franz von.**
Geognostiche Beobachtungen und Entdeckungen in der Gegend von Dresden. Weissenfels, F. Severin, 1798.
120p. 554.321/L62g

1386 **Liebknecht, Johann Georg,** 1679-1740.
Hassiae subterraneae specimen clarissima testimonia diluvii universalis heic et in locis vicinioribus occurrentia, ex priplici regno, animali, vegetabili et minerali, petita, figurisque aeneis exposita, imo omnis antiquitatis exempla certissima exhibens. Occasione arboris in mineram ferri mutatae, quae variis observationibus, et nova ferrum ex limo coquendi methodo illustratur. Cui accedit: I. Io. Geilfusii . . . Rariss. de terra sigillata Laubacensi tractatio . . . II. De serratis et bigatis numis, dissertatio epistolica . . . Giessae, E. H. Lammers, 1730.
490p., 14 plates. 551.791/L62h

1387 **Limbourg, P. de.**
Mémoire pour servir à l'histoire naturelle des fossiles des Pays-Bas. Par M. de Limbourg, le jeune. Lu à la séance du 7 février 1774. [Bruxelles, 1774].
[369]-418p. 549.9492/L62m

From Mémoires de l'Académie royale des sciences de Belgique, t.I.

1388 **Linden, Diederick Wessel.**
A treatise on the three medicinal mineral waters at Llandrindod, in Radnorshire, South Wales. With some remarks on mineral and fossil mixtures, in their native veins and beds; at least as far as respects their influence on water. London, Printed by J. Everingham and T. Reynolds, for the author, 1756.
336p., front. x613.38/L642t

1389 **Lindley, John,** 1799-1865.
The fossil flora of Great Britain; or Figures and descriptions of the vegetable remains found in a fossil state in this country.

¶ IOYFVLL NEVVES OVT OF

the newe founde worlde, wherein is *declared the rare and singuler vertues of diuerse* and sundrie Hearbes, Trees, Oyles, Plantes, and Stones, with *their aplications, aswell for Phisicke as Chirurgerie, the saied be-* yng well applied bryngeth suche present remedie for all deseases, as maie seme altogether incredible: notwithstandyng by practize founde out, to bee true: Also the portrature of the saied Hearbes, very aptly discribed: Englished by Jhon Frampton *Marchaunt.*

¶ Jmprinted at London in Poules Churche-yarde, by Willyam Norton.

Anno Domini.

1577.

Cat. No. 1601

By John Lindley . . . and William Hutton . . . London, J. Ridgway, 1831-37.
3v., illus., 230 plates. 561/L64f

1390 **Linné, Carl von,** 1707-1778, praeses.
. . . Specimen academicum de crystallorum generatione . . . Upsaliae, [1747].
30p., fold. plates. 548.5/L64s

Diss. — Upsala (M. Kähler, respondent).
Error in paging: p. 26 numbered 29.
Also in Linné's Amoenitates academicae, v. 1 [ed. 1], 1749, p.454-488; [ed. 2], 1749 and ed. 3, 1787, p.454-482. For other reprints and translations see J. M. Hulth, Bibliographia Linnaeana, [1907], p.55.

1391 **Lister, Martin,** 1638?-1712.
Historiae Animalium Angliae, tres tractatus — Unus de araneis. Alter de cochleis tum terrestribus, tum fluviatilibus. Tertius de cochleis marinis. Quibus adjectus est quartus de lapidibus ejusdem insulae ad cochlearum quandam imaginem figuratis . . . Londini, apud Joh. Martyn Regiae societatis typographum, 1678.
250p., 9 plates. uncat.

1392 **Lister.**
Appendicis ad historiae animalium Angliae, Tres Tractatus, &c. Altera editio, auctior & emendatior. Lóndini, S. Smith, 1685.
45p. 595.7/G55d/1685

Bound with Geodartius, Johannes, De insectis, in methodum redactus; cum notularum additione. Londini, S. Smith, 1685.

1393 **Lister.**
Historiae sive synopsis methodicae conchyliorum et tabularium anatomicarum. Editio altera. Recensuit et indicibus auxit Gulielmus Huddesford . . . Oxonii, e typographeo Clarendoniano, 1770.
77, 6p., 439 plates (8 fold.). xq594/L69h/1770

1394 **Lister.**
Historia sive synopsis methodica conchyliorum. Editio tertia. Recensuit et indice locupletissimo instruxit L. W. Dillwyn. Oxonii, e typographeo Clarendoniano, 1823.
48p. q594/L69h/1823

This ed. issued without the plates.

1395 **Loader, T. B.**
T. B. Loader's scientific and commercial map of England and Wales; in which are delineated the canals, rail-roads, & nav-

igable rivers. The extent of the navigation of each river, with the elevations of the canals, together with the geology, and principal situations of the mineral productions. [London?], T. B. Loader, 1831.

G5750/.C5/1831/.L8

Col. map, 134 x 108 cm. on 4 sheets.
Scale, 1 inch equals 10 miles.

1396 **Löhneyss, Georg Engelhardt von,** 1522-1622.
Bericht, vom Bergkwerck, wie man dieselben bawen, und in guten Wolstandt bringen soll, sampt allen darzu gehörigen Arbeiten, Ordnung und Rechtlichen Process. n.p., Zellerfeldt, 1617.
343p., illus. q uncat.

1397 **Lommer, Christian Hieronymus,** d. 1787.
Bergmännischer Beytrag zu der von der Königlichen Grossbrittanischen Societät der Wissenschaften, auf das Jahr 1781 ausgestellten Preissfrage: Wie waren die Bergwerke bey den Alten eigentlich beschaffen und eingerichtet? Und lässt sich nicht nach angestellter Vergleichung derselben, mit den unsrigen, zum Vortheil des Bergbaues, und Hüttenwerke in unsern Zeiten, etwas von den Alten lernen? Freyberg, Carl Craz, 1785.
44p., illus. 622.0943/L838b

The library owns a second copy which is bound with Flade, Christian Gottlob, De re Metallica . . . [1791]. (549/F59).

1398 **Long, Stephen Harriman,** 1784-1864.
See JAMES, EDWIN, Account of an expedition from Pittsburgh to the Rocky Mountains . . . 1823.

1399 **Longuemar, Alphonse le Touzé de,** b. 1803.
Étude géologique des terrains de la rive gauche de l'Yonne compris dans les arrondissements d'Auxerre et de Joigny. Auxerre, Impr. et lithographie de Perriquet, 1843.
251p., map, 9 plates. 554.441/L85e

Copy imperfect: map and plates listed in table of contents missing.

1400 **Löscher, Carl Immanuel.**
Uebergangsordnung bei der Kristallisation der Fossilien wie sie aus einander entspringen und in einander übergehen. Leipzig, S. L. Crusius, 1796.
61p., 6 fold. plates. 548.1/L89u

1401 **Lovell, Robert,** 1630?-1690.
Παυζωορυκτολογια, sive panzoologicomineralogia. Or a compleat history of animals and minerals, containing the summe of all authors, both ancient and modern, galenicall and chymicall, touching animals . . . Oxford, printed by Hen. Hall, for Jos. Godwin, 1661.
48, 519, 152p. 615/L94p

1402 **Lowry, Devalle (Varley).**
See VARLEY, DEVALLE (LOWRY).

1403 **Lucas, Jean André Henri,** 1780-1825.
Tableau méthodique des espèces minérales, présentant la série complète de leurs caractères, et la nomenclature de leurs variétés, extrait du Traité de minéralogie de M. Haüy, et augm. des nouvelles découvertes; auquel on a joint l'indication des gisemens de chaque espèce, et la description abrégée de la collection de minéraux du Muséum d'histoire naturelle. Paris, Levrault, Schoell et c^ie^; [etc., etc.], 1806-13.
2v., front. (port.), fold. plates. 549/L962t

Library owns v.1 only.

1404 **Luchet, Jean Pierre Louis de la Roche du Maine de,** 1740-1792.
Essais sur la minéralogie et la métallurgie. Maestricht, J. E. Dufour & P. Roux, 1779.
232p. 549/L96e

1405 **Ludwig, Christian Friedrich,** 1751-1823.
Handbuch der Mineralogie nach A. G. Werner; zu Vorlesungen entworfen von Christian Friedrich Ludwig. Leipzig, S. L. Crusius, 1803-04.
2v. in 1, illus. (part. col.). 549/L966h

Includes bibliographies.

1406 **Lyell, Charles,** 1797-1875.
Life, letters and journals of Sir Charles Lyell, bart. . . . Ed. by his sister-in-law, Mrs. Lyell. London, J. Murray, 1881.
2v., fronts. (ports.). B/L9841

"Geological papers and works by Sir Charles Lyell": v.2, p.479-482.

1407 **Lyell.**
Principles of geology, being an attempt to explain the former changes of the earth's surface, by reference to causes now in operation. London, J. Murray, 1830-35.
3v., fronts., illus., maps (part. fold.), diagrs. 550/L98p

HISTORIA UNIVERSAL
DOS
TERREMOTOS,
QUE TEM HAVIDO NO MUNDO, de que ha noticia, desde a sua creaçaõ até o seculo presente.

Com huma

NARRAÇAM INDIVIDUAL
Do Terremoto do primeiro de Novembro de 1755., e noticia verdadeira dos seus effeitos em Lisboa, todo Portugal, Algarves, e mais partes da Europa, Africa, e América, aonde se estendeu:

E huma

DISSERTAÇAÕ PHISICA
Sobre as causas geraes dos Terremotos, seus effeitos, differenças, e Prognosticos; e as particulares do ultimo.

POR
JOACHIM JOSEPH
MOREIRA DE MENDONÇA

LISBOA:
Na Offic. de ANTONIO VICENTE DA SILVA.

Anno de M.DCCLVIII.

Com todas as licenças necessarias.

Cat. No. 1608

"The author has found it impossible to compress into two volumes, according to his original plan, the wide range of subjects, which must be discussed . . . it will therefore be necessary to extend the 'Principles of geology' to three volumes." — v.2, pref.

Vol. 3 was issued for the first time in 1833, a little after the 2d edition of v.1-2, with which it is usually placed, as no 2d edition of v.3 ever appeared.

1408 **Lyell.**
Principles of geology, being an attempt to explain the former changes of the earth's surface, by reference to causes now in operation. The 2d ed. London, J. Murray, 1832-33.
3v., fronts., illus., plates, maps (part. fold.), diagrs. x550/L98p/1832

1409 **Lyell.**
Principles of geology: being an inquiry how far the former changes of the earth's surface are referable to causes now in operation. In four volumes . . . 4th ed. London, J. Murray, 1835.
4v., front., illus., col. plates, maps (part. fold.), diagrs.
550/L98p/1835

1410 **Lyell.**
Principles of geology, being an inquiry how far the former changes of the earth's surface are referable to causes now in operation. 5th ed. London, J. Murray, 1837.
4v., front., illus., plates, fold. maps. 550/L98p5

1411 **Lyell.**
Principles of geology; or, The modern changes of the earth and its inhabitants considered as illustrative of geology. 8th and entirely rev. ed. Illustrated with maps, plates, and woodcuts. London, J. Murray, 1850.
811p., front., illus., 11 plates (incl. 7 maps, 4 fold.).
550/L98p/1850

"Glossary of geological and other scientific terms": p. [775]-790.

1412 **Lyell.**
Principles of geology; or, The modern changes of the earth and its inhabitants considered as illustrative of geology. 9th and entirely rev. ed. Illustrated with maps, plates, and woodcuts. London, J. Murray, 1853.
835p., front., illus., 5 plates (incl. fold. map). 550/L98p/1853

"Glossary of geological and other scientific terms used in this work." p. 800-815.

1413 **Lyell.**
Principles of geology; or, The modern changes of the earth and its inhabitants considered as illustrative of geology. New

and entirely rev. ed. . . . New York, D. Appleton and company, 1854.
2v., fronts., illus., 6 plates (incl. fold. map), diagrs. 550/L98p9

"List of the dates of publication of successive editions of the 'Principles,' 'Elements,' and the 'Antiquity of man' ": v.1, p. ix-x.

1414 **Lyell.**
Principles of geology; or, The modern changes of the earth and its inhabitants considered as illustrative of geology. New and entirely rev. ed. . . . New York, D. Appleton & co., 1857.
834p., front., illus., 4 plates (incl. fold. map), diagrs. 550/L98p9a

"Dates of the successive editions of the 'Principles' and 'Elements' (or Manual) of geology, by the author": p. [vii].
"Glossary of geological and other scientific terms": p. [800]-816.

1415 **Lyell.**
Principles of geology: or The modern changes of the earth and its inhabitants considered as illustrative of geology. 10th and entirely rev. ed. London: J. Murray, 1867-1868.
2v., illus., maps. 550/L98p/1867

1416 **Lyell.**
Principles of geology; or, The modern changes of the earth and its inhabitants considered as illustrative of geology. 11th and entirely rev. ed. New York, D. Appleton and company, 1872.
2v., illus. 550/L98p11

1417 **Lyell.**
Principles of geology; or, The modern changes of the earth and its inhabitants considered as illustrative of geology. 12th ed. . . . London, J. Murray, 1875.
2v., fronts., illus., 6 plates (incl. fold. map), diagrs. 550/L98p/1875

"List of the dates of publication of successive editions of the 'Principles,' 'Elements,' and the 'Antiquity of man' ": v.1, p. xi-xii.

1418 **Lyell.**
Principes de géologie, ou Illustrations de cette science empruntées aux changements modernes que la terre et ses habitants ont subis. Ouvrage traduit de l'anglais sur la sixième edition, et sous les auspices de M. Arago, par Mme Tullia Meulien . . . Paris, Langlois et Leclercq; [etc., etc.], 1843-48.
4v., front., illus., 11 plates (incl. maps). 550/L98pFa

1419 **Lyell.**
Principes de géologie; ou, Illustrations de cette science empruntées aux changements modernes de la terre et de ses

habitants. Ouvrage traduit sur la dernière ed. anglaise, entièrement refondue, par M. J. Ginestou. Paris, Garnier frères, 1873.

2v., illus. 550/L98pFg

1420 **Lyell.**

Elements of geology. London, J. Murray, 1838.

543p., illus., maps, diagrs. 550/L98/1838

The 3d, 4th and 5th eds. have title: A manual of elementary geology.

1421 **Lyell.**

Elements of geology. 2d ed. London, J. Murray, 1841.

2v., illus., 7 plates (incl. 2 col. maps). 550/L98e/1841a

Copy imperfect: front. wanting.

1422 **Lyell.**

Elements of geology. Reprinted from the 2d English ed. . . . Boston, 1841.

2v., illus., 7 plates (incl. map), diagrs. 550/L98/1841

1423 **Lyell.**

A manual of elementary geology: or, The ancient changes of the earth and its inhabitants, as illustrated by geological monuments. 4th and entirely rev. ed. Illustrated with 500 woodcuts. London, J. Murray, 1852.

512p., front., illus. 550/L98m/1852

"Dates of the successive editions of the 'Principles' and 'Elements' (or Manual) of geology, by the author": p. xxxi.

1424 **Lyell.**

A manual of elementary geology; or The ancient changes of the earth and its inhabitants, as illustrated by the geological monuments. Reprinted from the fourth and entirely revised edition. New York, D. Appleton and Co., 1853.

512p., 500 woodcuts. 550/L98m

1425 **Lyell.**

A manual of elementary geology: or, The ancient changes of the earth and its inhabitants, as illustrated by geological monuments. Reprinted from the 4th and entirely rev. ed. Illustrated with five hundred woodcuts. New York, D. Appleton & Co., 1854.

512p., front., illus. 550/L98m/1854

"Dates of the successive editions of the 'Principles' and 'Elements' (or Manual) of geology, by the author": p. xxxii.

DE

CROSTACEI

E DEGLI ALTRI

MARINI CORPI

Che si truovano su' monti

LIBRI DUE

DI ANTON-LAZZARO MORO.

VENEZIA

APPRESSO STEFANO MONTI

CON LICENZA DE' SUPERIORI, E PRIVILEGIO.

Si vende presso Angiolo Geremia in Merceria, all'insegna della Minerva.

MDCCXL.

Cat. No. 1611

1426 **Lyell.**
A manual of elementary geology: or, The ancient changes of the earth and its inhabitants as illustrated by geological monuments. Reprinted from the 6th ed., greatly enl. Illus. with 750 woodcuts. New York, D. Appleton and Company, 1857.
647p., front., illus. 550/L98m/1857

1427 **Lyell.**
Elements of geology; or, The ancient changes of the earth and its inhabitants as illustrated by geological monuments. 6th ed., greatly enlarged and illustrated with 770 woodcuts. New York, D. Appleton and Company, 1868.
803p., illus. 550/L98

1428 **Lyell.**
The student's elements of geology. With more than 600 illustrations on wood. New York, Harper & brothers, 1891.
[25]-640p., front., illus., diagrs. 550/L98s/1891

On cover: The student's series.

1429 **Lyell.**
Eight lectures on geology, delivered at the Broadway Tabernacle in the city of New York. Reported for "The New-York tribune." New York, Greeley & McElrath, 1842.
56p., illus., diagrs. 550/L98ei

At head of title: Lyell's lectures on geology.

Contents. — I. Fresh water formation of Auvergne; extinct volcanoes of successive periods. — II. Structure of Aetna; origin of granitic rocks; changes in the organic world. — III. Upheaval and subsidence of the earth. — IV. Coral reefs. — V. Origin of coal. — VI. Fossil foot-prints. — VII. Recession of the falls of Niagara. — VIII. Boulders and icebergs.

1430 **Lyell.**
Lectures on geology, delivered at the Broadway Tabernacle, in the city of New York. 2d ed., rev., and corr., with a general introd., to which is added a sketch of a lecture on the different races of men, by J. Augustine Smith. Reported for the New-York Tribune by H. J. Raymond. New York, Greeley & McElrath, 1843.
55p., illus., front. 550/L98ei/1843

First published in 1842 under title: Eight lectures on geology.

1431 **Lyell.**
Travels in North America; with geological observations on

the United States, Canada, and Nova Scotia. London, J. Murray, 1845.
2v., col. fronts., illus., plates, maps, facsims. 917/L98t

1432 **Lyell.**
Travels in North America, in the years 1841-42; with geological observations on the United States, Canada, and Nova Scotia. New York, Wiley and Putnam, 1845.
2v. in 1., illus., fold. plates. x917/L98t/1845a

1433 **Lyell.**
Lyell's travels in North America in the years 1841-2; abridged and ed. by John P. Cushing. New York, C. E. Merrill co., 1909.
172p. (Maynard's English classic series).
SANDBURG/917/L98t/1909

Title of original edition "Travels in North America; with geological observations . . ." London, 1845. 2v.
"In the present edition the technical geological portions of the work have been omitted."

1434 **Lyell.**
A second visit to the United States of North America. London, J. Murray, 1849.
2v., illus. 917.3/L98

1435 **Lyell.**
A second visit to the United States of North America. New York, Harper & Bros., 1849.
2v., illus. 917.3/L98s/1849

1436 **Lyell.**
A second visit to the United States of North America. New York, Harper & Bros., 1855-68.
2v., illus. 917.3/L98s/1855

1437 **Lyell.**
Zweite Reise nach den Vereinigten Staaten von Nordamerika. Deutsch nach der 2.Ausg. des englischen Originals von Ernst Dieffenbach. Braunschweig, Bieweg, 1851.
2v., illus. 917.3/L98sGd

Translation of A second visit to the United States.

1438 **Lyell.**
The geological evidences of the antiquity of man, with remarks on theories of the origin of species by variation. London, J. Murray, 1863.
520p., front., illus. (incl. maps), plates. 573/L986g

1439 **Lyell.**
The geological evidences of the antiquity of man, with remarks on theories of the origin of species by variation. 2nd ed., revised. London, J. Murray, 1863.
528p., front., illus. (incl. maps), plates. 573/L986g/1863

1440 **Lyell.**
The geological evidences of the antiquity of man, with remarks on theories of the origin of species by variation. 3rd ed., revised. London, J. Murray, 1863.
551p., illus. (incl. maps), plates. 573/L986g/1863a

1441 **Lyell.**
The geological evidences of the antiquity of man, with an outline of glacial and post-tertiary geology and remarks on the origin of species, with special reference to man's first appearance on the earth. 4th ed., revised. London, J. Murray, 1873.
572p., illus. (incl. maps), plates. 573/L986g/1873

1442 **Lyell.**
The geological evidence of the antiquity of man. London, J. Dent and New York, E. P. Dutton, [1914].
407p., illus., plates. 573/L986g/1914

Introduction and notes by R. H. Rastall. Bibliography, p. xii.

1443 **Lyon, George Francis,** 1795-1832.
A narrative of travels in northern Africa, in the years 1818, 19, and 20; accompanied by geographical notices of Soudan, and of the course of the Niger. With a chart of the routes, and a variety of coloured plates, illustrative of the costumes of the several natives of northern Africa. By Captain G. F. Lyon . . . London, J. Murray, 1821.
383p., front. (fold. map), 17 col. plates. 916.1/L99n

Observations, p. [361]-369. "I here subjoin a short account, by Professor Buckland of Oxford, of a few specimens of the rocks and minerals of Tripoli and Fezzan, which I was enabled to collect in the course of my tour," p. [361].

Imperfect: t.p. and 1 plate wanting; title supplied from Library of Congress Catalog.

1444 **Lyon, J.**
Explanation of the accompanying geological map of England and Wales; with a geological indication of the Isle of Wight. London, Printed for the author by Miller and Field, [n.d.].
12p., fold. col. map. 554.2/L99e

DISCOVRS ADMIRABLES, DE LA NATVRE DES EAVX ET FONTEINES, TANT NATVRELLES QV'ARTIFICIELLES, des metaux, des sels & salines, des pierres, des terres, du feu & des emaux.

AVEC PLVSIEVRS AVTRES EXCELLENS secrets des choses naturelles.

PLVS VN TRAITE' DE LA MARNE, FORT vtile & necessaire, pour ceux qui se mellent de l'agriculture.

LE TOVT DRESSE' PAR DIALOGVES, Esquels sont introduits la theorique & la practique.

Par M. BERNARD PALISSY, *inuenteur des rustiques figulines du Roy, & de la Royne sa mere.*

A TRESHAVT, ET TRESPVISSANT sieur le sire Anthoine de Ponts, Cheualier des ordres du Roy, Capitaine des cents gentils-hommes, & conseiller tresfidele de sa maiesté.

A PARIS,

Chez Martin le Ieune, à l'enseigne du Serpent, deuant le college de Cambray.

1580.

AVEC PRIVILEGE DV ROY.

Cat. No. 1717

1445 **Macculloch, John Ramsey,** 1773-1835.
A description of the Western Islands of Scotland, including the Isle of Man; comprising an account of their geological structure, with remarks on their agriculture, scenery and antiquities. London, Printed for A. Constable, Edinburgh, 1819.
3v., 33 plates, maps (2 fold., part. col.). 554.117/M13d

1446 **Macculloch.**
A geological classification of rocks, with descriptive synopses of the species and varieties, comprising the elements of practical geology. London, 1821.
655p. 552/M13g

1447 **Macedo, Francisco de,** d. 1681.
Theatrum meteorologicum in quo aetherea, aerea, ignea, aquea, terrestria, subterranea, ac ex his mista meteora spectantur Authore Francisco à S. Augustina Macedo . . . Romae, Typis I. Dragondelli, 1660.
211p. x551/M15t

1448 **Maclure, William,** 1763-1864.
Observations on the geology of the United States, explanatory of a geological map, read January 20th, 1809, *in* Transactions American Philosophical Society, v.6, 1809.
p. 411-428, map. 506/APT

Translated and published in Paris (1811) with geology this time on the Volney base map. Title: Suite des observations sur la géologie des Etats-Unis servant à l'explication de la carte ci-jointe, *in* Journal de physique, de chimie, d'histoire naturelle et des arts, v.72, 1811, p. 137-165 (505/JOUR).

1449 **Maclure.**
Observations on the geology of the United States of America; with some remarks on the effect produced on the nature and fertility of soils, by the decomposition of the different classes of rocks; and an application to the fertility of every state in the Union, in reference to the accompanying geological map. Philadelphia, Printed for the author by Abraham Small . . . ; . . . sold by him and J. Melish, 1817.
[10]-127p., 2 fold. leaves of plates; col. maps (engravings).
x557.3/M226o

"Read as a memoir before the American Philosophical Society, and inserted in the 1st vol. of their Transactions, New series."

References: G. W. White, William Maclure's maps . . . in Journ. soc. bibl. nat. hist. 8 (1977), p. 268.

Library owns 3 copies. Copy 1 contains the state B map as described by White; copy 2 contains the state A map on thin paper; copy 3 contains the state C map.

1450 **Maclure.**
Essay on the formation of rocks; or, An inquiry into the probable origin of their present form and structure. New-Harmony, Indiana, Printed for the author, 1832.
53p. x552/M22e/1832

Mutilated copy.

1451 **Maclure.**
Essay on the formation of rocks; or, An inquiry into the probable origin of their present form and structure. Philadelphia, Printed by J. Wilbank, 1838.
32p. 552/M22e

1452 **Macquart, Louis Charles Henri,** 1745-1818.
Essais, ou, Recueil de mémoires sur plusieurs points de minéralogie, avec la description des pièces déposées chez le roi, la figure, & l'analyse chimique de celles qui sont les plus intéressantes & la topographie de Moscou. Après un voyage fait au nord par ordre du gouvernement. Paris, Cuchet, 1789.
580p., 7 plates. 549/M24e

1453 **Macrinus, Joseph.**
. . . De Vesuvio. Item Poetica opuscula eiusdem. Neapoli, Typis Hieronymi Fasuli, 1693.
156p. 551.21/M248d

1454 **Maffei, Giovanni Camillo.**
Scala naturale, ovvero fantasia dolcissima intorno alle cose occulte, e desiderate nella filosofia. Venezia, Lucio Spineda, 1601.
126p. uncat.

This work mentions the origin of stones, of valleys, of caves and mountains, of springs and rivers, of volcanoes, and of the salinity of the ocean.

1455 **Magnus, Olaus,** Abp. of Uppsala, 1490-1558.
Historiae . . . variis conditionibus statibusque, & de morum, rituum, superstitionum, exercitiorum, regiminis, disciplinae, uictusque mirabili diversitate. Item de billis, structuris, instrumentisque mirabilibus. Item de mineris metallicis, & varijs animalium generibus, in illis regionibus degentium. Opus, omnibus cuiuscunque conditionis hominibus magno usui futurum, rerum cognitione delectationesque refertum, expres-

sis figuris pictis illustratum . . . Basileae, Ex officina Henric. Petrina [1567].
854p., illus., map. uncat.

1456 **Maillet, Benoît de,** 1656-1738.
See LE MASCRIER, JEAN BAPTISTE, Description de l'Egypte . . . 1735.

1457 **Maillet.**
Telliamed; ou, Entretiens d'un philosophe indien avec un missionnaire françois sur la diminution de la mer, la formation de la terre, l'origine de l'homme, &c. Mis en ordre sur les mémoires de feu M. de Maillet, par J. A. G*** [i.e. Jean Antoine Guer]. Amsterdam, L'Honoré, 1748.
2v. in 1. 500/M282t

1458 **Maillet.**
Telliamed, ou, Entretiens d'un philosophe indien avec un missionnaire françois sur la diminution de la mer, la formation de la terre, l'origine de l'homme, &c. Mis en ordre sur les mémoires de feu M. de Maillet, par J. A. G***. À Basle, Chez les Libraires associés, 1749.
407p. x500/M282t/1749

1459 **Maillet.**
Telliamed; ou, Entretiens d'un philosophe indien avec un missionnaire françois sur la diminution de la mer. Nouvelle édition, revue, corrigée & augmentée sur les originaux de l'auteur. Avec une vie de M. de Maillet. La Haye, P. Gosse, 1755.
2v. 500/M282t/1755a

1460 **Maillet.**
Telliamed; or, Discourses between an Indian philosopher and a French missionary, on the diminution of the sea, the formation of the earth, the origin of men and animals, and other curious subjects relating to natural history and philosophy. Being a translation of the French original of Mr. Maillet . . . London, Printed for J. Loyseau, 1750.
284p. x500/M282tE/1750

1461 **Maillet.**
Telliamed; or the world explain'd: containing discourses between an Indian philosopher and a missionary, on the diminution of the sea, the formation of the earth, the origin of men & animals and other singular subjects relating to Natural History & Philosophy. A very curious work. Baltimore: Printed

LE MOYEN DE DEVENIR RICHE, ET LA MANIERE VERITABLE,

par laquelle tous les hommes de la France pourront apprendre à multiplier & augmenter leurs threſors & poſſeſsions.

AVEC PLVSIEVRS AVTRES excellens ſecrets des choſes naturelles, deſquels iuſques à preſent l'on n'a ouy parler.

Par Maiſtre BERNARD PALISSY de Xaintes Ouurier de terre & Inuenteur des Ruſtiques Figulines du Roy.

A PARIS,
Chez ROBERT FOÜET, ruë S. Iacques à l'Occaſion deuant les Mathurins.

M. DC. XXXVI.
AVEC PRIVILEGE DV ROY.

Cat. No. 1718

by W. Pechin, No. 15, Market-street for D. Porter, at the Observatory, Federal-Hill, 1797.
268p. uncat.

1462 **Maine. Geological Survey.**
. . . First report on the geology of the public lands in the state of Maine. By Charles T. Jackson. Boston, Dutton and Wentworth, printers to the state, 1837.
47p. x557/M38h/cop.1

Bound with Hitchcock, Edward, Report on the geology, mineralogy, botany, and zoology of Massachusetts . . . 1833.

1463 **Maine. Geological Survey.**
Second annual report on the geology of the public lands, belonging to the two states of Massachusetts and Maine. By C. T. Jackson. Boston, Dutton and Wentworth, state printers, 1838.
93p., illus. x557/M38h/cop.1

House doc. no. 70, Massachusetts General Court, 1838.
Printed also in Augusta, 1838.
Bound with Hitchcock, Edward, Report on the geology . . . of Massachusetts . . . 1833.

1464 **Maine. Geological Survey.**
Second [third annual] report on the geology of the State of Maine, by Charles T. Jackson. Augusta, L. Severance, printer, 1838-39.
2v. in 1, illus. 557/M28g

Third report has imprint: Augusta, Smith & Robinson, Printers to the State, 1839.

1465 **Maiolo, Simone,** 1520-1597?
Dies caniculares hoc est colloquia tria et viginti physica, nova et penitus admiranda ac summa iucunditate concinnata . . . Quibus pleraque naturae admiranda, quae aut in aethere fiunt, aut in Europa, Asia, atque Africa, quin etiam in ipso orbe novo, & apud omnes antipodes sunt, item mirabilia arte hominum confecta recesentur, ordine, quem sequens pagina tertia indicabit. Ursellis, apud Cornelium Sutorium, impensis Ioannis Theobaldi Schönwetteri, 1600.
824p. uncat.

1466 **Maiolo.**
Dies caniculares . . . hoc est colloquia physica nova & admiranda, tum lectu iucunda & suprafidem recreabilia, tum cognitu insignia & penitus necessaria. Quibus pleraque naturae admiranda, quae aut in aethere fiunt, aut in Europa, Asia,

atque Africa, quin etiam in ipso orbe novo, & apud omnes antipodes sunt, item mirabilia arte hominum confecta recensentur, ordine, quem sequens pagina 4 indicabit. Moguntiae impensis Ioannis Theobaldi Schönwetteri, 1615.
7v. in 1. Baldwin q uncat.

1467 **Mairan, Jean Jacques Dortous de,** 1678-1771.
Dissertation sur la glace, ou, Explication physique de la formation de la glace, & de ses divers phénomènes. Paris, Imprimerie royale, 1749.
384p., front., 4 fold. plates, fold. diagr. x551.31/M28d

1468 **Mairan.**
Abhandlung von dem Eisse, oder, Physikalische Erklärung der Entstehung des Eisses, und der dabey vorkommenden verschiedenen Erscheinungen. Aus dem Französischen ins Deutsche übers. Leipzig, F. Landkischens Erben, 1752.
298p., front., 4 fold. plates. 551.31/M28d9G

1469 **Maironi da Ponte, Giovanni,** 1748-1833.
Osservazioni sopra alcune particolari petrificazioni nel Monte-Misma, dipartimento di Serio. Bergamo, Tipografia di L. Sonzogni, 1812.
51p. 552/M28o

Dedication by Il professore Maironi Daponte.

1470 **Maironi da Ponte.**
Fontane intermittenti della provincia bergamasca; memoria di Gio. Maironi Daponte. Bergamo, Stamperia Mazzoleni, 1825.
33p. 551.49/M28f

1471 **Maironi da Ponte.**
Sulla geologia della provincia bergamasca; memoria di Giovanni Maironi da Ponte presentata all'Imperiale regio istituto di scienze ed arti. Bergamo, Stamperia Mazzoleni, 1825.
200p. 554.5/M28s

Bibliographical footnote.

1472 **Malesherbes, Chrétien Guillaume de Lamoignon de,** 1721-1794.
Observations sur l'Histoire naturelle, générale et particulière de Buffon et Daubenton. Paris, C. Pougens, 1798.
2v., port. 500/B86hYm

Published with an introduction and notes by Louis Paul Abelle. Cf., Cat. gén. de la Bibl. nat.

1473 **Mallet, Fredrik,** 1728-1797.
Werlds-beskrifning. [Upsala, Tryckt hos J. Edman, 1772-74].
3v. in 4, fronts., fold. plates. x910/M288w/[v.1]

Title from spine. Each vol. in the set has special t.p.

Contents. — 1. del. Allmänn eller mathematisk beskrifning om jordklotet, af F. Mallet. 1772. — 2. del. Physisk beskrifning öfver jordklotet, på T. Bergman. 2v. 1773-74. — 3. del. Om folkslagens seder och lynne, af S. Insulin, 1772.

1474 **Mallet.**
Allmänn eller mathematisk beskrifnung om jordklotet, med cosmographiska sällskapets bifall författad. Upsala, J. Edman, 1772.
411p., 7 fold. plates. x910/M288w/[v.1]

1475 **Mallet, Robert,** 1810-1881.
On a hitherto unobserved structure discovered in certain trap rocks in the county of Galway. Dublin, Printed by R. Graisberry, 1838.
7p. 554.174/M29o

1476 **Mallet.**
On the dynamics of earthquakes; being an attempt to reduce their observed phenomena to the known laws of wave motion in solids and fluids, *in* Transactions of the Royal Irish Academy, Dublin, 1848, (read in 1846).
v.21, p.51-105, 3 plates. 062/ROA

1477 **Malte-Brun, Conrad,** 1775-1826.
A system of universal geography; or, A description of all the parts of the world, on a new plan, according to the great natural divisions of the globe, accompanied with analytical, synoptical, and elementary tables. With additions and corrections by James G. Percival. Embellished with a complete atlas, and a series of beautiful engravings. Boston, S. Walker, 1834.
3v., plates, maps, diagrs. 910/M29gE/1834

1478 **Mammatt, Edward,** 1807-1860.
A collection of geological facts and practical observations, intended to elucidate the formation of the Ashby coal-field, in the parish of Ashby-de-la-Zouch and the neighboring district; being the result of forty years' experience and research. Illustrated by a map and profiles, coloured sections of the stratification and one hundred and two plates of vegetable fossils, after drawings taken from nature, by Robert

IL MARE
ESSAMINATO
DEL SIGNOR
DOMENICO PANAROLO
ROMANO
FILOSOFO, E MEDICO.

In Roma, Per il Moneta. 1656. *Con lic. de Superiori.*

Cat. No. 1726

Ironmonger. Ashby-de-la-Zouch, Printed and sold by W. Hextall; London, G. Lawford, 1834.
101p., front. (fold. map), plates (part. col.). q551.2/M31c

1479 **Manni, Domenico Maria,** 1690-1788.
Vita del letteratissimo Monsig. Niccolò Stenone de Danimarca, vescovo di Titopoli e vicario apostolico. Firenze, Stamperia de G. Vanni, 1775.
335p. B/S8263m

1480 **Mantell, Gideon Algernon,** 1790-1852.
The fossils of the South Downs; or, Illustrations of the geology of Sussex. London, L. Relfe, 1822.
327p., illus., 62 plates (part. fold. col.), col. map. q554.225/M31f

"The engravings executed by Mrs. Mantell, from drawings by the author."

1481 **Mantell.**
Notice on the iguanodon, a newly discovered fossil reptile, from the sandstone of Tilgate forest, in Sussex, *in* Philosophical Transactions of the Royal Society of London, 1825.
v.115, p.179-186, 1 plate. 506/RO/1825

1482 **Mantell.**
Illustrations of the geology of Sussex; containing a general view of the geological relations of the south-eastern part of England, with figures and descriptions of the fossils of Tilgate Forest. London, Lupton Relfe, 1827.
92p., front., plates (incl. col. map). q554.225/M31i

1483 **Mantell.**
The geology of the south-east of England. London, Longman, Rees, Orme, Brown, Green & Longman, 1833.
415p., illus., fold. col. map. 554.2/M31ge

1484 **Mantell.**
The wonders of geology; or, A familiar exposition of geological phenomena, being the substance of a course of lectures delivered at Brighton. In two volumes . . . First American, from the third London edition. Newhaven, Conn., A. H. Maltby; London, Relfe and Fletcher, 1839.
2v., fronts., illus., 10 plates (part. col.). 550/M31w/1839

1485 **Mantell.**
The wonders of geology; or A familiar exposition of geological phenomena. 8th ed., rev. and augm. by T. Rupert Jones . . . London, H. G. Bohn, 1864-66.

2v., front., illus., 6 plates (incl. map), 2 fold. tables. 550/M31

Paged continuously.

On t.-p. of v.2: 7th ed., rev. and augm. by T. Rupert Jones . . . London, Bell & Daldy, 1866.

"Obituary notice of the late Dr. G. A. Mantell": v.1, p. [vi]-xvii.

"Introduction to the American edition, by Professor Silliman": v.1, p. 1-24.

1486 **Mantell.**
The medals of creation; or First lessons in geology, and in the study of organic remains. London, H. G. Bohn, 1844.
2v., col. fronts., illus., plates. 560/M31m

"A geological excursion to Tilgate forest; A.D. 2000. By Thomas Hood": p. 982-985.

1487 **Mantell.**
The medals of creation; or, First lessons in geology, and in the study of organic remains. Second edition, entirely re-written. London, H. G. Bohn, [1853].
2v., col. fronts., illus., plates. 560/M31m2

1488 **Mantell.**
Thoughts on a pebble, or: A first lesson in geology. 8th ed., with thirty-two illustrations. London, Reeve, Benham and Reeve, 1849.
102p., illus. 550/M31t/1849

1489 **Mantell.**
See BARBER, THOMAS, Barber's picturesque guide to the Isle of Wight . . . 1850.

1490 **Mantell.**
A pictorial atlas of fossil remains, consisting of coloured illustrations selected from Parkinson's "Organic remains of a former world," and Artis's "Antediluvian phytology." With descriptions by Gideon Algernon Mantell . . . London, H. G. Bohn, 1850.
207p., front., illus., 74 col. plates. q560/M31p

1491 **Mantell.**
Petrifactions and their teachings; or, A hand-book to the gallery of organic remains of the British museum. London, H. G. Bohn, 1851.
496p., illus., diagrs. 560/M31pe

1492 **Mantell.**
Geological excursions round the Isle of Wight, and along the adjacent coast of Dorsetshire; illustrative of the most inter-

esting geological phenomena and organic remains. 3d ed. London, H. G. Bohn, 1854.
356p., illus., plates, diagrs., fold. map. 554.2/M31g

Bibliography: p. 300-304.

1493 **Maravigna, Carmelo,** 1782-1851.
Istoria dell'incendio dell'Etna del mese di maggio 1819. Catania, Torchi della R. Università, F. Pastore, tipografo, 1819.
102p., 2 plates. 551.21/M32i

1494 **Maravigna.**
Mémoires pour servir à l'histoire naturelle de la Sicile. Paris, Chez J. B. Baillière, libraire de l'Académie royale de médecine, 1838.
86p., 6 fold. plates. 570.9458/M33m

Contents. — 1. Abrégé d'orictognosie etnéenne; ou, Description abrégée des minéraux simples qui se trouvent à Etna, avec l'indication des principales roches auxquelles ils appartiennent. — 2. Monographie des formes diverses qu'offre le souffre cristallisé en Sicile. Rapport à l'Institut de France, par M. Cordier. — 3. Monographie de la célestine de la Sicile. — 4. Catalogue méthodique des mollusques qu'on trouve en Sicile. — 5. Examen sur la question suivante: "Peut-on prouver les rapports qui existent entre le basalte et la téphrine d'après l'inspection de leurs caractères orictognostiques, etc."

1495 **Marbode,** fl. 1067-1101.
De lapidibus pretiosis Encheridion, cum scholijs pictorij villingensis. Eiusdem pictorii de lapide molari carmen. 1531.
55p. uncat.

1496 **Marcellini, Silvestro.**
Trattato compendioso orittologico. Camerino, Dalli Torchi Goriani, 1801.
278p. 549/M665t

1497 **Mariotte, Edmé,** ca. 1620-1684.
Oeuvres de M. Mariotte, de l'Académie Royale des Sciences; comprenant tous les traitez de cet auteur, tant ceux qui avoient déja paru séparément, que ceux qui n'avoient pas encore été publiés; imprimées sur les exemplaires les plus exacts & les plus complets; revûes & corrigées de nouveau. Nouvelle édition. La Haye, Jean Neaulme, 1740.
2v., 26 plates. 530/M330/1740

1498 **Mariotte.**
Traité du mouvement des eaux et des autres corps fluides. Divisé en v. parties. Paris, Estienne Michallet, 1686.
408p., illus. uncat.

RAISONNEMENS
PHILOSOPHIQVES

Touchant la Salûre, Flux & reflux de la Mer, & l'origine des Sources, tant des Fleuues que des Fontaines.

Par NICOLAS PAPIN, Medecin de la Ville de Blois.

Auſquels eſt adiouſté vn Traicté de la Lumiere de la Mer, Composé par le meſme Autheur.

A BLOIS,
Par FRANÇOIS DE LA SAVGERE,
Imprimeur du Roy, de ſon Alteſſe
Royalle, & de la Ville, 1647.

Cat. No. 1728

1499 **Mariotte.**
Traité du mouvement des eaux et des autres corps fluides . . . Par feu M. Mariotte . . . Mis en lumière par les soins de M. de La Hire . . . Nouvelle éd. corrigée. À Paris, Chez J. Jombert, 1700.
390p., diagrs. x532.5/M334t/1700

1500 **Mariotte.**
The motion of water, and other fluids. Being a treatise of hydrostaticks. Written originally in French, by the late Monsieur Mariotte . . . Divided into five parts, and translated into English. Together with a little treatise of the same author, giving practical rules for fountains, or jets d'eau. By J. T. Desaguliers . . . By whom are added, several annotations for explaining the doubtful places. London, Printed for J. Senex and W. Taylor, 1718.
290p., 17 fold. plates. x532.5/M335tEd/1718

1501 **Mariotte.**
Tractatus de motu aquarum aliorumque corporum fluidorum. Viennae, Typis J. J. Kürner, 1728-29.
2v. in 1, 9 plates. 532.5/M335tL

Translation of Traité du mouvement des eaux et des autres corps fluides.

Vol.2 has imprint: Viennae, Typis M. T. Voigtin, 1729.

1502 **Marmocchi, Francesco Costantino,** 1805-1858.
Prodromo della storia naturale, generale e comparata d'Italia. Firenze, Società editrice fiorentina, 1844.
2v., fold. map. 551.4/M345p

Half-title: Biblioteca dell'italiano. v.1, pt. 1-2.
Paged continuously.

1503 **Marsigli, Luigi Ferdinando,** 1658-1730.
Brieve ristretto del saggio fisico intorno alla storia del mare; scritto alla Regia Accademia delle scienze di Parigi. Ora esposto in una lettera all'eccellentiss. signor Cristino Martinelli, nobile veneto. Venezia, Presso A. Poletti, 1711.
72p., 3 col. plates. q551.46/M35b

Half title: Osservazioni naturali intorno al mare, ed alla grana detta kermes.

1504 **Marsigli.**
Histoire physique de la mer. Ouvrage enrichi de figures dessinées d'après le naturel. Amsterdam, Aux dépens de la Compagnie, 1725.

173p., front., 15 plates, 2 fold. maps, profiles, 6 tables. xq551.46/M35h

Latin and French in parallel columns.

In four parts; a projected fifth part upon fishes and insects was not published.

1505 **Marsigli.**

Danubius Pannonicus-Mysicus, observationibus geographicis, astronomicis, hydrographicis, historicis, physicis perlustratus et in sex tomus digestus . . . Hagae Comitum, Apud P. Gosse et al.; Amstelodami, Apud H. Uytwerf & F. Changuion, 1726.

6v. in 3, illus. xf914.39/M36d

Library owns v.3 & 4 (in 1); v.3, 35 plates; v.4, 32 plates.

Each vol. has separate t.p. and paging.

In v.3, most plates are numbered in Roman numerals, some in Arabic. In v.4, all plates are numbered in Arabic numerals.

Set imperfect: 6 fronts. wanting.

1506 **Märter, Franz Joseph.**

[Letters], *in* Physikalische Arbeiten der Einträchtigen Freunde in Wien. Edited by Ignaz von Born. Wien, Christian Friederich Wappler, 1785.

v.1, pt. 3, p.53-65, 66-67, 82-87; pt. 4, p.20-39. x500/PS692

Parts 3 & 4 of the Physikalische Arbeiten include a detailed account of the voyage to America of a Professor Märter, who describes, among other geological features, the rocks near Philadelphia and New Jersey.

1507 **Martin, Benjamin,** 1704-1782.

The philosophical grammar; being a view of the present state of experimental physiology, or natural philosophy. . . 3d ed., corr. and improved. London, Printed for J. Noon, 1748.

362p., 25 fold. plates (incl. maps), fold. table. x500/M363ph/1748

Contents. — pt.I. Somatology — pt.II. Cosmology. — pt.III. Aerology. — pt.IV. Geology.

1508 **Martin.**

The philosophical grammar; being a view of the present state of experimental physiology or natural philosophy. 7th ed. with alterations, corrections, and very large additions by way of notes. London, Printed for J. and F. Rivington, 1769.

362p., 26 fold. plates (incl. maps), 2 fold. tables. x500/M363ph/1769

Contents. — pt.I. Somatology. — pt.II. Cosmology. — pt.III. Aerology. — pt.IV. Geology.

1509 **Martin, Martin,** d. 1719.

A description of the Western Islands of Scotland, containing a full account of their situation, extent, soils, product, har-

bours, bays, tides, anchoring places, and fisheries . . . a brief description of the Isles of Orkney and Schetland. Second edition, very much corrected. London, Andrew Bell, 1716.
392p. uncat.

1510 **Martin, Peter John,** 1786-1860.
A geological memoir on a part of western Sussex; with some observations upon chalk-basins, the weald-denudation, and outliers-by-protrusion. London, Sold by J. Booth, 1828.
100p., 2 col. maps (1 fold.), 2 col. profiles, fold. table.
q554.225/M36g

1511 **Martin, William,** 1767-1810.
Figures and descriptions of petrifactions, collected in Derbyshire: to which are added, a systematical list of the minerals, which have been found constituting the substance of extraneous fossils in that country, and a brief introduction to the knowledge of petrifactions in general. Wigan, printed for the author, by Lyon and Atkinson, 1793.
[52]p., [2] folded leaves, 28 leaves of plates, col. illus. x560/M36f

Includes on p.[52] a description of missing plate 29; ms. note on p.[52]: Not published. Some plates bound out of sequence.

"The plates and descriptions now offered to the public, formed part of a general History of Derbyshire fossils in which I was engaged a few years since, in conjunction with Mr. Watson of Bakewell. Unforseen circumstances at first retarded, and have, at length prevented (at least for the present) the publication of the original work." —Pref., p.iii.

A rev. and expanded version, with 52 plates and with no reference to Watson, was published in 1809 under title: Petrificata derbiensia.

Laid in: typescript (2 leaves) on the publication of this work, by Trevor Ford, with a few notes by G. W. White.

References: Watson, White. A delineation of the strata of Derbyshire. 1811. p.ii-iii (2nd group). An Account of the life and writings of the late William Martin, in the Monthly magazine, or, British register. v.32, p.556-565.

1512 **Martin.**
Outlines of an attempt to establish a knowledge of extraneous fossils on scientific principles; in two parts. Macclesfield, Printed by J. Wilson, 1809.
250p. 560/M36o

1513 **Martin.**
Petrificata derbiensia; or, Figures and descriptions of petrifactions collected in Derbyshire. [vol. I]. Wigan, Printed by D. Lyon, 1809.

FLETA MINOR.

THE

LAVVS

OF

ART and NATURE,

IN

Knowing, Judging, Aſſaying, Fining, Refining and Inlarging the BODIES of confin'd

METALS.

In Two Parts.

The *Firſt* contains *ASSAYS* of *Lazarus Erckern*, Chief *Prover* (or *Aſſay-Maſter General* of the Empire of *Germany*) in V. BOOKS: originally written by him in the *Teutonick* Language, and now tranſlated into *Engliſh*.

The *Second* contains *ESSAYS* on *Metallick Words*, as a DICTIONARY to many pleaſing DISCOURSES.

By Sir *John Pettus*, of *Suffolk*, K^t. Of the *Society* for the *MINES ROYAL*.

Illuſtrated with 44 *Sculptures*.

Mal. 3. 3. *Numb.* 31. 31.
Jehovah Chimiſta Supremus.
Carolus D. G. Secundus.

LONDON,
Printed for and ſold by *Stephen Bateman* at the Sign of the Bible over againſt *Furnivals*-Inn Gate in *Holbourn*. MDCLXXXVI.

Cat. No. 1755

[100], 28p., 52 plates (partly col.). 560/M36p

"No more published." — Brit. Mus. Catalogue.

1514 **Martinenghi, Giovanni.**
Distribuzione ragionata del Museo mineralogico dell'Università di Pavia; opera del citt. Gio. Martinenghi dedicata alla nazione bresciana. Pavia, Stamperia Bolzani, 1801.
142p., fold. plan. 549/P28d

Supplimento di Giovanni Martinenghi alla sua Distribuzione ragionata . . . ovvero, Adattazione della nuova nomenclatura del chiariss. Ab. Haüy ai pezzi del medesimo, coll'aggiunta in fine d'un Progetto dello stesso Martinenghi circa il metodo . . . diffondere nella Rep. italiana il genio per la più utile coltivazione delle miniere. Pavia, Tipografia di G. Capelli, 1802.
58p. 549/P28d

The library owns another copy without the "Supplimento" (549/P28d/cop.2).

1515 **Martinenghi.**
Lettera di Giovanni Martinenghi . . . a Monsieur I. A. S., in cui egli I. si difende contro gli attachi dell'autore anonimo dell'Ombra vendicata di Lazzaro Spallanzani. II. Difende l'onore del suo maestro il celebre Werner, la cui fama il medesimo anonimo cerca diminuire. III. Difende la sua attuale sistemazione del Museo mineralogico de Pavia contro le dicerìe di alcuni spiriti amanti di novità. IV. Dà un'idea ragionata d'una sua nuova distribuzione de'corpi fossili, in cui combina le nuove teorie chimiche francesi coi principj mineralogici dei tedeschi. Pavia, Tipografia Capelli, 1803.
54p. 549/M36ℓ

1516 **Marzari-Pencati, Giuseppe,** 1779-1836.
Corsa pel bacino del Rodano e per la Liguria d'occidente divisa in sei sezioni, di cui la principale, cioè quella che diede motivo all'opera contiene la orittografia del monte Coiron, situato nella dianzi provincia vivarese, ora dipartimento de l'Ardeche . . . Vicenza, Paroni, 1806.
174p., fold. plates. 554.482/M36c

1517 **Mather, William Williams,** 1804-1859.
Elements of geology, for the use of schools. Norwich, [Conn.], W. Lester, jr., printed by J. Dunham, 1833.
139p., illus. x550/M421e

1518 **Mather.**
See OHIO. GEOLOGICAL SURVEY, First and second annual report . . . 1838.

1519 **Mather.**
Geology of New York, Part 1, comprising the geology of the First Geological District. Albany, Carroll and Cook, 1843.
653p., 46 plates (incl. geol. maps, sections, fossils).
q508.742/N48/v.4, pt.1

Natural History of New York, v.4, pt.1.

1520 **Maupertuis, Pierre Louis Moreau de,** 1698-1759.
See KEILL, JOHN, An examination of Dr. Burnet's theory... with a summary of exposition of the Cartesian and Newtonian systems [by] Mons. de Maupertuis ... 1734.

1521 **Maupertuis.**
La figure de la terre, déterminée par les observations de Messieurs de Maupertuis, Clairaut, Camus, Le Monnier... et de M. l'Abbé Outhier... accompagnés de M. Celsius ... faites par ordre du roi au cercle polaire. Amsterdam, J. Catuffe, 1738.
216p., 10 plates (incl. map). 526/M44f/1738

1522 **Maupertuis.**
Examen desinteressé des différens ouvrages qui ont été faits pour déterminer la figure de la terre. 2.éd., augm. de l'histoire du livre. A Amsterdam, 1741.
160p. 526.1/M44e/1741

1523 **Maupertuis.**
Eléments de géographie. Nouv. ed. Paris, G. Martin, 1742.
109p., tables. 525.14/M45e/1742

1524 **Mauro, of Florence, Servite,** 1495-1556.
Annotationi sopra la lettione della spera del Sacro Bosco. Dove si dichiarano tutti e principii mathematici & naturali, che in quella si possan' desiderare. Con alcune quistioni notabili a detta spera necessarie, & altri notandi & rari segreti, che in quella son ascosti. Con le infrascritte cose, cio, e Una nuova & fedele (ad verbum) traduttione de detta spera. Una spera theologica divina & christiana. Una spera platonica, con alcune eccitationi mathematiche, theologiche & divine. Una nuova inventione, & astronomico instrumento ... [Firenze, 1550].
219p., illus., diagrs. x523/Sal4sYm/1550

1525 **Maximilian, Fridrich Ludewig Ferdinand.**
Dissertatio inauguralis de alcali vegetabili quatenus sit eductum et productum. Halae ad Salam, Stanno Beyeriano, 1775.

23p. 549/F59

Bound with Flade, Christian Gottlob, De re Metallica . . . [1791].

1526 **Mawe, John,** 1764-1829.
The mineralogy of Derbyshire; with a description of the most interesting mines in the north of England, in Scotland, and in Wales; and an analysis of Mr. William's work, intitled "The mineral kingdom." Subjoined is a glossary of the terms and phrases used by miners in Derbyshire. London, Printed and sold by W. Phillips, 1802.
211p., 2 plates, map. 549.942/M44m

1527 **Mawe.**
A new descriptive catalogue of minerals, consisting of more varieties than heretofore published, and intended for the use of students, with which they may arrange the specimens they collect. 3d. ed., entirely rewritten, and generally following the system of Werner, with an explanation of peculiar phrases, and a description of the blow-pipe and lapidaries' apparatus. London, Printed for and sold by the author, and Longman, Hurst, Rees, Orme, and Brown, 1818.
96p., front. 549/M441n/1818

1528 **Mawe.**
Familiar lessons on mineralogy and geology; explaining the easiest methods of discriminating minerals, and the earthy substances, generally called rocks, which compose the primitive, secondary, floetz or flat, and alluvial formations: to which is added, a description of the lapidaries' apparatus, etc. 2d ed. London, Printed for, and sold by, the author, 1820.
96p., illus., col. plates. 549/M441f

1529 **Mayr, Christoph.**
See WERNER, ABRAHAM GOTTLOB, Herrn Bergrath Werner's Leztes Mineral-System . . . 1820.

1530 **Meade, William.**
An experimental enquiry into the chemical properties and medicinal qualities of the principal mineral waters of Ballston and Saratoga, in the state of New-York. With directions for the use of those waters in the various diseases to which they are applicable; and observations on diet and regimen. To which is added an appendix, containing a chemical analysis of the Lebanon spring in the state of New-York. Philadelphia, H. Hall, 1817.
195p., illus., plates. x615.79/M46e

THE

Natural Hiſtory

OF

STAFFORD-SHIRE.

BY

ROBERT PLOT. LLD.

Keeper of the

ASHMOLEAN MUSÆUM

And

PROFESSOR of CHYMISTRY

in the

UNIVERSITY

of

OXFORD.

Ye ſhall Deſcribe *the Land, and bring the* Deſcription *hither to Me.* Joſhua 8. *v.* 6.

OXFORD

Printed at the THEATER, *Anno* M. DC. LXXXVI.

Cat. No. 1804

1531 **Mease, James,** 1771-1846.
See SMITH, THOMAS, The wonders of nature and art . . . Rev., corr., and improved by James Mease . . . 1806-07.

1532 **Mease.**
A geological account of the United States; comprehending a short description of their animal, vegetable and mineral productions, antiquities, and curiosities. Philadelphia, Birch & Small, R. Carr, Printer, 1807.
496p., illus. 570.973/M46g and uncat.

1533 **Mecatti, Giuseppe Maria.**
Racconto storico-filosofico del Vesuvio e particolarmente de quanto è occorso in quest' ultima eruzione principiata il dì 25.ottobre 1751, e cessata il dì 25.febbrajo 1752, al luogo detto l'Atrio del Cavallo. Napoli, G. di Simone, 1752.
5 fold. plates. 551.21/M46r

1534 **Megenberg, Konrad von,** 14th cent.
Naturbuch, von nutz, eigenschafft wunderwirckung und Gebrauch aller Geschöpff, element un Creaturn. Dem menschen zu güt beschaffen. Nit allein den ärtzten und kunstliebern. Sonder einem ieden hauszvatter in seinem hause nützlich und lustig zu haben, zulesen und zu wissen. Franckfurt am Meyn, Bei L. Egenolff, [1540].
66p., illus. xq831/K833/Ob1540

The ninth book is on precious stones and stones.

1535 **Mella. Commissione di Beneficenza ai Danneggiati.**
Delle inondazioni del Mella e de'suoi confluenti nella notte del 14 al 15 agosto 1850; memoria storica e osservazioni geologiche. Brescia, Pietro di Lorenzo Gilberti, 1851.
60p., front. 551.48/C73d

1536 **Melle, Jacob von,** 1659-1743.
De lapidibus figuratis agri litorisque lubecensis, ad v. cl. Josephum Monti, commentatio epistolica. Lubecae, Typis Struckianis, 1720.
44p., 4 plates. 560.9431/M487d

1537 **Melleville, Maximilien,** 1807-1872.
Du diluvium; recherches sur les dépôts auxquels on doit donner ce nom, et sur la cause qui les a produits. Paris, Langlois et Leclercq, 1842.
86p. 551.31/M48d

1538 Memoirs concerning Herculaneum, the subterranean city,

lately discovered at the foot of Mount Vesuvius, giving a particular account of the most remarkable buildings, statues, paintings, medals, and other curiosities found there to the present time. Tr. from the Italian of a memorial drawn by the secretary of the Marquis d'Hospital, . . . and illustrated with notes, by William Fordyce . . . To which are added an extract from Xiphilinus's Epitome of Dion and Pliny the younger's two letters, giving an account of the overthrow of this city, and the eruption of Vesuvius that occasioned it. London, Printed for D. Wilson, 1750.
68p. x913.377/M519

1539 **Mendes da Costa, Emanuel,** b. 1717.
A natural history of fossils. Vol. I, pt. 1. London, Printed for L. Davis and C. Reymers, 1757.
294p., plates. x560/M52n

No more published.

1540 **Mendes da Costa.**
Elements of Conchology: or An Introduction to the knowledge of shells. London, Printed for B. White, 1776.
318p., 2 fold. tables, 7 fold. plates. x594/M52e

1541 **Mercati, Michele,** 1541-1593.
Metallotheca. Opus posthumum, auctoritate, & munificentiâ Clementis Undecimi Pontificis Maximi. E' tenebris in lucem eductum; opera autem, & studio Joannis Mariae Lancisii, archiatri pontificii illustratum. Cui accessit appendix cum XIX. recens inventis iconibus. Romae, Apud J. M. Salvioni Typographum Vaticanum, 1719.
378, 53p., illus. q553/M53m

Colophon dated 1717.
Half title: Metallotheca Vaticana.
Appendix has special t.p. and separate paging.

1542 **Merian, Peter,** 1795-1883.
Beiträge zur Geognosie. Basel, In der Schweighauser'schen Buchhandlung, 1821-1831.
2v. in 1, 3 col. illus. x551.13/M54b

Contents. — Bd.1. Uebersicht der Beschaffenheit der Gebirgsbildungen in den Umgebungen von Basel. — Bd.2. Geognostische Uebersicht des südlichen Schwarzwaldes.

1543 **Merian.**
Ueber die Theorie der Gletscher. [Basel, Berichte Verlag, 1843].
52p. 551.31/M54u

1544 **Merz, Karl Friedrich,** d. 1813, jt. author.
See LEONHARD, KARL CÄSAR VON, Systematisch-tabellarische . . . 1806.

1545 **Metz, G. H.**
Handbuch zur allgemeinen, das ist mathematischen und physikalischen Erdbeschreibung. Leipzig, J. S. Heinsius, 1799.
2v. in 1, 8 plates. 551.4/M57h

1546 **Michelin, Hardouin,** 1786-1867.
Iconographie zoophytologique, description par localités et terrains des polypiers fossiles de France et pays environnants. Accompagnée de figures lithographiées par Ludovic Michelin et Delarue fils. Paris, P. Bertrand, 1840-47.
2v., 348p., and atlas (79 plates). uncat.

1547 **Michell, John,** 1724?-1793.
Conjectures concerning the cause, and observations upon the phenomena of earthquakes; particularly of that great earthquake of the first of november, 1755, which proved so fatal to the city of Lisbon, and whose effects were felt as far as Africa, and more or less throughout almost all Europe, *in* Philosophical Transactions of the Royal Society, London, 1761.
v.51, pt.II, p.566-634. 506/RO

1548 **Michelotti, Giovanni,** 1812-1898.
Specimen zoophytologiae diluvianae. Aug. Taurinorum, Heredes S. Botta, 1838.
227p., 7 leaves of plates. 593.5/M582s

From the library of Louis Agassiz.

1549 **Michelotti.**
Description des fossiles des terrains miocènes de l'Italie septentrionale. Leide: A. Arnz, 1847.
408p., 17 leaves of plates, illus. q551.781/M582d

1550 **Michigan. Geological Survey.**
Annual report of the state geologist. [1839-40. Detroit, 1840-41].
124p. x557/M58a/1839

Fold. map in 1839 report.
The State Geologist was Douglas Houghton.

1551 **Milano, Michele,** 1778-1843.
Cenni geologici sulla provincia di terra d'Otranto. Livorno, Presso G. Masi, 1820.
56p. 554.5/M58c

IOANNIS QVIRINI

De Teſtaceis Foſſilibus Muſæi Septalliani

ET

IACOBI GRANDII

De Veritate Diluuij Vniuerſalis, & Teſtaceorum, quæ procul à Mari reperiuntur Generatione.

EPISTOLÆ.

VENETIIS, Typis Valuaſenſis, M.DC LXXVI.

SVPERIORVM PERMISSV.

Cat. No. 1823

1552 **Miller, Hugh,** 1802-1856.
Essays, historical & biographical, political, social, literary & scientific, edited by Peter Bayne. Boston, 1866.
501p. 824/M61

1553 **Miller.**
[Works]. Boston, Gould and Lincoln, 1869-70.
10v. 828/M61

1554 **Miller.**
Leading articles on various subjects. Edited by his son-in-law, The Rev. John Davidson. Edinburgh, W. P. Nimmo, 1870.
453p., port. 828/M61ℓ

1555 **Miller.**
The old red sandstone; or, New walks in an old field. Edinburgh, J. Johnstone, 1841.
275p., fold. col. front., 9 plates. 551.7/M61o/1841

1556 **Miller.**
The old red sandstone; or, New walks in an old field. Illustrated with numerous engravings. From the fourth London edition. Boston, Gould and Lincoln, 1851.
288 (i.e. 264)p., col. fold. front., 10 plates. 551.7/M61o/1851

1557 **Miller.**
The old red sandstone; or, New walks in an old field. To which is appended a series of geological papers, read before the Royal physical society of Edinburg; a new, improved, and enl. ed. Boston, Gould and Lincoln, [1858].
403p., illus. 551.7/M61o/1858

Last page misnumbered as 427.

1558 **Miller.**
The old red sandstone; or, New walks in an old field. To which is appended a series of geological papers, read before the Royal Society of Edinbugh. [7th ed.] Edinburgh, A. and C. Black, 1861.
385p., illus., 14 plates, fold. col. diagrs. 551.7/M61o/1861

1559 **Miller.**
First impressions of England and its people. London, J. Johnstone, 1847.
407p. 914.2/M61f/1847

1560 **Miller.**
First impressions of England and its people. Boston, Gould & Lincoln, 1866.
430p., port. 914.2/M61

1561 **Miller.**
The foot-prints of the Creator; or, The Asterolepis of Stromness. From the 3d London ed. With a memoir of the author, by Louis Agassiz. Boston, Gould and Lincoln, 1850.
[25]-337p., illus. 560/M61f/1850

1562 **Miller.**
The foot-prints of the Creator; or, the Asterolepis of Stromness. With a sketch of the author's life and an estimate of his writings by Sir David Brewster, Bart. From the 3d London ed. Cincinnati, W. H. Moore, 1851.
308p., illus. 560/M61f/1851

1563 **Miller.**
The foot-prints of the Creator: or, The Asterolepis of Stromness. From the 3d London ed. With a memoir of the author, by Louis Agassiz. Boston, Gould and Lincoln, 1866.
[25]-337p., illus., port. 560/M61f/1866

1564 **Miller.**
An autobiography. My schools and schoolmasters; or The story of my education. Boston, Gould and Lincoln, 1854.
537p., port. B/M648m1/1854

1565 **Miller.**
An autobiography. My schools and schoolmasters; or The story of my education. Boston, Gould and Lincoln, 1860.
537p., port. B/M648m1/1860

1566 **Miller.**
The testimony of the rocks; or, Geology in its bearings on the two theologies, natural and revealed. Edinburgh, Shepherd & Elliot, 1857.
500p., illus., port. 550.1/M61/1857

Also a second copy published in Boston by Gould and Lincoln.

1567 **Miller.**
The testimony of the rocks; or, Geology in its bearings on the two theologies, natural and revealed. With memorials of the death and character of the author. Boston, Gould and Lincoln, 1860.
502p., illus. 550.1/M61

1568 **Miller.**
The testimony of the rocks; or, Geology in its bearings on the two theologies, natural and revealed. Edinburgh, A. and C. Black, 1862.
454p., illus. 550.1/M61/1862

1569 **Miller.**
The testimony of the rocks; or, Geology in its bearings on the two theologies, natural and revealed . . . with memorials of the death and character of the author. Boston, Gould and Lincoln, 1867.
502p., illus. 550.1/M61

1570 **Miller.**
The cruise of the Betsey; or, A summer ramble among the fossiliferous deposits of the Hebrides. With rambles of a geologist; or, Ten thousand miles over the fossiliferous deposits of Scotland. Edinburgh, T. Constable, 1858.
486p. 554.1/M61/1858

Edited by W. S. Symonds.

1571 **Miller.**
The cruise of the Betsey; or, A summer ramble among the fossiliferous deposits of the Hebrides. With rambles of a geologist; or, Ten thousand miles over the fossiliferous deposits of Scotland. Boston, Gould and Lincoln, 1860.
524p. 554.1/M61

1572 **Miller.**
The cruise of the Betsey; or, A summer ramble among the fossiliferous deposits of the Hebrides. With rambles of a geologist; or, Ten thousand miles over the fossiliferous deposits of Scotland. Boston, Gould and Lincoln, 1865.
524p. 554.1/M61

1573 **Miller.**
The cruise of the Betsey; or, A summer ramble among the fossiliferous deposits of the Hebrides. With rambles of a geologist; or, Ten thousand miles over the fossiliferous deposits of Scotland. New York, Hurst, [1899].
524p. 554.1/M61/1899

1574 **Miller.**
Sketch-book of popular geology; being a series of lectures delivered before the Philosophical Institution of Edinburgh. With an introductory preface, giving a résumé of the progress of geological science within the last two years. By Mrs. Miller. Edinburgh, T. Constable, 1859.
358p. 550.4/M61s/1859

1575 **Miller.**
Popular geology: a series of lectures read before the Philosophical institution of Edinburgh, with descriptive sketches

DE FONTIVM MVTINENSIVM ADMIRANDA SCATVRIGINE TRACTATVS PHYSICO-HYDROSTATICVS

BERNARDINI RAMAZZINI

In Mutinensi Lyceo Medicinæ Professoris.

AD SERENISSIMVM

FRANCISCVM II.

MVTINÆ, REGII, &c.

DVCEM X.

MVTINAE,

Typis Hæredum Suliani Impressorum Ducalium. 1691.

Superiorum Permissu.

Cat. No. 1827

from a geologist's portfolio. With an introductory résumé of the progress of geological science within the last two years, by Mrs. Miller. Boston, Gould and Lincoln, 1860.
423p. 550.4/M61p/1860

1576 **Miller.**
Popular geology; a series of lectures read before the Philosophical Institution of Edinburgh with descriptive sketches from a geologist's portfolio. With an introductory résumé of the progress of geological science within the last two years by Mrs. Miller. Boston, Gould and Lincoln, 1867.
423p. 550.4/M61p

Head of title: Sketch book of popular geology.

1577 **Miller.**
Sketch-book of popular geology. 4th ed. Edinburgh, W. P. Nimmo, 1869.
356p. 550.4/M61s/1869

1578 **Miller.**
See DAVIES, THOMAS ALFRED, Answer to Hugh Miller and theoretic geologists . . . 1860.

1579 **Miller.**
Edinburgh and its neighborhood, geological and historical; with the geology of the Bass rock. 4th ed. Edinburgh, W. P. Nimmo, 1870.
344p., illus. 554.1145/M61e/1870

1580 **Miller, William Hallows,** 1801-1880.
See PHILLIPS, WILLIAM, An elementary introduction to mineralogy . . . with extensive alterations and additions, by . . . W. H. Miller . . . 1852.

1581 **Millin, Aubin Louis,** 1759-1818.
Minéralogie homérique; ou, Essai sur les minéraux, dont il est fait mention dans les poëmes d'Homère. Paris, Garnery, 1790.
118p. 549/M62m

Bound with Bernard, Mémoire sur les avantages et les inconvéniens de l'emploi du charbon, 1780; Faujas de Saint-Fond, Barthélemy, Mémoire sur la manière de reconnoitre les différentes especes de pouzzolane, 1780; and Faujas de Saint-Fond, Barthélemy, Recherches sur la pouzzolane . . . , 1778.

1582 **Millin.**
Mineralogie des Homer. Aus dem Französischen mit Anmerkungen und Berichtigungen von Friedrich Theodor Rink.

Königsberg und Leipzig, Bey Friedrich Nicolovius, 1793.
126p. 549/M62mGr

1583 **Mirabaud, Jean-Baptiste.**
See HOLBACH, PAUL HENRI THIRY D', Systême de la nature . . . 1770 and 1771.

1584 **Mitchell, Elisha,** 1793-1857.
Elements of geology, with an outline of the geology of North Carolina; for the use of the students of the University . . . [n.p.], 1842.
141p., front. (fold. col. map). uncat.

1585 **Mitchill, Samuel Latham,** 1764-1831.
A sketch of the mineralogical history of the State of New York, *in* Medical Repository (N.Y.).
v.1, 1798, p. 293-314, 445-452; v.3, 1800, p. 325-335; v.5, 1801, p. 212-214.
x610.5/MERE

1586 **Mitchill,** supposed author.
See CALDWELL, JOSEPH, A tour through part of Virginia . . . 1809.

1587 **Mitchill.**
(Circular) American zoology and geology, New York Institution, March 11, 1817. Samuel L. Mitchill, Chairman. New-York, 1817.
[4]p. x590.742747/C496

Caption title.

Summary: A circular, issued by the New York Historical Society, soliciting help in establishing a zoological museum.

This copy signed by Mitchill and addressed to Parker Cleaveland, Esq., Professor, &c, &c, Brunswicke, Maine.

Includes on p.[3]-[4] a holograph letter, dated 19th March 1817, from Mitchill to Cleaveland, concerning rock samples and fossils in Mitchill's possession.

1588 **Mitchill.**
See CUVIER, GEORGES, Baron, Essay on the theory of the earth . . . To which are now added, observations on the geology of North America; illustrated by the description of various organic remains, found in that part of the world, by Samuel L. Mitchill . . . 1818.

1589 **Mitchill.**
See PHILLIPS, WILLIAM, An elementary introduction to the knowledge of mineralogy . . . with notes and additions on American articles by Samuel L. Mitchill . . . 1818.

1590 **Mitchill.**
Catalogue of the organic remains, which, with other geological and some mineral articles, were presented to the New York lyceum of natural history, in August, 1826, by their associate, Samuel L. Mitchill . . . New York, Printed by J. Seymour, 1826.
40p. 560/M69c

1591 **Mohs, Friedrich,** 1773-1839.
Des Herrn Jac. Fried. von der Null Mineralien Kabinet, nach einem, durchaus auf äussere Kennzeichen gegründeten Systeme geordnet, beschrieben, und durch Hinzuthuung vieler, dem gegenwärtigen Zustande der Mineralogie angemessener, erläuternder Anmerkungen und nöthiger Berichtigungen als Handbuch der Oryctognosie brauchbar gemacht. Wien, Auf Kosten des Besitzers, und in Commission der Camesinaischen Buchhandlung, 1804.
3v. in 2. 549/M72h

1592 **Mohs.**
Die Charaktere der Klassen, Ordnungen, Geschlechter und Arten; oder, Die Charakteristik des Naturhistorischen Mineral-Systemes. Zweite, verbesserte auflage. Mit 3 Kupfertafeln. Dresden, Arnoldische Buchhandlung, 1821.
226p., 3 fold. plates. 549/M72c/1821

1593 **Mohs.**
The characters of the classes, orders, genera, and species; or, The characteristic of the natural history system of mineralogy. Intended to enable students to discriminate minerals on principles similar to those of botany and zoology. Edinburgh, W. and C. Tait, 1820.
109p. 549/M72cE

1594 **Mohs.**
Treatise on mineralogy; or, The natural history of the mineral kingdom. Tr. from the German, with considerable additions, by William Haidinger . . . Edinburgh; Printed for A. Constable and Co., Edinburgh, and Hurst, Robinson, and co., London, 1825.
3v., plates. 549/M72gEh

Translation of: Grundriss der Mineralogie. v.1, v.2, v.3.

1595 **Mohs.**
Leichtfassliche Anfangsgründe der Naturgeschichte des Mineralreiches. Zum Gebrauche bei seinen Vorlesungen über

Three *Physico-Theological* DISCOURSES,

CONCERNING

I. The Primitive CHAOS, and Creation of the World.

II. The General DELUGE, its Causes and Effects.

III. The Dissolution of the WORLD, and Future Conflagration.

Wherein are largely discussed,

The Production and Use of Mountains; the Original of Fountains, of Formed Stones, and Sea-Fishes Bones and Shells found in the Earth; the Effects of particular Floods, and Inundations of the Sea; the Eruptions of *Vulcano*'s; the Nature and Causes of Earthquakes.

Also an Historical Account of those Two late remarkable Ones in *Jamaica* and *England*.

With PRACTICAL INFERENCES.

By *JOHN RAY*, late Fellow of the *Royal Society*.

The Third Edition, Illustrated with Copper-Plates, and much more Enlarged than the former Editions, from the Author's own MSS.

London: Printed for WILLIAM INNYS, at the *Prince's Arms* in S. *Paul*'s Church yard, 1713.

Cat. No. 1844

die Mineralogie . . . nebst einem Anhange welcher Gleichungen zur Berechnung einfacher und zusammengesetzter Krystallgestalten und Beispiele der letztern enthält. Wien, J. B. Wallishausser, 1832.
643p., plates. 549/M72ℓ

1596 **Mohs.**
Leichtfassliche Anfangsgründe der Naturgeschichte des Mineralreiches. Zum Gebrauche bei seinen Vorlesungen über die Mineralogie. 2. verm. und verb. Aufl. . . . Wien, Gedruckt und im Verlage bei C. Gerold, 1836-39.
2v., plates. 549/M72t/1836

1597 **Mohs.**
Die ersten Begriffe der Mineralogie und Geognosie für junge praktische Bergleute der K. k. Österreichischen Staaten. Im Auftrag der K. k. Hofkammer in Münz- und Bergwesen. Herausgegeben nach seinem Tode . . . Wien, C. Gerold, 1842.
2v., 34 plates. 549/M72e

Contents. — v. 1. Mineralogie. — v. 2. Geognosie.

1598 **Moiseienko, Fedor Petrovych,** 1754-1781.
Mineralogische Abhandlung von dem Zinnsteine. Leipzig, J. G. I. Breitkopf, 1779.
91p. 549.2/M72m

1599 **Molyneux, Thomas,** 1661-1733.
See A natural history of Ireland . . . 1726.

1600 **Monardes, Nicolás,** ca. 1512-1588.
Dos libros, el uno que trata de todas las cosas que traen de nuestras Indias Occidentales, que sirven al uso de la medicina, y el otro que trata de la piedra bezaar, y de la Yerva escuerçonera. Sevilla, en casa de Hernando Diaz, 1569.
[n.p.], port. uncat.

1601 **Monardes.**
Joyfull newes out of the newe founde worlde, wherein is declared the rare and singuler vertues of diverse and sundrie hearbes, trees, oyles, plantes, and stones . . . Englished [from the Spanish] by John Frampton . . . London, Willyam Norton, 1577.
foll. 109 [in 3 pts.]. uncat.

1602 **Monardes.**
Joyfull newes out of the newe founde worlde, wherein is declared the rare and singuler vertues of diverse and sundrie

hearbes, trees, oyles, plantes, and stones... Englished by John Frampton ... London, Printed by E. Allde, by the assigne of Bonham Norton, 1596.
foll. 109 [in 3 pts.]. uncat.

Bound with three other books, including the author's, "The dialogue of iron, which treateth of the greatness thereof, and how it is the most excellent metall of all others, and the thing most necessarie for the service of man; and of the greate medicinall vertues which it hath." foll. 129-172.

1603 **Monti, Giuseppe,** 1682-1760.
De monumento diluviano nuper in agro Bononiensi detecto dissertatio. In qua permultae ipsius inundationis vindicae ... Bononiae studiorum, apud Rossi & Socios, 1719.
50p. 627.5/M76m

1604 **Monticelli, Teodoro,** 1759-1845.
Storia de'fenomeni del Vesuvio, avvenuti negli anni 1821, 1822, e parte del 1823, con osservazioni e sperimenti de T. Monticelli e N. Covelli ... letta nella Reale accademia delle scienze. Napoli, Torchi del Gabinetto bibliografico e tipografico, 1823.
208p., tables (1 fold.), 4 fold. plates. 551.21/M76s

1605 **Mont-Rond, P. E. de.**
Du Rhône et de ses affluents des Alpes. Paris, Carilian Goeury et Ve Dalmont, 1847.
252p. 551.48/M76d

1606 **Moore, Nathaniel Fish,** 1782-1872.
Ancient mineralogy; or, An inquiry respecting mineral substances mentioned by the ancients; with occasional remarks on the uses to which they were applied. New York, G. & C. Carvill, 1834.
192p. x549.09/M78a

1607 **Morand, Jean François Clément,** 1726-1784.
L'art d'exploiter les mines de charbon de terre. [Paris, Saillant et Nyon], 1768-
f622.33/M79a

Library owns v.1-v.2$^{1-2}$ in 1v. and v.2$^{3-4}$ in 1v.

1608 **Moreira de Mendonça, Joaquim José.**
Historia universal dos terremotos, que tem havido no mundo, de que ha noticia, desde a sua creação até o seculo presente. Com huma narraçam individual do terremoto do primeiro de novembro de 1755, e noticia verdadeira dos seus effeitos

em Lisboa, todo Portugal, Algarves, e mais partes da Európa, Africa, e América, aonde se estendeu: e huma dissertação phisica sobre as causas geraes dos terremotos, seus effeitos, differenças, e prognosticos; e as particulares do ultimo. Lisboa, na offic. de A. V. da Silva, 1758.
272p. 551.22/M82h

1609 **Morel de Vindé, Charles Gilbert Terray,** 1759-1842.
Des révolutions du globe; conjecture formée d'après les découvertes de Lavoisier sur la décomposition et recomposition de l'eau. Paris, Du Pont, 1797.
30p. 551/M81d

1610 **Morin, Jean Baptiste,** 1583-1656.
Nova mundi sublunaris anatomia . . . Parisiis, Apud Nicolaum Du Fosse, 1619.
144p. uncat.

1611 **Moro, Antonio Lazzaro,** 1687-1764.
De' crostacei e degli altri marini corpi che si truovano su' monti; libri due. Venezia, S. Monti, 1740.
452p., 8 fold. plates. 565.3/M82c

1612 **Moro.**
Neue Untersuchung der Veränderungen des Erdbodens, nach Anleitung der Spuren von Meerthieren und Meergewächsen, die auf Bergen, und in trockener Erde gefunden werden. Aus dem italienischen übersetzet. Leipzig, B. C. Breitkopf, 1751.
465p., 9 fold. plates. x550/M82nG

1613 **Moro.**
See ANONYMOUS, Philosophische Ergötzungen . . . 1765.

1614 **Morris, John,** 1810-1886.
A catalogue of British fossils. Comprising all the genera and species hitherto described; with references to their geological distribution and to the localities in which they have been found. London, J. Van Voorst, 1843.
222p. 560.942/M83c

1615 **Morris.**
A catalogue of British fossils; comprising all the genera and species hitherto described, with references to their geological distribution and to the localities in which they have been found. Second edition, considerably enlarged. London, Printed for and published by the author, 1854.
372p. 560.942/M83c/1854

THESAURUS
IMAGINUM
PISCIUM TESTACEORUM;

Quales sunt
Cancri, *Echini*, *Echinometra*, *Stellæ Marinæ*, &c.
Ut &

COCHLEARUM;

Inter quas numerantur
Lunares; *Laciniatæ*; *Trochi*; *Valvatæ*, sive *Semilunares*; *Valvatæ striatæ*; *Cassides tuberosæ*, *verrucosæ*, *læves* & *Murices*; *Globosæ*; *Buccina*; *Strombi*; *Volutæ*; *Alatæ*; *Porcellanæ majores* & *minores*; *Cylindri*, &c.
Quibus accedunt

CONCHYLIA,

Ut
Nautilus, *Cornu Ammonis*, &c.

CONCHÆ UNIVALVIÆ & BIVALVIÆ;

Quarum species sunt
Solenes univalvii, *Chamæ asperæ*, *Chamæ læves*, *Pectines*, *Pectunculi*, *Tellinæ*, *Solenes bivalvii*, *Musculi*, *Pinnæ*, *Ostrea*, &c.
Denique

MINERALIA;

Uti
Metalla, *Lapides* & *Argillæ*, variis in locis reperta.
Quorum omnium maximam partem

GEORGIUS EVERHARDUS RUMPHIUS M. D.

Et *Academiæ Cesareæ Naturæ Curiosorum Collega*, dictus PLINIUS INDICUS, collegit;
Jam vero Naturæ Amator & Curiosus quidam in hunc ordinem digessit,
& nitidissimè æri incidi curavit.

LUGDUNI BATAVORUM,
Apud PETRUM vander Aa, Bibliopolam. M D CC XI.

Cat. No. 1935

1616 **Morse, Jedidiah,** 1761-1826.
The American geography; or, A view of the present situation of the United States of America. Containing astronomical geography — geographical definitions, discovery, and general description . . . with a particular description of Kentucky, the western territory and Vermont . . . to which is added, an abridgment of the geography of the British, Spanish, French and Dutch dominions in America and the West Indies — of Europe, Asia and Africa. 2d ed. London, Printed for J. Stockdale, 1792.
536p., 2 fold. maps (incl. front.). 917.3/M83a

1617 **Mousson, Albert,** 1805-1890.
Geologische Skizze der Umgebungen von Baden im Canton Aargau. Zürich, Orell, Füssli und Compagnie, 1840.
122p., fold. tables, 4 fold. plates, fold. map. 554.94/M86g

1618 **Müller, Johannes,** 1801-1858.
Über die Fossilen Reste der Zeuglodonten von Nord-America mit Rücksicht auf die europäischen Reste aus dieser Familie. Berlin, G. Reimer, 1849.
38p., 27 plates (partly col.). f567.3/M91u

1619 **Murchison, Charles,** 1830-1879 ed.
See FALCONER, HUGH, Palaeontological memoirs . . . 1868.

1620 **Murchison, Roderick Impey,** 1792-1871.
The Silurian system, founded on geological researches in the counties of Salop, Hereford, Radnor, Montgomery, Caermarthen, Brecon, Pembroke, Monmouth, Gloucester, Worcester, and Stafford; with descriptions of the coalfields and overlying formations. London, J. Murray, 1839.
2v., illus., plates (partly fold., partly col.), maps, and portfolio of 2 fold. maps. xq551.73/M93si

1621 **Murchison,** jt. author.
See SEDGWICK, ADAM, On the physical structure of Devonshire . . . 1840.

1622 **Murchison,** jt. author.
See SEDGWICK, ADAM, On the distribution and classification of the older or Palaeozoic deposits of the North of Germany and Belgium . . . 1842.

1623 **Murchison.**
The geology of Russia in Europe and the Ural mountains. By Roderick Impey Murchison . . . Édouard de Verneuil . . . and

Count Alexander von Keyserling . . . London, J. Murray; [etc., etc.], 1845.
2v., front. (v.1) illus., plates, fold. maps, fold. col. profiles, diagrs. 554.7/M93g

Vol. 2 published in French with title: Géologie de la Russie d'Europe et des montagnes de l'Oural . . .
Contents. — v.1. Geology. — v.2. Paléontologie.

1624 **Murchison.**
Outline of the geology of the neighborhood of Cheltenham. A new ed., augm, and rev. by James Buckman . . . and H. E. Strickland . . . London, J. Murray; Cheltenham, H. Davies, 1854.
109p., front. (fold. map), 14 plates (part. col.). 554.241/M93o/1845

1625 **Murchison.**
Ueber den Gebirgsbau in den Alpen, Apenninen und Karpathen, namentlich um einen Uebergang aus sekundären Gebilden in tertiäre darzuthun, und über die Entwicklung eocener Ablagerungen im südlichen Europa. Bearb. von Gustav Leonhard . . . Stuttgart, J. B. Müller, 1850.
162p., fold. plates. uncat.

Bound with his Ueber die älteren vulkanischen Gebilde . . . 1851.

1626 **Murchison.**
Ueber die älteren vulkanischen Gebilde im Kirchenstaate und über die Spalten, welchen in Toscana heisse Dämpfe entsteigen, und deren Beziehungen zu alten Eruptions- und Bruch-Linien. Stuttgart, J. B. Müller's Verlagshandlung, 1851.
65p., illus. uncat.

Bound with his Ueber den Gebirgsbau in den Alpen . . . 1850.

1627 **Murchison.**
Siluria. The history of the oldest known rocks containing organic remains, with a brief sketch of the distribution of gold over the earth. London, J. Murray, 1854.
523p., illus., 37 plates, fold. maps. 551.73/M93s

1628 **Murchison.**
Siluria. The history of the oldest fossiliferous rocks and their foundations, with a brief sketch of the distribution of gold over the earth. Third edition, including "The Silurian System," with maps and many additional illustrations. London, J. Murray, 1859.
592p., 41 plates, illus., front. 551.73/M93s3

1629 **Murchison.**
Geological map of Europe exhibiting the different systems of rocks according to the most recent research and inedited materials, by Sir Roderick I. Murchison and James Nicol; constructed by A. Keith Johnston. Edinburgh, W. Blackwood & Sons and W. & A. K. Johnston, 1856.
G5700/.C5/1856/.M9

Scale, 1:4,800,000.

1630 **Murchison.**
First sketch of a new geological map of Scotland, with explanatory notes, by Sir Roderick I. Murchison and Archibald Geikie. Constructed by A. Keith Johnston, geographer to the Queen. Edinburgh, 1861.
22p., fold. col. map. G5770/.C5/1861/.M8

Scale not given.
Bibliographical footnotes.

1631 **Murchison.**
First sketch of a new geological map of Scotland, with explanatory notes, by Sir Roderick I. Murchison and Archibald Geikie. Constructed by A. Keith Johnston, Edinburgh, W. & A. K. Johnston, 1862.
23p., fold. col. map. 554.14/M93f

Scale not given.
Error in binding: p. 23 follows t.p. Murchison introduction is not dated (as in the 1861 edition).
Geological map accompanying this edition is revised to January 1865.

1632 **Murray, John,** d. 1820.
A comparative view of the Huttonian and Neptunian systems of geology: in answer to the Illustrations of the Huttonian theory of the earth, by Professor Playfair. Edinburgh, Printed for Ross and Blackwood; [etc., etc.], 1802.
256p. 551.7/M96c

1633 **Murray, John,** 1786?-1851.
A memoir on the diamond; including its economical and political history. 2d ed. London, Relfe & Fletcher, 1839.
84p., 2 plates. 553.8/M96m2

1634 **Mylius, Gottlieb Friedrich,** 1675-1726.
Memorabilium Saxonie subterraneae. Mit vielen Kupffern gezieret. Leipzig, In Verlegung des Autoris; zu finden bey F. Groschuffen, 1709-18.
2v. in 1, illus. 561/M99m

D'AMBOINSCHE RARITEITKAMER,

Behelzende eene BESCHRYVINGE van allerhande zoo weeke als harde

SCHAALVISSCHEN,

te weete raare

KRABBEN, KREEFTEN,

en diergelyke Zeedieren,

als mede allerhande

HOORNTJES en SCHULPEN,

die men in d'Amboinſche Zee vindt:

Daar benevens zommige

MINERAALEN, GESTEENTEN,

en ſoorten van AARDE, die in d'Amboinſche, en zommige omleggende Eilanden gevonden worden.

Verdeelt in drie Boeken,

En met nodige PRINTVERBEELDINGEN, alle naar 't leven getekent; voorzien

Beſchreven door

GEORGIUS EVERHARDUS RUMPHIUS;

van Hanauw, Koopman en Raad in Amboina, mitsgaders Lid van het Kyzerlyke kweekſchool der onderzoekers van de Natuurkunde in 't Duitſche Roomſche Ryk opgerecht onder den naam van

PLINIUS INDICUS.

T'AMSTERDAM,

By JAN ROMAN DE JONGE, Boekverkoper, 1741.

Cat. No. 1937

1635 A natural history of Ireland; in three parts. By several hands . . . Dublin, Printed by and for George Grierson, 1726.
213p., plates (partly fold.). x508.415/N219

Contributions mainly by Gerard Boate (1604-1650) and by Thomas Molyneux (1661-1733).

1636 **Naumann, Carl Friedrich,** 1797-1873.
Grundriss der Krystallographie. Leipzig, J. A. Barth, 1826.
408p., 3 fold. plates. 548/N22g

1637 **Naumann.**
Entwurf der Lithurgik; oder, Ökonomischen Mineralogie; ein Leitfaden für Vorlesungen. Leipzig, Bei Adolph Wienbrack, 1826.
300p. 549/N22en

1638 **Naumann.**
Lehrbuch der reinen und angewandten Krystallographie. Leipzig, F. A. Brockhaus, 1830.
2v., 905 diagrams on 39 fold. plates. 548/N22ℓ

1639 **Naumann.**
Anfangsgründe der Krystallographie. Leipzig, Arnold, 1854.
292p., 26 plates. 548/N22

1640 **Naumann.**
Elemente der theoretischen Krystallographie. Leipzig, W. Engelmann, 1856.
383p., illus. 548/N22e

1641 **Naumann.**
Lehrbuch der Geognosie. Erster-[dritter] Band . . . Zweite, vermehrte und verbesserte Auflage. Leipzig, W. Englemann, 1858-[1872].
3v. and atlas. 551/N22ℓ2

Vol. 3 ends with p.576; it was never completed, and has no title page.

1642 **Necker de Saussure, Louis Albert,** 1786-1861.
Voyage en Écosse et aux îles Hebrides, Genève, J. J. Paschoud, 1821.
3v., illus. 914.1/N28v

1643 **Necker de Saussure.**
Travels in Scotland; descriptive of the state of manners, literature, and science. Translated from the French. London, Printed for R. Phillips, 1821.
112p. uncat.

1644 **Needham, B.,** jt. author.
See OWEN, DAVID DALE, Geological reports . . . lands of D. Kingsbury . . . [1857?].

1645 **Needham,** jt. author.
See OWEN, DAVID DALE, Reports on the geology of the Tunungwant Coal Field, M'Kean County, Pa. . . . [1857].

1646 **Nesi, Lorenzo.**
Storia fisica della terra compilata sulle traccie della Geografia fisica di Kant e sulle più recenti scoperte, ed ultime transazioni politiche d'Europa . . . Milano, F. Baret, 1816-17.
2v., fold. map. 551.4/N369s

1647 **New Brunswick. Geological Survey.**
Report on the Geological survey of the province of New-Brunswick, [1838]-1842. By Abraham Gesner, F. G. S., provincial geologist, &c. Saint John, Printed by H. Chubb, 1839-43.
5v. in 1, illus., diagrs. 557.15/N42r

[1838-1841] title begins: First-fourth report. 1842 has subtitle; with a topographical account of the public lands, and the districts explored in 1842.

1648 **New Hampshire. Geological Survey.**
First annual report [1840/1] on the geology of the state of New Hampshire. By Charles T. Jackson, state geologist. Concord, N.H., C. Barton, state printer, 1841.
[5]-164p., illus. x557.42/N45

1649 **New Hampshire. Geological Survey.**
Final report on the geology and mineralogy of the state of New Hampshire; with contributions towards the improvement of agriculture and metallurgy. By Charles T. Jackson, M.D. Published by order of the Legislature. Concord, N.H., Caroll & Baker, state printers, 1844.
376p., front., illus., plates (part. fold. and col.), fold. map, tables.
557.42/N45f

1650 **New Jersey. Geological Survey.**
Report on the Geological Survey of the state of New Jersey, by Henry D. Rogers . . . Philadelphia, Desilver, Thomas & co., 1836.
174p., front. (profiles). 557/N46fr

1651 **New York. Natural History Survey.**
. . . Communication from the governor, transmitting several reports relative to the geological survey of the state [1837. Albany, 1838].

2v., illus., fold. tables and atlas of plates. 508.747/N48c

Atlas accompanies 2d report, 1837.
Library owns v.2 only.
The communication is signed W.L. Marcy. It contains the following reports: Mineralogy and Chemistry by L. D. Beck, Paleontology by T.A. Conrad, Geology of First District by W.W. Mather, Geology of Second District by E. Emmons, Geology of Third District by L. Vanuxem, Geology of Fourth District by J. Hall.

1652 **Newte, Thomas,** pseud.
See THOMSON, WILLIAM, 1746-1817.

1653 **Nicol, James,** 1801-1879.
Guide to the geology of Scotland; containing an account of the character, distribution and more interesting appearances of its rocks and minerals. Edinburgh, Oliver & Boyd; London, Simpkin, Marshall, 1844.
272p., [11] leaves of plates, illus. (some col.), folded col. map.
554.1/N54g

1654 **Nicol.**
See MURCHISON, RODERICK IMPEY, Geological map of Europe . . . 1856.

1655 **Nicollet, Joseph Nicolas,** 1768-1843.
Report intended to illustrate a map of the hydrographical basin of the Upper Mississippi River. Washington, Blair and Rives, 1843.
170p., map. x581.977/N54r

26th Cong., 2nd Sess., Senate Doc. 237, Serial 380.
Library copy lacks the original map. Report and map, as part of the Congressional Set, reproduced on microcard in Rare Book Room, and the Map and Geography Library owns a facsimile of the map (G4042/.M5/1843/.N5/1970).
The report was also printed by the House, 1845, as H.R. Doc. 52, 28th Cong., 2nd Sess. with a smaller map. Rare Book Room owns this version as Congressional Documents Serial 464.

1656 **Nicols, Thomas,** fl. 1659.
Arcula gemmea; or, A cabinet of jewels. Discovering the nature, vertue, value of pretious stones, with infallible rules to escape the deceit of all such as are adulterate and counterfeit. London, Printed for Nath. Brooke, 1653.
239., fold. tab. 549/N548ℓ/1653

Published in 1652 with title: A lapidary; or, The history of pretious stones; and in 1659 with title: Gemmarius fidelius; or, The faithful lapidary.

VOYAGES
DANS LES ALPES,
PRÉCÉDÉS

D'UN ESSAI

SUR L'HISTOIRE NATURELLE

DES ENVIRONS

DE GENEVE,

Par HORACE-BÉNEDICT DE SAUSSURE, *Professeur de Philosophie dans l'Académie de Geneve.*

TOME PREMIER.

Carl Hackert del.

Nec species sua cuique manet, rerumque novatrix,
Ex aliis alias reparat Natura figuras.

Ovid.

A NEUCHATEL,

CHEZ SAMUEL FAUCHE, IMPRIMEUR ET LIBRAIRE DU ROI.

M. DCC. LXXIX.

Cat. No. 1954

1657 **Nilsson, Sven,** 1787-1883.
Petrificata suecana formationis cretaceae, descripta et iconibus illustrata a S. Nilsson . . . Pars prior, vertebrata et mollusca sistens. Londini Gothorum, ex officina Berlingiana, 1827.
39p., 10 plates. q560.176/N59p

No more published?

1658 **Nogaret, de.**
Essais sur les montagnes. Amsterdam, 1785.
2v. 551.43/N68e

1659 **Nöggerath, Johann Jacob,** 1788-1877, ed.
Das Gebirge in Rheinland-Westphalen nach mineralogischen und chemischen Bezüge. Bonn, E. Weber, 1822-26.
4v., plates (incl. col. maps, col. profiles). 552/N68g/1822

1660 **North Carolina. Geological Survey.**
. . . Report of Professor Emmons, on his geological survey of North Carolina. Raleigh, S. Gales, printer to the legislature, 1852.
181p. 567/N81re

At head of title: Executive document no. 13.
"First annual report."
No more published.

1661 **Norwood, Joseph Granville,** 1807-1895.
Report. [Washington, Gov't. Printing Office], 1848.
p. 75-134. Meine/557/Ow2r

Bound with and part of Owen, David Dale, 1807-1860, Report of a geological reconnoissance of the Chippewa land district of Wisconsin . . . 1844.

1662 **Norwood.**
Report. [Washington, Gov't. Printing Office], 1848.
p. 75-129. x557/W75o

Bound with and part of Owen, David Dale, Letter of the Secretary of the Treasury, communicating a report of a geological reconnoissance of the Chippewa land district of Wisconsin . . . 1848.

1663 **Norwood.**
See U.S. GEOLOGICAL CORPS. Geological map of Wisconsin, Iowa and Minnesota . . . 1851.

1664 **Nose, Karl Wilhelm,** 1753-1835.
Orographische Briefe über das Siebengebirge und die benachbarten zum Theil vulkanischen Gegenden bey der Ufer des Nieder-Rheins an Herrn Joseph Paul Edeln von Cobres

. . . Frankfurt am Mayn, Gebbard und Körber, 1789-90.
2v. in 1, illus. x549.943423/N84o

1665 **Nose.**
Ueber einige Ereignisse in der mineralogische Literatur unserer Tage. Frankfurt am Mayn, 1793.
56p. 549/V54u

Bound with Veltheim, August Ferdinand, Graf von, Ueber der Herren Werner und Karsten Reformen in der Mineralogie . . . 1793.

1666 **Nose.**
Beschreibung einer Sammlung von meist vulcanisirten Fossilien, die Deodat-Dolomieu im Jahre 1791 von Maltha aus nach Augsburg und Berlin versandte; mit verschiedenen Aufsätzen. Frankfurt am Main, In der Gebhard- und Körberschen Buchhandlung, 1797.
82p. q560/N84b

1667 **Nota, Alberto,** 1775-1847.
Del terremoto avvenuto nella città e provincia de S. Remo l'anno 1831. Pinerolo, Tipografia di P. Ghighetti, 1832.
46p. 551.22/N84d

1668 **Null, Jacques Friedrich von der.**
See MOHS, FRIEDRICH, Des Herrn Jac. Fried. von der Null Mineralien Kabinet . . . 1804.

1669 **Nuttall, Thomas,** 1786-1859.
A journal of travels into the Arkansas territory, during the year 1819. With occasional observations on the manners of the aborigines. Illustrated by a map and other engravings. Philadelphia, T. H. Palmer, 1821.
[9]-296p., front. (fold. map), plates, tables. 917.67/N96j

A journey from Philadelphia, down the Ohio and Mississippi rivers to the Arkansas; thence across Arkansas to the interior of the modern Oklahoma; returning via the Arkansas and Mississippi rivers, and then to New Orleans.

1670 **Oberlin, Henri Gottfried.**
Propositions géologiques, pour servir d'introduction à un ouvrage sur les élémens de la chorographie, avec l'exposé de leur plan, et leur application à la description géognostique, oeconomique et médicale du Ban de la Roche. Strasbourg, Impr. de Levrault, impr. de l'École de médecine, 1806.
261p., illus., 5 plates (incl. maps). 554.438/Ob2p

1671 **Ohio. Geological Survey.**
First [and] second annual report on the Geological survey of the state of Ohio. 1837-38. By W. W. Mather, principal geologist, and the several assistants. Columbus, 1838.
2v. in 1, illus., plates. 557/Oh31

1672 **Oken, Lorenz,** 1779-1851.
Okens Lehrbuch der Naturgeschichte. 1-3. th. Leipzig, C. H. Reclam, 1813-26.
3v. in 5, and atlas of 2v., 58 plates. 500/Ok2ℓ
Library owns v.1 only (Mineralogie).

1673 **Oken.**
Allgemeine Naturgeschichte für alle Stände. Stuttgart, Hoffmann, 1833-41.
7v. in 13, illus. 500/Ok2a

1674 **Oken.**
Abbildungen zu Oken's Allgemeiner Naturgeschichte für alle Stände. Stuttgart, Hoffman'sche Verlags-Buchhandlung, [1843?].
[50], 22, [24], 24, [64]p., 164 plates (part. col.). q500/Ok2a/plates

1675 **Olafsson, Eggert,** 1726-1768.
Travels in Iceland, performed by order of His Danish Majesty, containing observations on the manners and customs of the inhabitants, a description of the lakes, rivers, glaciers, hot-springs, and volcanoes; of the various kinds of earths, stones, fossils, and petrifactions; as well as of the animals, insects, fishes, &c., by Messrs. Olafsen & Povelsen. Translated from the Danish. London, Printed for R. Phillips by Barnard & Sultzer, 1805.
162p., 4 plates, fold. map. 914.91/Oℓ1vE

1676 **Oliger, Jacob,** 1650-1701.
Muséum Regium seu catalogus rerum tam naturalium, quàm artificialium, quae in Basilica Bibliothecae . . . Christiani Quinti Hafniae asservantur, descriptus ab Oligero Jacobaeo. Hafniae, Joachim Schmetgen, 1696.
201p., illus. uncat.
Catalog of the Royal Museum of Natural History of Copenhagen. Presentation copy of Johann Christian Ziege.

1677 **Omalius d'Halloy, Jean Baptiste Julien d',** 1783-1875.
Mémoires pour servir à la description géologique des Pays-Bas, de la France et de quelques contrées voisines. Namur, Impr. de D. Gérard, 1828.
307p., fold. col. map, fold. profiles. 554/Om1m

ΟΥΡΕΣΙΦΟΙΤΗΣ HELVETICUS,

SIVE

ITINERA ALPINA

Tria: In Quibus

Incolæ, Animalia, Plantæ, Montium Altitudines Barometricæ, Cœli & Soli Temperies, Aquæ Medicatæ, Mineralia, Metalla, Lapides Figurati, aliaque Fossilia; & quicquid insuper in Natura, Artibus, & Antiquitate, per *Alpes Helveticas* & *Rhæticas*, rarum sit, & notatu dignum, exponitur, & Iconibus illustratur.

Authore

JOHANNE JACOBO SCHEUCHZERO,
Med. Doct. *Tigurino*, & Societatis Regiæ *Londinensis* Socio.

IMPRIMATUR,

IS. NEWTON, P. R. S.

LONDINI:
Impensis *Henrici Clements*, ad Insigne *Lunæ Falcatæ* in Cœmeterio D. *Pauli*, Anno MDCCVIII.

Cat. No. 1966

1678 **Omalius d'Halloy.**
Introduction à la géologie, ou Première partie des éléments d'histoire naturelle inorganique, contenant des notions d'astronomie, de météorologie et de minéralogie avec un atlas de 3 tableaux et 17 planches. 3.éd. Bruxelles, Société belge de librairie, 1838.
571p., and atlas (16 plates). 549/Om1i/1838

1679 **Omalius d'Halloy.**
Éléments de géologie; seconde partie des Éléments d'histoire naturelle inorganique. 3.éd. Bruxelles, Société belge de librairie, 1838.
414p., illus. 550/Om1e/1838
Later editions published under titles Précis élémentaire de géologie and Abrégé de géologie. Cf. Biog. nat., v.16 (1901).

1680 **Omallius d'Halloy.**
Précis élémentaire de géologie. Paris, A. Bertrand, 1843.
790p., fold. plate, fold. table. 550/Om1p

1681 **Omalius d'Halloy.**
Abrégé de géologie. 7.éd. Bruxelles et Leipzig, A. Schnée, 1862.
626p., illus. 550/Om1a/1862
"Y compris celles publiées sous les titres d'Éléments et de Précis de géologie.

1682 **Omalius d'Halloy.**
Géologie. Paris, F.-L. Mathias, 1863.
612p., illus. 550/Om1a/1863
Previous editions published under title: Abrégé de géologie.

1683 **Orbigny, Alcide Dessalines d',** 1802-1857.
Foraminifères fossiles du bassin tertiaire de Vienne (Autriche), découverts par... Joseph de Hauer... et décrits par Alcide d'Orbigny... Ouvrage publié sous les auspices de sa Majesté l'empereur d'Autriche. Paris, Gide et comp^{e}, 1846.
312p., 21 plates. 563.1/Or1f
Added t.-p.: Die fossilen Foraminiferen des tertiaeren Beckens von Wien ...
Text in German and French.

1684 **Orbigny.**
Cours élémentaire de paléontologie et de géologie stratigraphiques. Paris, Victor Masson, 1849-52.
3v. and atlas. 560/Or1c
560/Or1c/atlas
Published as v.1; v.2, pt. 1, v.2, pt. 2; atlas.

1685 **Orbigny.**
Prodrome de paléontologie stratigraphique universelle des animaux mollusques & rayonnés, faisant suite au Cours élémentaire de paléontologie et de géologie stratigraphiques. Paris, V. Masson, 1850-1852.
3v. 560/Or1p

1686 **Orbigny.**
Paléontologie des coquilles & des mollusques. Paris, Adolphe Delahayes, 1855.
392p. 564/Or1p

1687 **Orbigny.**
Monographie d'un nouveau genre de mollusque gastéropode de la famille de trochoïdes. [n.p., n.d.].
p. 340-352, plate. 564/Or1m

1688 **Orbigny, Charles Dessalines d',** 1806-1876.
Géologie appliquée aux arts et à l'agriculture, comprenant l'ensemble des révolutions du globe; ouvrage orné de vignettes intercalées dans le texte, et d'un tableau gravé sur acier, représentant, par ordre chronologique, les terrains stratifiés et les principaux fossiles qui les caractérisent; suivi d'un vocabulaire donnant la définition des termes scientifiques employés dans le cours de l'ouvrage. Par MM. C. d'Orbigny et A. Gente . . . Paris, A. Gente, 1851.
534p., fold. col. plate. 550.Or12g

1689 **Ordinaire, Claude Nicolas,** 1736-1808.
Histoire naturelle des volcans, comprenant les volcans sous-marins, ceux de boue, et autres phénomènes analogues. Paris, Levrault frères, 1802.
342p., fold. map. 551.21/Or21h

1690 **Otter, William,** Bp. of Chichester, 1768-1840.
The life and remains of Edward Daniel Clarke, professor of mineralogy in the University of Cambridge. London, G. Cowie and Co., 1825.
2v. B/C597o

First edition, 1824.

1691 **Oviedo y Valdés, Gonzalo Fernández de,** 1478-1557.
Libro secondo delle Indie Occidentali: Summario de la naturale et generale historia de l'Indie Occidentali, *in* Summario de la generale historia de l'Indie Occidentali . . . [Vinegia, 1534].
x970/Su63

Description of placer gold and of the fact that it originates from veins in the nearby mountains. See also translation into English as part of EDEN, RICHARD, History of trauayle in the West and East Indies . . . London, 1577.

1692 **Owen, David Dale,** 1807-1860.
Report of a Geological Reconnoissance of the State of Indiana, made in the year 1837, in conformity to an order of the Legislature. Indianapolis, Indiana, J. W. Osborn & J. S. Willets, 1838.
49p. uncat.

1693 **Owen.**
See INDIANA. STATE GEOLOGIST . . . 1839 and 1859.

1694 **Owen.**
. . . Report of a geological exploration of part of Iowa, Wisconsin, and Illinois, made under instructions from the secretary of the Treasury of the United States, in the autumn of the year 1839; with charts and illustrations. By David Dale Owen, M.D., principal agent to explore the mineral lands of the United States. [Washington, Gov't print. off.], 1844.
191p., plates (partly fold.), fold. maps. x557/Io9o/1844

[U.S.] 28th Cong., 1st sess. Senate doc. no. 407.
Rev. ed. of the author's Mineral lands of the United States.

1695 **Owen.**
Report of a geological exploration of part of Iowa, Wisconsin, and Illinois, made under instructions from the Secretary of the Treasury of the United States, in the autumn of the year 1839. [Washington, Gov't. Printing Office], 1844.
191p., 25 plates (part. fold.). Meine/557/Ow2r

Bound with his Report of a geological reconnoissance of the Chippewa land district of Wisconsin . . . 1844.

1696 **Owen.**
Report of a geological reconnoissance of the Chippewa land district of Wisconsin, and, incidentally of a portion of the Kickapoo country, and of a part of Iowa and of the Minnesota Territory, made under instructions from the United States Treasury Department. [Washington, Gov't. Printing Office], 1848.
p.1-72, 23 plates, 6 numbered sections, 6 unnumbered sections, map. Meine/557/Ow2r

Bound with his Report of a geological exploration of part of Iowa, Wisconsin, and Illinois . . . 1844.

ΣΥΝΘΕΩ.
HOMO
DILUVII TESTIS
Et
ΘΕΟΣΚΟΠΟΣ
Publicæ συζητήσει expositus

à

JOH. JACOBO SCHEUCH-ZERO,

Med. D. Math. P. Acad. Imperialis Carolinæ & Societ. Regg. Anglicæ ac Prussicæ Membro.

Respp.

Pro Examine Philosophico consequendo.

(In Classe Prima.)	(In Classe Secunda.)
HENRICO WOLPHIO,	CONRADO FUESLINO,
HENR. KILCHSPERGERO,	JOHANNE FEERIO,
JACOBO WASERO,	CASPARO TOMANNO,
FELICE AMMIANO,	HENRICO WAEBERO,
JOHANNE GROBIO,	JOHANNE SUTERO,
CASPARO GROBIO,	
JACOBO Laub.	
HULDRICO GROBIO, Togg.	

In Auditorio Æstivo.

Ad diem Junij H. L. Q. S.

TIGURI

Typis JOH. HENRICI BYRGKLINI,
Anno MDCCXXVI.

Cat. No. 1973

1697 **Owen.**

Letter of the Secretary of the Treasury, communicating a report of a geological reconnoissance, of the Chippewa land district of Wisconsin, and the northern part of Iowa. [Washington, 1848].

134p., 37 plates (partly fold., incl. diagrs.), fold. map. x557/W75o

Report submitted to the commissioner of the General land office, by David Dale Owen, U.S. geologist for Wisconsin.

Senate ex. doc. 57, 30th Cong., 1st sess.

Half-title [p. 3]: Report of a geological reconnoissance of the Chippewa land district of Wisconsin; and incidentally, of a portion of the Kickapoo country, and of a part of Iowa and of the Minnesota territory . . .

Bound with Norwood, Joseph Granville, Report . . . 1848.

1698 **Owen.**

See U.S. GEOLOGICAL CORPS. Geological map of Wisconsin, Iowa and Minnesota . . . 1851.

1699 **Owen.**

Report of a geological survey of Wisconsin, Iowa, and Minnesota; and incidentally of a portion of Nebraska Territory. Made under instructions from the United States Treasury department. Philadelphia, Lippincott, Grambo & co., 1852.

[39]-638p., illus., and atlas of plates, maps and sections.

q557.3/Ow2
q557.3/Ow2/atlas

Final report, addressed to the commissioner of the U. S. General land office.

Atlas has title: Illustrations to the geological report of Wisconsin, Iowa and Minnesota . . .

In the library's copy 1, the atlas illustrations are bound with the text.

1700 **Owen.**

See SALINE COAL & MANUFACTURING CO., Extracts . . . 1856?

1701 **Owen.**

Geological reports on the mineral & timber lands of D. Kingsbury in the counties of Clearfield, Elk, and Jefferson, State of Pennsylvania, by David Dale Owen and B. Needham. Buffalo, Commercial Advertiser Steam Press, [1857?].

39p., profiles, tables. 553/Ow2g

Extract from Dr. D. D. Owens' geological report on the State of Kentucky: p.[37]-39.

1702 **Owen.**
Reports on the geology of the Tunungwant Coal Field, M'Kean County, Pa., by Dr. D. D. Owen, and Prof. B. Needham. Buffalo, Steam Press of E. R. Jewett & Co., [1857].
54p., fold. plates (1 col.), map. uncat.

1703 **Owen.**
See ARKANSAS. GEOLOGICAL SURVEY, First [and Second] report of a geological reconnoissance of the northern [middle and southern] counties of Arkansas . . . 1858 and 1859.

1704 **Owen.**
Report of a geological reconnoissance of the state of Indiana made in the year 1837, in conformity to an order of the legislature. Indianapolis, John C. Walker, 1859.
63p., 69p. x557/In24r

1705 **Owen.**
Report of a geological reconnoissance . . . 1859, *See* INDIANA. STATE GEOLOGIST, Report . . . 1859.

1706 **Owen, Edward.**
Observations on the earths, rocks, stones and minerals, for some miles about Bristol, and on the nature of the hot-well and the virtues of its water. London, Printed and sold by W. Johnston, 1754.
250p. uncat.

1707 **Owen, Richard,** 1804-1892.
Description of the skeleton of an extinct gigantic sloth, *Mylodon robustus,* Owen, with observations on the osteology, natural affinities, and probable habits of the megatherioid quadrupeds in general. By Richard Owen . . . conservator of the Museum of the Royal college of surgeons in London. Pub. by direction of the Council. London, Printed by R. and J. E. Taylor, 1842.
176p., 24 plates (4 fold.). q569.3/Ow2d

1708 **Owen.**
A history of British fossil mammals, and birds. London, J. Van Voorst, 1846.
560p., illus., table. 566/Ow2

1709 **Owen.**
A history of British fossil reptiles. London, 1849-84.
4v. q568/Ow2h

CONTENTS. — v.1, 3, Text. — v.2, 4, Plates.

1710 **Owen.**
Palaeontology; or, A systematic summary of extinct animals and their geological relations. Edinburgh, A. and C. Black, 1860.
420p., illus. 560/Ow2p

1711 **Owen.**
Palaeontology; or, A systematic summary of extinct animals and their geological relations. 2nd ed. Edinburgh, A. and C. Black, 1861.
463p., illus. 560/Ow2

1712 **Owen, Richard,** 1810-1890.
Key to the geology of the globe: an essay, designed to show that the present geographical, hydrographical, and geological structures, observed on the earth's crust, were the result of forces acting according to fixed, demonstrable laws, analogous to those governing the development of organic bodies. Nashville, Tenn., Stevenson & Owen; New York, A. S. Barnes & Co.; [etc., etc.], 1857.
[9]-256p., 3 fold. plates, 2 fold. maps, fold. chart, fold. table, fold. diagr. 550/Ow2k

1713 **Packe, Christopher,** 1686-1749.
'Ανκολραφια; sive, Convallium descriptio. In which are briefly but fully expounded the origine, course and insertion; extent, elevation and congruity of all the valleys and hills, brooks and rivers (as an explanation of a new philosophico-chronological chart) of East-Kent. Occasionally are interspers'd some transient remarks that relate to the natural history of the country; and to the military marks and signs: of Caesar's route thro it, to his decisive battle in Kent. Canterbury, Printed and sold by J. Abre, for the author, 1743.
110p. x914.223/P126a

1714 **Palassou, Pierre Bernard,** 1745-1830.
Essai sur la minéralogie des Monts-Pyrénées. Paris, Didot jeune, 1784.
330p., 12 plates, 8 fold. maps. 554.6/P17e

1715 **Palissy, Bernard,** 1510?-1590.
Oeuvres de . . . revues sur les exemplaires de la Bibliothèque du Roi, avec des notes par MM. Faujas de Saint-Fond, et Gobet. Paris, Ruault, 1777.
734p. x504/P17o

Kupfer-Bibel,

In welcher

Die

PHYSICA SACRA,

Oder

Geheiligte

Natur-Wissenschafft

Derer

In Heil. Schrifft vorkommenden

Natürlichen Sachen,

Deutlich erklärt und bewährt

Von

JOH. JACOB SCHEUCHZER,

Med. D. Prof. in Lyceo Tigurino, Academiæ Imper. Naturæ Curiosor. LEOPOLDINO-CAROLINÆ Adjuncto, & Socc. Regg. Anglicæ & Prussicæ Membro.

Anbey

Zur Erläuterung und Zierde des Wercks

In

Künstlichen Kupfer-Tafeln

Ausgegeben und verlegt

Durch

Johann Andreas Pfeffel,

Kayserlichen Hof-Kupferstecher in Augspurg.

Erste Abtheilung

Von

TAB. I — CLXXIV.

Samt Vorbericht und Verzeichnis der Auctorum, &c.

Augspurg und Ulm,

Gedruckt bey Christian Ulrich Wagner, 1731.

Cat. No. 1975

1716 **Palissy.**
Oeuvres complètes, . . . , édition conforme aux textes originaux imprimés du vivant de l'auteur; avec des notes et une notice historique par Paul-Antoine Cap. Paris, J.-J. Dubochet, 1844.
437p. 504/P17o/1844

"Note sur les précédentes éditions des oeuvres de Bernard Palissy": p.[xxxvi]-xxxix.

1717 **Palissy.**
Discours admirables, de la nature des eaux et fontaines, tant naturelles qu'artificielles, des metaux, des sels & salines, des pierres, des terres, du feu & des emaux. Avec plusieurs autres excellens secrets des choses naturelles. Plus un traité de la marne, fort utile & necessaire pour ceux qui se mellent de l'agriculture. Le tout dressé par dialogues, esquels sont introduits la theorique & la practique. Paris, chez Martin le Ieune, 1580.
361p. uncat.

Some pages incorrectly numbered (160 = 360)

1718 **Palissy.**
Le moyen de devenir riche, et la maniere veritable, par laquelle tous les hommes de la France pourront apprendre à multiplier & augmenter leurs thresors & possessions. Avec plusieurs autres excellens secrets des choses naturelles, desquels iusques à présent l'on n'a ouy parler. . . Paris, chez Robert Foüet, 1636.
1st part 255p., 2nd part 526p. uncat.

Part 2 has the following title-page: Seconde partie du moyen de devenir riche, contenant les discours admirables de la nature des eaux & fontaines, tant naturelles qu'artificielles des fleuves, puits, cisternes, estãgs, marez & autres eaux douces de leur origine, bonté & autres qualitez. De l'alchimie des metaux, de l'or potable, du mitridat, des glaces, des sels vegetatifs ou generatifs, du sel commun. Description des marez salans. Des pierres tant communes, que precieuses. Des causes de leur generation, formes, couleur, pesanteur & qualités d'icelles, des terres d'argille, de l'art, de la terre, de son utilité, & du feu, de la marne, & le moyen de les cognoistre.

1719 **Palissy.**
Resources: a treatise on "Waters and springs." Written in 1557. Translated by E. E. Willett, 1876. Brighton, W. J. Smith, [1876].
39p. x551.49/P17r

"Fifty copies only."

1720 **Pallas, Peter Simon,** 1741-1811.
Elenchus zoophytorum sistens generum adumbrationes generaliores et specierum cognitarum succinctas descriptiones cum selectis auctorum synonymis. Hagae Comitum, P. V. Cleef, 1766.
[17]-451p. 593.3/P17e

1721 **Pallas.**
Betrachtungen über die Beschaffenheit der Gebürge und Veränderungen der Erdkugel, besonders in Beziehung auf das Russische Reich. Vorgelesen in der öffentlichen Versammlung der Russisch-Kaiserl. Akademie der Wissenschaften den 23ten Junius 1777, da dieselbe mit der hohen Gegenwart des Herrn Grafen von Gothland beehret wurde. Frankfurt, 1778.
87p. 551.43/P17b

1722 **Pallas.**
Observations sur la formation des montagnes, et les changemens arrivés à notre globe, pour servir à l'Histoire Naturelle de M. le Comte de Buffon. St. Pétersbourg [etc.], Méquignon, 1782.
93p. 551.43/P17bF/1782

Translation of Betrachtungen über die Beschaffenheit der Gebürge und Veränderungen der Erdkugel.

1723 **Pallas.**
Voyages du Professeur Pallas, dans plusieurs provinces de l'Empire de Russie et dans l'Asie septentrionale; traduits de l'allemand par le C. Gauthier de la Peyronie. Nouvelle édition, revue et enrichie de notes par les C.C. Lamarck . . . Langlès . . . Paris, Maradan, 1789-1799.
9 vols. 914.7/P17rFg

Translation of: Reise durch verschiedene Provinzen des Russischen Reichs. Vol. 9 has t.p. Tableau physique et topographique de la Tauride, Suivi d'Observations sur la formation des Montagnes, et les changements arrivés à notre Globe, Paris, Guillaume, Gide, 1799.

1724 **Pallas.**
Travels through the southern provinces of the Russian empire, in the years 1793 and 1794. Tr. from the German of P. S. Pallas . . . London, T. N. Longman & O. Rees [etc.], 1802-03.
2v., col. illus., 52 plates, maps (part. fold.), fold. plan. 914.7/P17rEb

Translation of Bemerkungen auf einer Reise in die südlichen Stotthaltershaften des Russischen Reichs in den Jahren 1793-1794.

1725 **Pallas.**
Travels through the southern provinces of the Russian empire in the years 1793 and 1794. Translated from the German of P. S. Pallas . . . Second edition . . . London, Printed for J. Stockdale, 1812.
2v., col. fronts., illus. (part. col.), plates (part. col.), maps (part. fold.). q914.7/P17rEb/1812

1726 **Panarolo, Domenico,** 1587-1657.
Il Mare essaminato . . . In Roma, Per il Moneta, 1656.
184p., port. uncat.

1727 **Pander, Christian Heinrich,** 1794-1865.
Beiträge zur Geognosie des russischen Reiches. St. Petersburg, Gedruckt bei K. Kray, 1830.
165p., 31 (i.e. 34) plates, fold. map. q560.947/P19b

1728 **Papin, Nicolas,** d. 1653.
Raisonnemens philosophiques touchant la salûre, flux & reflux de la mer, & l'origine des sources, tant des fleuves que des fontaines. Ausquels est adiousté un traicté de la lumiere de la mer, composé par le mesme autheur. A Blois, par François de La Saugère, 1647.
156p. x551.46/P19r

1729 **Paragallo, Gaspare.**
Istoria naturale del monte Vesuvio. Divisata in due libri. Napoli, Stamperia di G. Raillard, 1705.
429p. 551.21/P21i

1730 **Parker, Martin.**
Supposed author. *See* ANONYMOUS, Dreadfull newes; or A true relation of the great, violent and late earthquake . . . 1638.

1731 **Parker, Samuel,** 1779-1866.
Journal of an exploring tour beyond the Rocky Mountains, under the direction of the A.B.C.F.M. performed in the years 1835, '36, and '37; containing a description of the geography, geology, climate, and productions; and the number, manners, and customs of the natives. With a map of Oregon Territory. By Rev. Samuel Parker, A.M. Ithaca, N.Y., The author, 1838.
[13]- 371p., incl. tables, front. (fold. map), plates. x917.95/P22j

1732 **Parker.**
Journal of an exploring tour beyond the Rocky Mountains, under the direction of the American board of commissions for foreign missions, in the years 1835, '36, and '37; containing a description of the geography, geology, climate, productions

Beyträge

zur mineralogischen Kenntniß des östlichen Theils von

Nordamerika

und seiner Gebürge

von

D. Johann David Schöpf

Hochfürstl. Brandenb. Onolzb. und Culmb. Hof- und Garni-
sonmedicus, auch Landphysikus und des med. Collegiums
zu Bayreuth Beysizer.

Cat. No. 1991

of the country, and the number, manners, and customs of the natives; with a map of Oregon Territory. By Rev. Samuel Parker... 2d ed. Ithaca, N.Y., Pub. by the author, Mack, Andrus, & Woodruff, printers, 1840.

400p., plates, fold. map. x917.95/P22j/1840

Map is dated 1838.

1733 **Parker.**

Journal of an exploring tour beyond the Rocky Mountains, under the direction of the A.B.C.F.M., in the years 1835, '36, and '37; containing a description of the geography, geology, climate, productions of the country, and the number, manners, and customs of the natives... By Rev. Samuel Parker, A.M. 3d ed. Ithaca, N.Y., Mack, Andrus, & Woodruff; Boston, Crocker & Brewster; [etc., etc.], 1842.

[17]-408p., front. (fold. map). x917.95/P22j3

Map lacking.

1734 **Parker.**

Journal of an exploring tour beyond the Rocky Mountains, under the direction of the A.B.C.F.M., containing a description of the geography, geology, climate, productions of the country, and the number, manners, and customs of the natives; with a map of Oregon Territory. By Rev. Samuel Parker, A.M. 4th ed. Ithaca, N.Y., Andrus, Woodruff, & Gauntlett; Boston, Crocker & Brewster; [etc., etc.], 1844.

[17]-416p., front. (fold. map), plates. x917.95/P22j/1844

1735 **Parkinson, James,** 1755-1824.

Organic remains of a former world. An examination of the mineralized remains of the vegetables and animals of the antediluvian world; generally termed extraneous fossils. London, J. Robson [etc.], 1804-11.

3v., fronts., col. plates. 560/P24or

Contents. — The vegetable kingdom. — The fossil zoophytes. — The fossil starfish, echini, shells, insects, amphibia, mammalia, &c.

1736 **Parkinson.**

Outlines of oryctology. An introduction to the study of fossil organic remains, especially of those found in the British strata; intended to aid the student in his enquiries respecting the nature of fossils, and their connection with the formation of the earth... London, The author, 1822.

346p., 10 plates. 560/P24o

1737 **Parkinson.**
Outlines of oryctology. An introduction to the study of fossil organic remains; especially of those found in the British strata: intended to aid the student in his inquiries respecting the nature of fossils, and their connection with the formation of the earth. 3d ed. With the author's latest corrections. London, M. A. Nattali, [183-].
350p., 10 plates. 560/P24o/183-

1738 **Parkinson.**
See MANTELL, GIDEON ALGERNON, A pictorial atlas of fossil remains . . . 1850.

1739 **Parrot, Georg Friedrich,** 1767-1852.
Grundriss der Physik der Erde und Geologie, zum Gebrauche für akademische Vorlesungen. Riga, F. Meinshausen, 1815.
718p., 2 fold. plates. 551/P24g

1740 **Passy, Antoine François,** 1792-1873.
Description géologique du département de la Seine-Inférieure. Ouvrage imprimé par ordre de l'Académie royale des sciences, belles-lettres et arts de Rouen. Rouen, Imp. de Nicétas Periaux, 1832.
371p., and atlas of 15p., 20 plates (part. col.), front (fold. map).
554.425/P26d

Atlas bound with main work, following p. 371.
"Table alphabétique des auteurs et des autorités": p. [345]-352.

1741 **Patin, Charles,** 1633-1693.
Traité des tourbes combustibles. A Paris, Chez Iean Du Bray & Pierre Variquet, 1663.
122p., port. uncat.

1742 **Patrin, Eugène Melchior Louis,** 1742-1815.
Histoire naturelle des minéraux, contenant leur description, celle de leur gîte, la théorie de leur formation, leurs rapports avec la géologie ou histoire de la terre, le détail de leurs propriétés et de leurs usages, leur analyse chimique, &c., avec figures dessinées d'après nature. 2.éd. Paris, Deterville, 1803.
5v., illus. 549/P27h/1803

Half-title: Histoire naturelle de Buffon, à laquelle on adjoint les Observations et les Découvertes des plus célèbres Naturalistes modernes sur la Minéralogie.

1743 **Patrin.**
Histoire naturelle des minéraux, contenant leur description, celle de leur gîte, la théorie de leur formation, leurs rapports avec la géologie ou histoire de la terre, le détail de leurs propriétés et de leurs usages, leur analyse chimique, &c., avec figures dessinées d'après nature. 2 éd. . . . Paris, Chez Raynal, 1830.
5v., plates. 549/P27h/1830

1744 **Peale, Rembrandt,** 1778-1860.
An historical disquisition on the mammoth, or great American incognitum, an extinct, immense, carnivorous animal, whose fossil remains have been found in North America. London, Printed for E. Lawrence by C. Mercier, 1803.
91p., illus. 569.6/P31h

1745 **Penn, Granville,** 1761-1844.
Conversations on geology; comprising a familiar explanation of the Huttonian and Wernerian systems; the Mosaic geology as explained by Mr. Granville Penn; and the late discoveries of Professor Buckland, Humboldt, Dr. MacCulloch and others. London, Printed for S. Maunder, 1828.
371p., illus., 12 plates. 550/P382c

1746 **Pennsylvania. Geological Survey.**
First-sixth annual report(s) on the Geological Survey of Pennsylvania. By Henry D. Rogers, state geologist . . . [1836-1841]. Harrisburg, Boas & Caplan, 1836-42.
6v., fold. fronts. (2d, 5th), fold. tables. 557/P38gea

Title varies: 1st, 2d, 5th: annual report of the state geologist . . . 3d-4th, 6th: annual report on the Geological survey. . . Library owns v.3, 1839. 118p.

1747 **Percival, James Gates,** 1795-1856.
See MALTE-BRUN, CONRAD, A system of universal geography. . . 1834.

1748 **Percival.**
See CONNECTICUT. GEOLOGICAL SURVEY, Report . . . 1842.

1749 **Perrault, Pierre,** ca. 1611-1680.
De l'origine des fontaines. Paris, 1674.
352p., front. uncat.

1750 **Perrault.**
De l'origine des fontaines. Paris, P. Le Petit, 1674.
353p., front. x551.49/P42d

Letter to Christiaan Huygens: p.[3]-[6].
Imperfect: front. wanting.

P. GASPARIS SCHOTTI

Regiscuriani è Societate Jesu

ANATOMIA

Physico-Hydrostatica

FONTIUM AC FLUMINUM

L I B R I S VI.

explicata,

ET FIGURIS ÆRI INCISIS

exornata;

Quibus, præmissâ fontium ac fluminum historiâ, primò eorundem principium & matrix accurato scrutinio inquiritur; deinde modus quo è matrice præcipua (quæ Mare est) oriuntur, ex Sacræ Scripturæ, Philosophiæ naturalis, & Hydrostaticæ principiis detegitur; postremò variæ proprietates ac veluti miracula in variis fontibus elucescentia discutiuntur.

Accedit in fine

Appendix de vera origine Nili.

HERBIPOLI,

Sumptibus

Viduæ Jo. Godofr. Schönwetteri,

Bibliopolæ Francofurtensis,

Excudit JOBUS HERTZ *Bibliopola, & Typogr. Herbipolensis.*

M. DC. LXIII.

Cat. No. 1993

1751 **Perrey, Alexis,** 1807-1882.
Mémoire sur les tremblements de terre ressentis dans le bassin du Rhône. [n.p., 1845?].
[p.265-370], illus. 551.22/P42m

Caption title.

1752 **Petermann, Augustus,** 1882-1878.
The atlas of physical geography, constructed by Augustus Petermann; with descriptive letter-press, embracing a general view of the physical phenomena of the globe by Thomas Milner. London, W. S. Orr, 1850.
142p., 16 leaves of plates (incl. col. maps). q912/P44a

1753 **Petiver, James,** ca. 1663-1718.
Opera, Historiam naturalem spectantia; or, Gazophylacium. Containing several 1000 figures of birds, beasts, reptiles, insects, fish, beetles, moths, flies, shells, corals, fossils, minerals, stones, fungusses, mosses, herbs, plants, etc. from all nations, on 156 copper plates, with Latin and English names. The shells etc. have English, Latin and native names. London, printed for John Millan, 1764.
2v. (chiefly plates). xq570/P4450

1754 **Petrini, Giovanni Vincenzo.**
Gabinetto mineralogico del Collegio Nazareno descritto secondo i caratteri esterni e distribuito a norma de' principj costitutivi. Roma, Presso i Lazzarini, [1791]-92.
2v. 549/P45g

1755 **Pettus, John,** 1613-1690.
Fleta minor. The laws of art and nature in knowing, judging, assaying, fining, refining . . . metals . . . London, Printed for and sold by Stephen Bateman, 1686.
345p., illus., port. q uncat.

Part 2 has separate title page: Fleta minor, spagyrick laws, the second part. Containing essays on metallick words . . . as a dictionary to Lazarus Erckern . . . London, Printed for the author by Thomas Dawks, 1683. 133p.

1756 **Phillips, John,** 1800-1874.
See SMITH, WILLIAM, Syllabus of lectures on geology in the Town-Hall, Scarborough . . . including three lectures on organic fossils by J. Phillips . . . 1824.

1757 **Phillips.**
Illustrations of the geology of Yorkshire; or, A description of the strata and organic remains of the Yorkshire coast; accom-

panied by a geological map, sections and plates of the fossil plants and animals. York, Printed for the author by T. Wilson and Sons, 1829.
192p., plates (partly col.), map. x554.27/P54i/1829

1758 **Phillips.**
Illustrations of the geology of Yorkshire; or, A description of the strata and organic remains, accompanied by a geological map, sections, and plates of the fossil plants and animals . . . London, Printed for J. Murray, 1835-36.
2v., front (v.1, map), plates (part. col.), fold. map, diagrs.
554.27/P54i

Vol. I is 2d edition.
Titles of two volumes vary slightly.
Contents. — pt. I. The Yorkshire coast. — pt. II. The mountain limestone district.

1759 **Phillips.**
A guide to geology. London; Longman, Rees, Orme, Brown, Green and Longman; 1834.
139p., 2 leaves of plates, illus. 550/P54g

1760 **Phillips.**
A guide to geology. 3d ed. London, Printed for Longman, Rees, Orme, Brown, Green, and Longman, 1836.
186p., illus., map. 550/P54g/1836

Bibliography: p. 176-182.

1761 **Phillips.**
An index geological map of the British Isles; constructed from published documents, communications of eminent geologists, and personal investigation by John Phillips. [2nd ed.]. London, John Weale, [1838].
C5740/.C5/1838/.P4

Col. map.
Scale not given.

1762 **Phillips.**
An index geological map of the British Isles; constructed from published documents, communications of eminent geologists, and personal investigation by John Phillips. [3rd ed.] London, John Weale, [1838].
G5740/.C5/1838/.P41

Col. map.
Scale, 1 inch equals ca. 25 miles.

1763 **Phillips.**
New index geological map of the British Isles, and adjacent coast of France, constructed from published documents, communications of eminent geologists and personal investigation by John Phillips. [5th ed.] London, John Weale, [1840].
G5740/.C5/1840/.P4

Col. map.
Scale, ca. 1 inch equals 25 miles.

1764 **Phillips.**
A treatise on geology, forming the article under that head in the seventh edition of the Encyclopaedia Britannica. Edinburgh, A. and C. Black, 1838.
295p., illus., plates, tables. 550/P54/1838

1765 **Phillips.**
. . . Treatise on geology. A new ed. London, Printed for Longman, Brown, Green & Longmans [etc.], 1852.
2v., illus. 550/P54

The cabinet cyclopaedia. Conducted by . . . D. Lardner . . . Natural history. [v. 94-95].

1766 **Phillips.**
Memoirs of William Smith, LL.D., author of the Map of the strata of England and Wales. London, J. Murray, 1844.
150p., 5 leaves of plates, illus., ports. x550.924/Sm6p and B/S6641p

Library owns two copies. Front. (port.) missing in B/S6641p.

1767 **Phillips.**
A map of the principal features of the Geology of Yorkshire. Published for the author [by] York, W. Monkhouse, 1853.
G5753/.Y2/1853/.P4

Col. map.
Scale, 1 inch equals ca. 4.5 British Statute miles.

1768 **Phillips.**
The rivers, mountains, and sea-coast of Yorkshire. With essays on the climate, scenery, and ancient inhabitants of the country. London, J. Murray, 1853.
302p., 36 plates. 914.274/P53r

1769 **Phillips.**
Vesuvius . . . Oxford, The Clarendon press, 1869.
355p., front., illus., plates, maps. 551.21/P54

1770 **Phillips.**
Geology of Oxford and the valley of the Thames. Oxford, Clarendon press, 1871.

523p., illus., 17 plates (partly col., incl. front., maps), diagrs. 554.257/P54g

"Works of authors consulted or referred to in the catalogues of fossils": p. [xxi]-xxiv.

1771 **Phillips, William,** 1775-1828.
An outline of mineralogy and geology, intended for the use of those who may desire to become acquainted with the elements of those sciences; especially of young persons . . . New York, Printed and sold by Collins and Co., 1816.
192p., 4 plates (part. col.). x549/P53o

1772 **Phillips.**
An elementary introduction to the knowledge of mineralogy; including some account of mineral elements and constituents; explanations of terms in common use; brief accounts of minerals, and of the places and circumstances in which they are found. Designed for the use of the student. With notes and additions on American articles, by Samuel L. Mitchill . . . New York, Collins and Co., 1818.
246p. x549/P53e

1773 **Phillips.**
An elementary introduction to mineralogy. New ed., with extensive alterations and additions, by H. J. Brooke . . . and W. H. Miller . . . London, Longman, Brown, Green, and Longmans [etc.], 1852.
700p., illus., diagrs. 549/P53e/1852

1774 **Piccolomini, Alessandro,** 1508-1578.
Della grandezza della terra et dell'acqua, trattato di M. Alessandro Piccolomini, nuovamente mandato in luce . . . Venetia, G. Ziletti, 1558.
4p., 41 (i.e. 43) numb. leaves, diagrs. x551.4/P58d

Printer's mark on t.-p. and on verso of last leaf.
Errors in foliation.
Bound with 8 other parts, paged separately, and not applicable to geology.

1775 **Picot de Lapeyrouse, Philippe Isidore,** 1744-1818.
Description de plusieurs nouvelles espèces d'orthocératites et d'ostracites . . . avec figures. À Erlang, aux dépens de Wolfgang Walther, libraire; À Paris, chez Didot le jeune [etc.], 1781.
45p., 13 plates. q564.5/P58d

In French and Latin.

1776 **Picquet, J. P.**
Voyage dans les Pyrénées françoises. Paris, Le Jay fils, 1789.
327p. 914.489/P58v

"Par Picquet, d'après Barbier. Attribué aussi à L. P. Berenger." — Cat. gén. de la Bibliothèque nat.

1777 **Pictet, François Jules,** 1809-1872.
Traité élémentaire de paléontologie; ou, Histoire naturelle des animaux fossiles considérés dans leurs rapports zoologiques et géologiques. Paris, Langlois et Leclerq, 1844-46.
4v., plates. 560/P58t

1778 **Pictet.**
Traité de paléontologie; ou, Histoire naturelle des animaux fossiles considérés dans leurs rapports zoologiques et géologiques. 2.éd., rev., cor., considérablement augm., accompagnée d'un atlas de 110 planches grand in-4° . . . Paris, J. B. Baillière; [etc., etc.], 1853-57.
4v., and atlas of 77p., 110 plates. 560/P58t/1853
q 560/P58t/1853/atlas

1779 **Pictet.**
Description des mollusques fossiles qui se trouvent dans les grès verts des environs de Genève, par F. J. Pictet et William Roux. Genève, impr. de J. G. Fick, 1847-1853.
457p. in various pagings, 51 leaves of plates. 564/P58d/1847

1780 **Pictet.**
Description des mollusques fossiles qui se trouvent dans les grès verts des environs de Genève, par F. J. Pictet et W. Roux. Quatrième livraison . . . Genève, J. Cherbuliez et J. Kessmann, 1853.
p.489-558, plates 41-51. 564/P58d/pt.4

Cover title.

1781 **Pictet.**
Description des fossiles du terrain aptien de la Perte du Rhône et des environs de Ste-Croix, par F.-J. Pictet et Eugène Renevier. Genève, J. Kessman, 1858.
184p. 23 plates. q560.9494/P58d

Half title: Matériaux pour la paléontologie suisse.
With joint author's autograph.

1782 **Pitot, Henri,** 1695-1771.
Description d'une machine pour mesurer la vitesse des eaux courantes, & le sillage des vaisseaux, *in* Mémoires de l'Académie Royale des Sciences, Amsterdam, 1732.
p.504-519, fold. plate. 506 PAH/1732

1783 **Pilla, Leopoldo,** 1805-1848.
Breve cenno sulla ricchezza minerale della Toscana. Pisa, R. Vannucchi, 1845.
224p. 549.9455/P64b

Bibliographical footnotes.

1784 **Pilla, Nicola.**
Saggio litologico sui vulcani estinti di Rocca Monfina, di Sessa, e di Tiano, dedicato all'illustrissimo signore D. Giuseppe Vairo. Napoli, V. Orsino, 1795.
74p. 552.1/P64s

1785 **Pini, Ermenegildo,** 1739-1825.
Addizioni al saggio di una nuova teoria della terra; in risposta all'esame fattone del Sig. De Luc. [Milano, G. Marelli, 1742].
52p., diagrs. 550.1/P65a

Caption title.
Interleaved, with some manuscript additions.

1786 **Pini.**
Osservazioni mineralogiche su la miniera de ferro di Rio ed altre parti dell'isola d'Elba. Milano, G. Marelli, 1777.
110p., fold. plate, fold. map. x549/P65o

"Catalogus fossilium": p.87-108.
Bibliographical footnotes.

1787 **Pini.**
Mémoire sur des nouvelles cristallisations des feldspaths et autres singularités renfermées dans les granites des environs de Baveno. Milan, J. Marelli, 1779.
62p., 2 fold. plates. 549.61/P65m

1788 **Pini.**
Della maniera di preparare la torba, e di usarla a fuoco più vantaggioso dell'ordinario. Istruzione pubblicata per ordine del R. Governo della Lombardia austriaca. Milano, Stamperia de G. Marelli, 1785.
138p., 2 fold. plates. 553.21/P65d

1789 **Pini.**
Della maniera de osservare nei monti la disposizione degli strati con uno stromento comodissimo a tal fine. [n.p., 1790?].
13p., illus. 551.43/P65d

1790 **Pini.**
De alcuni fossili singolari della Lombardia austriaca, e di altre parti dell'Italia; memoria . . . nella quale trattasi pure de

un vulcano supposto nella Lombardia medesima. Milano, G. Marelli, 1790.

48p. 552/P65d

1791 **Pinkerton, John,** 1758-1826.

Petrology. A treatise on rocks. London, Printed for White, Cochrane & Co. by S. Hamilton & Co., 1811.

2v., illus., 2 plates. 552/P652p

1792 **Plattes, Gabriel,** fl. 1638.

A discovery of subterraneall treasure; viz. of all manner of mines and mineralls, from the gold to the coale; with plaine directions and rules for the finding of them in all kingdomes and countries. And also the art of melting, refining, and assaying of them is plainly declared, so that every ordinary man, that is indifferently capacious, may with small charge presently try the value of such oares as shall be found either by rule or by accident. Whereunto is added a real experiment whereby every ignorant man may presently try whether any peece of gold that shall come to his hands be true or counterfeit, without defacing or altering the form thereof, and more certainly than any goldsmith or refiner could formerly discerne. Also a perfect way to try what colour any berry, leaf, flower, stalke, root, fruit, seed, barke, or wood will give: with a perfect way to make colours, that they shall not stayne nor fade like ordinary colours. Very necessary for every one to know, whether he be travailer by land or sea, or in what country, dominion, or plantation soever shall inhabit. London, by I. Okes for Iasper Emery, 1639.

60p. uncat.

The library owns several editions with slightly different titles and different editors: London, 1639 (uncat.); London, 1679 (uncat.); London, 1738 (x669/B23aEs/1739, bound with Barba, A collection of scarce and valuable treatises upon metals . . . 1739); Philadelphia, 1784 (x549/P69d/1784); Philadelphia, 1792 (x549/P69d/1792).

1793 **Plattner, Karl Friedrich,** 1800-1858

. . . Probirkunst mit dem Löthrohre oder vollständige Anleitung zu qualitativen und quantitativen Löthrohruntersuchungen. Vierte Auflage, neu bearbeitet und vermehrt von Theodor Richter. . . Leipzig, J. A. Barth, 1865.

679p., illus. 544.3/P69

The library has two copies of this edition.

DE
CORPORIBUS MARINIS
LAPIDESCENTIBUS

QUÆ DEFOSSA REPERIUNTUR

AUCTORE AUGUSTINO SCILLA

ADDITA DISSERTATIONE

FABII COLUMNÆ

DE GLOSSOPETRIS

EDITIO ALTERA EMENDATIOR.

ROMÆ, MDCCLII.

Sumptibus Venantii Monaldini Bibliopolæ in via Cursus.

EX TYPOGRAPHIA LINGUARUM ORIENTALIUM
ANGELI ROTILII, ET PHILIPPI BACCHELLI
IN ÆDIBUS MAXIMORUM.

SUPERIORUM PERMISSU.

Cat. No. 2005

1794 **Plattner.**
The use of the blowpipe, in the examination of minerals, ores, furnace-products, and other metallic combinations. Translated from the German, with notes, by James Sheridan Muspratt. With a pref. by Professor Liebig. London, Taylor and Walton, 1845.
364p., illus., tables. 544.3/P69pEm

Translation of Probirkunst mit dem Löthrohre.

1795 **Plattner.**
The use of the blowpipe in the qualitative and quantitative examination of minerals, ores, furnace products, and other metallic combinations. By Professor Plattner . . . and Dr. Sheridan Muspratt . . . Illustrated by numerous diagrams . . . 3d. ed., rev. and further enl. London, J. Churchill, 1854.
405p., illus., tables. 544.3/P69pEm/1854

Translation of Probirkunst mit dem Löthrohre.

1796 **Playfair, John,** 1748-1819.
The works . . . with a memoir of the author. Edinburgh, Printed for A. Constable, and London, Hurst Robinson & Co., 1822.
4v. 508/P69/1822

1797 **Playfair.**
Illustrations of the Huttonian theory of the earth. Edinburgh, Printed for W. Creech; [etc., etc.], 1802.
528p. 551.7/H979tYp

Title page in photocopy.

1798 **Playfair.**
Explication de Playfair sur la théorie de la terre par Hutton; et examen comparatif des systèmes géologiques fondés sur le feu et sur l'eau, par M. Murray; en réponse à l'explication de Playfair: tr. de l'anglais, et accompagnés de notes et de planches, par C. A. Basset . . . Paris, Bossange et Masson; [etc., etc.], 1815.
2v. in 1, 2 plates. 551.7/H979tYpFb

1799 **Playfair.**
See MURRAY, JOHN, A comparative view of the Huttonian and Neptunian systems of geology . . . 1802.

1800 **Plot, Robert,** 1640-1696.
Enquiries to be propounded to the most ingenious of each

county in my travels through England and Wales, in order to their history of nature and arts. [n.p., n.d.].
[4]p. xq570.942/P72e

Caption title.

Dated London? 1670? in the Brit. Mus. Cat. and Feb. 1678/79 by Falconer Madan in his Oxford books. v.3 (1651-1680), p.374, item 3220.

Perhaps an enlarged issue of the author's "Quaer's to be propounded to the most ingenious of each county in my travels through England." [Oxford? 1674?] Wing P 2584.

Devised by the author for his studies of the counties of England, notably Oxfordshire and Staffordshire.

1801 **Plot.**
The natural history of Oxford-shire, being an essay toward the natural history of England. Oxford, Printed at the theater; [etc., etc.], 1677.
358p., 16 plates, fold. map. xq570.942/P72n

Imperfect: map wanting.

1802 **Plot.**
The natural history of Oxford-shire, being an essay towards the natural history of England. 2d ed., with large additions and corrections: to which is prefix'd a short account of the author, &c. Oxford, Printed by L. Lichfield for C. Brome [etc.]. London, 1705.
366p., 16 plates, fold. map. xq570.942/P72n/1705

1803 **Plot.**
De origine fontium, tentamen philosophicum. In praelectione habita coram Societate Philosophica nuper Oxonii instituta ad Scientiam naturalem promovendam. Oxonii, Hen. Clements, 1685.
187p., front. uncat.

1804 **Plot.**
The natural history of Stafford-shire. Oxford, Printed at the Theater, 1686.
450p., 38 plates (partly fold.), fold. map. xq570.942/P72na

Imperfect: plates 9 and 19 wanting.
Error in paging: 417 numbered 407.
Error in binding: plate 15 bound after p.228.
This copy does not contain the rare leaf with the coats of arms omitted in the fold. map.

1805 **Plot.**
The natural history of Stafford-shire. Oxford, Printed at the Theater, 1686.

450p., 37 plates (part. fold.), fold. map. xq570.942/P72na/1686a

Error in paging: 417 numbered 407.
Error in binding: plates 14 and 15, 23 and 24, reversed.
This copy has the rare leaf with the coats of arms omitted in the fold. map.

1806 **Pluche, Noël-Antoine,** 1688-1761.
Spectacle de la nature: or Nature display'd being discourses on such particulars of natural history as were thought most proper to excite the curiosity and form the minds of youth. Illustrated by copper plates. Translated from the original French by Mr. Humphreys. London, Francklin et al., 1740-1753.
5v. x580/P72s Eh

V. 1 the seventh edition, v.2-3 the fifth, and v.4-5 the third edition.
Le spectacle de la nature was the second most popular book in France after Buffon's Histoire naturelle. Volume 3 contains a physical description of the earth and the sea, an interpretation of inorganic and organic fossils, and a description of mines.

1807 **Pontoppidan, Erik,** 1698-1764.
The natural history of Norway: containing a particular and accurate account of the temperature of the air, the different soils, waters, vegetables, metals, minerals, stones, beasts, birds, and fishes; together with the dispositions, customs, and manner of living of the inhabitants; interspersed with physiological notes from the eminent writers, and transactions of academies. Translated from the Danish original of the Right Revd. Erich Pontoppidan. London, A. Linde, 1755.
2v. in 1, illus., map. xq570.9481/P77vEb

1808 **Poppe, Johann Heinrich Moritz von,** 1776-1854.
Die artesischen Brunnen; ihre Beschaffenheit, die Art ihrer Verfertigung und ihre Benutzung mit allen dazu gehörigen Instrumenten und Maschinen zum Bohren, Ausfüttern und gutem Einrichten dieser Brunnen als Spring-, Lauf- und Pumpbrunnen. 2.verb. und verm. Aufl. Tübingen, C. F. Osiander, 1831.
73p., 4 fold. plates. 551.49/P81a/1831

1809 **Portlock, Joseph Ellison,** 1794-1864.
Report on the geology of the county of Londonderry and of parts of Tyrone and Fermanagh. Dublin, A. Milliken, Hodges and Smith, 1843.
784p., illus., fold. col. map, plates, col. profiles. 554.15/P83r

1810 **Portlock.**
A rudimentary treatise on geology; for the use of beginners. By Lieut.-Colonel Portlock . . . London, J. Weale, 1849.
182p., front., illus. 550/P83r

Weale's rudimentary series, no. 3.
Annotated leaves inserted.

1811 **Pötzsch, C. G.**
Ausführliche mineralogische Beschreibung der Gegend um Meissen. Dresden, Waltherischen Hofbuchhandlung, 1779.
138p., illus. 549.9/P85

1812 **Pott, Georgii Evstaf'evich,** d. 1858.
Geschichte und wissenschaftliche Beschäftigungen der in St. Petersburg gestifteten Russisch-Kaiserlichen Gesellschaft für die Gesammte Mineralogie von 1817 bis 1842, vom Mitstifter der Gesellschaft H. A. G. von Pott. St. Petersburg, 1842.
20, [188p.], 8 fold. plates (6 col., with mica adhesions). 549.06/VS2p

Pt.1 of Schriften der in St. Petersburg gestifteten Russisch-Kaiserlichen Gesellschaft für die Gesammte Mineralogie.

1813 **Pott, Johann Heinrich,** 1692-1777.
Lithogéognosie; ou, Examen chymique des pierres et des terres en général, et du talc, de la topaze & de la stéatite en particulier, avec une dissertation sur le feu & sur la lumière. Ouvrages tr. de l'allemand. Paris, J. T. Herissant, 1753.
2v., plates. 552/P85cFa

Vol. 2 has title: Continuation de la Lithogéognosie pyrotechnique, où l'on traite plus particulièrement de la connoissance des terres & des pierres, & de la manière d'en faire l'examen.
Tr. by Didier d'Arclais de Montamy.
"Table des effets des mélanges différents des terres de la lithogéognosie, de M. Pott": ciii p. at end of v.2.

1814 **Pownall, Thomas,** 1722-1805.
A topographical description of such parts of North America as are contained in the (annexed) map of the middle British colonies, &c. in North America. London, J. Almon, 1776.
46, 16p., front. (fold. map). xf912.73/Ev1

In case with Evans, Lewis, 1700?-1756, Geographical, historical, political, philosophical and mechanical essays. Philadelphia . . . 1755. See Evans for contents of case.
The original map was published at Philadelphia in 1755 in connection with a pamphlet by Evans having the following title: Geographical, historical, political, philosophical and mechanical essays. The first, containing An analysis of A general map of the middle British colonies in America . . . The "Topographical description," in

addition to the map, includes numerous extracts from this essay of Evans.

Appendix: I. The account of Capt. Anthony Van Schaick of the ground between the entrance of lake Champlain at Crown point, and the mouth of Otter creek. — II. Captain Anthony Van Schaick's journal, 1756. — III. Captain Hobb's account of the way from No. 4, in New Hampshire, to the mouth of Otter creek. — IV. Extracts from the journal of Captain Harry Gordon . . . who was sent from Fort Pitt . . . to Illinois, in 1766. — V. Extract from Mr. Lewis Evans' journal. 1743. — VI. A journal of Christopher Gist's journey, began from Col. Cresap's at the old town on Potomack river, Maryland, October 31, 1750, continued down the Ohio, within 15 miles of the falls thereof; and from thence to Roanoak river in North Carolina, where he arrived May 19, 1751; undertaken on account of the Ohio company.

1815 **Prestwich, Joseph,** 1812-1896.

A geological inquiry respecting the water-bearing strata of the country around London, with reference especially to the water-supply of the metropolis; and including some remarks on springs. London, J. Van Voorst, 1851.

240p., fold. map, diagrs. 554.21/P77g

Copy signed by author.

1816 **Prévost, Constant,** 1787-1856.

Coupe des terrains tertiaires du bassin de Paris, faite en suivant le cours de la Seine, de Moret à Mantes. Strasbourg, F. G. Levrault, [n.d.].

1 sheet. col. map and profiles. 551.781/P92c

21 x 95 cm. fold. to 21 x 13 cm.

1817 **Priest, Josiah,** 1788-1851.

American antiquities, and discoveries in the West: being an exhibition of the evidence that an ancient population of partially civilized nations, differing entirely from those of the present Indians, peopled America, many centuries before its discovery by Columbus. And inquiries into their origin, with a copious of many of their stupendous works, now in ruins. With conjectures concerning what may have become of them. Comp. from travels, authentic sources, and the researches of antiquarian societies. Albany, Printed by Packard, Hoffman and White, 1833.

400p., fold. front. 913.73/P93a

"Draining of the western lakes," p.367-373; "Lake Ontario formed by a volcanoe," p.376 ff. C. S. Rafinesque contributed to this part of the book, but it is uncertain which geological articles were by Rafinesque and which by Priest.

JOANNIS ANTONII SCOPOLI,

S. C. R. ET APOST. MAJESTATIS IN MONTANISTICIS ET MONETARIIS CONSILIARII, SUPREMI HUNGARIÆ INFERIORIS CAMERGRAFIATUS OFFICII ASSESSORIS, MINERALOGIÆ PROFESSORIS, CÆSAREÆ REGIÆ AGRARIÆ SOCIETATIS STIRIÆ, CARNIOLIÆ, GORIZIÆ ET GRADISCÆ, OECONOMICÆ BERNENSIS, ET APIARIÆ LUSATIÆ SUPERIORIS SOCII.

PRINCIPIA MINERALOGIÆ SYSTEMATICÆ ET PRACTICÆ

SUCCINCTE EXHIBENTIA

STRUCTURAM TELLURIS, SYSTEMATA MINERALOGICA, LAPIDUM CLASSES, GENERA, SPECIES, CUM PRÆCIPUIS VARIETATIBUS, EORUMQUE CHARACTERIBUS, SYNONYMIS, ANALYSI ET USU, NEC NON REGULIS NONNULLIS GENERALIBUS, AD DOCIMASIAM ET PYROTECHNIAM METALLURGICAM PERTINENTIBUS.

VETERO-PRAGÆ

APUD WOLFGANGUM GERLE 1772.

Cat. No. 2009

1818 **Priest.**

American antiquities and discoveries in the West: being an exhibition of the evidence that an ancient population of partially civilized nations differing entirely from those of the present Indians peopled America many centuries before its discovery by Columbus. 4th ed. Albany, Printed by Hoffman & White, 1834.

400p., illus., 2 plates (incl. fold. front.), fold. plan. 913.73/P93a4

1819 **Priest.**

American antiquities and discoveries in the West: being an exhibition of the evidence that an ancient population of partially civilized nations differing entirely from those of the present Indians peopled America many centuries before its discovery by Columbus . . . 5th ed. Albany, Hoffman & White, 1838.

400p., illus., 2 plates (incl. fold. front., fold. plan). 913.73/P93a5

1820 **Pryce, William,** 1725?-1790.

Mineralogia cornubiensis: a treatise on minerals, mines, and mining; containing the theory and natural history of strata, fissures, and lodes, with the methods of discovering and working of tin, copper, and lead mines, and of cleansing and metalizing their products; shewing each particular process for dressing, assaying and smelting of ores. To which is added, an explanation of the terms and idioms of miners. London, Printed and sold for the author by J. Phillips [etc.], 1778.

331p., front. (port.), 7 plates (incl. maps, 2 fold.), 2 fold. tables. xq622/P95

1821 **Puton, Ernest.**

Des métamorphoses et des modifications survenues dans certaines roches des Vosges. Paris, etc., J. B. Baillière, etc., 1838.

54p. uncat.

Signed by author.

Extrait des Mémoires de la 5e session du Congrès Scientifique de France, 1837.

Bound with Hogard, Henri, Description minéralogique et géologique des regions granitique et arénacée du système des Vosges, 1837.

1822 **Quad, Matthias,** 1557-1613.

Compendium universi complectens geographicarum enarrationum libros sex, ex quibus totus terrarum orbis situs . . . regionum & aquarum qualitates . . . gentium mores, ferarum genera . . . percipi possunt . . . Ex optimus ut pluri-

mum tam veteris quam huius aeuri scriptoribus excerpta, in ordinem concinnum digesta, & in epitomen hanc redacta. Coloniae Agrippinae, Sumptibus Wilhelm Lutzenkirchij, 1600.
714p., fold. maps. x910/Q22c

1823 **Quirini, Giovanni.**
. . . De testaceis fossilibus Musaei Septalliani et Iacobi Grandi. De veritate diluvij universalis, & testaceorum quae procul à Mari reperiuntur generatione. Epistolae. Venetiis, Typis Valuasensis, 1676.
76p. uncat.

1824 **Rafinesque, Constantine Samuel,** 1783-1840.
See PRIEST, JOSIAH, American antiquities, and discoveries in the West . . . 1833, 1834, and 1838.

1825 **Rafinesque.**
A life of travels and researches in North America and south Europe, or Outlines of the life, travels, and researches of C. S. Rafinesque . . . Containing his travels in North America and the south of Europe; the Atlantic Ocean, Mediterranean, Sicily, Azores, &c., from 1802 to 1835 — with sketches of his scientific and historical researches, &c. . . . Philadelphia, Printed for the author by F. Turner, 1836.
148p. 910.4/R117ℓ

1826 **Rafinesque.**
The good book, and amenities of nature. Or Annals of historical and natural sciences. Containing selections, of observations, researches and novelties in all the branches of physical and historical knowledge, with letters of eminent authors—chiefly on zoology, botany, agronomy, geognosy, ethnography . . . or organized beings, and fossils, nations and languages. Philadelphia, Printed for the Eleutherium of knowledge, 1840.
84p. 570/R11g

1827 **Ramazzini, Bernardino,** 1633-1714.
De fontium mutinensium admiranda scaturigine. Tractatus physico-hydrostaticus . . . Mutinae, Typis Haeredum Suliani Impressorum Ducalium, 1691.
87p., fold. plate. uncat.

1828 **Ramazzini.**
The Abyssinian philosophy confuted; or, Telluris theoria neither sacred, nor agreable to reason. Being for the most part, a translation of Petrus Ramazzini, of the wonderful springs

of Modena. Illustrated with many curious remarks and experiments by the author and translator. To which is added, a new hypothesis deduced from Scripture, and the observation of nature. With an addition of some miscellany experiments. By Robert St. Clair, M.D. . . . London, Printed for the author, 1697.
208p., fold. front., illus. uncat.

1829 **Rammelsberg, Karl Friedrich,** 1813-1899.
Handwörterbuch des chemischen Theils der Mineralogie. Erste-zweite Abteilung. Berlin, C. G. Lüderitz, 1841.
2v. in 1. 549.13/R14h

1830 **Rammelsberg.**
Handwörterbuch des chemischen Theils der Mineralogie. Erstes-fünftes Supplement zu dem Handwörterbuch des chemischen Theils der Mineralogie. Berlin, 1843-1853.
5v. in 2. 549.13/R14h/sup.

1831 **Ramond de Carbonnières, Louis François Elisabeth,** 1755-1827.
Observations faites dans les Pyrénées, pour servir de suite à des observations sur les Alpes, insérées dans une traduction des lettres de W. Coxe, sur la Suisse. Paris, Belin, 1789.
284p., fold. map, profile. 914/R14o

1832 **Randall, S. S.**
Incentives to the cultivation of the science of geology: designed for the use of the young. By S. S. Randall, Dep. Superintendent of the Common Schools of the State of New York . . . New York, Greenley & McElrath, 1846.
189p., illus. 551/R152i

1833 **Rashleigh, Philip,** 1729-1811.
Specimens of British minerals, selected from the cabinet of Philip Rashleigh . . . with general descriptions of each article. London, Printed by W. Bulmer and co., 1797-1802.
2v. in 1, col. plates. q549.942/R17s

1834 **Raspe, Rudolf Erich,** 1737-1794.
Specimen historiae naturalis globi terraquei, praecipue de novis e mari natis insulis, et ex his exactiùs descriptis & observatis, ulteriùs confirmandâ, Hookiana telluris hypothesi, de origine montium et corporum petrefactorum . . . Amstelodami & Lipsiae, sumptibus J. Schreuder & P. Mortier, 1763.
191p., 3 plates (2 fold.). 551.42/R18s

1835 **Raspe.**
Beytrag zur allerältesten und natürlichen Historie von Hessen; oder, Beschreibung des Habichwaldes und verschiedner andern Niederhessischen alten Vulcane in der Nachbarschaft von Cassel. Cassel, J. J. Cramer, 1774.
76p., fold. plate. 554.35/R18b

1836 **Raspe.**
An account of some German volcanos, and their productions. With a new hypothesis of the prismatical basaltes; established upon facts. Being an essay of physical geography for philosophers and miners. Published as supplementary to Sir William Hamilton's observations on the Italian volcanos. London, Printed for L. Davis, 1776.
140p., 2 fold. plates. x551.21/R18a

1837 **Raspe.**
See FERBER, JOHANN JACOB, Travels through Italy . . . translated . . . with notes . . . by R. E. Raspe . . . 1776.

1838 **Raspe.**
See BORN, IGNAZ, Baron Inigo Born's new process of amalgamation of gold and silver ores . . . 1791.

1839 **Raumer, Karl Georg von,** 1783-1865.
Geognostische Fragmente. Nürnberg, J. L. Schrag, 1811.
78p., fold. map. 554.3/R19g

"Diese Fragmente enthalten einen Theil der Resultate aus den geognostischen Untersuchungen, welche ich in Gemeinschaft mit Herrn Moritz von Engelhardt gemacht."

Contents. — Ueber die Sienitformation. Nach Beobachtungen im Erzgebirge. — Fragmente von einem Aufsatze über die Flötzgebirge.

1840 **Raumer.**
Versuch eines Abc Buchs der Krystallkunde. 1.Theil. Berlin, E. Reimer, 1820.
264p., diagrs. 548/R19v

No more published?

"Nachträge . . . Berlin, G. Reimer, 1821" (44p.) inserted after p.xl.

1841 **Ravergie.**
See RENOU, ÉMILIEN JEAN, Géologie de l'Algérie . . . 1848.

1842 **Ray, John,** 1627-1705.
Observations topographical, moral & physiological; made in a journey through part of the Low-countries, Germany, Italy and France: with a catalogue of plants not native of England,

found spontaneously growing in those parts, and their virtues . . . whereunto is added a brief account of Francis Willughby Esq; his voyage through a great part of Spain. p.466-499. London, Printed for John Martyn, 1673.
499p. uncat.

1843 **Ray.**
Miscellaneous discourses concerning the dissolution and changes of the world. Wherein the primitive chaos and creation, the general deluge, fountains, formed stones, sea-shells found in the earth, subterraneous trees, mountains, earthquakes, vulcanoes, the universal conflagration, and future state, are largely discussed and examined. London, for S. Smith, 1692.
259p. uncat.

1844 **Ray.**
Three physico-theological discourses. Concerning I. The primitive chaos, and creation of the world; II. The general deluge, its causes and effects; III. The dissolution of the world, and future conflagration. Wherein are largely discussed the production and use of mountains; the original of fountains, of formed stones, and sea-fishes bones and shells found in the earth; the effects of particular floods and inundations of the sea; the eruptions of vulcano's; the nature and causes of earthquakes: with an historical account of those two late remarkable ones in Jamaica and England. 2d ed. corrected, very much enlarged, and illustrated with copper-plates. London, printed for Sam. Smith, 1693.
406p., illus. uncat.

The library also owns the third edition, 1713, printed for William Innys.

1845 **Ray.**
The wisdom of God manifested in the works of the creation. In two parts. Viz. The heavenly bodies, elements, meteors, fossils, vegetables, animals . . . more particularly in the body of the earth . . . and in the admirable structure of the bodies of man and other animals; as also in their generation, &c. With answers to some objections . . . The 5th ed., corr., and very much enl. London, Printed by J. B. for Benj. Walford, 1709.
464p., front. (port.). 215/R21w/1709

First edition published 1691.

HISTORIA
SVCCINORVM
CORPORA ALIENA

INVOLVENTIVM

ET

NATVRAE OPERE

PICTORVM ET CAELATORVM

EX

REGIIS
AVGVSTORVM
CIMELIIS

DRESDAE CONDITIS

AERI INSCVLPTORVM

CONSCRIPTA

A

NATHANAELE SENDELIO D.

MEDICO REGIO ET PHYSICO

ELBINGENSI ORDINARIO.

LIPSIAE

Apud IO. FRIDERICVM GLEDITSCHIVM

A. O. R. MDCC XLII.

Cat. No. 2023

1846 **Ray.**
The wisdom of God manifested in the works of the creation. In two parts. 8th ed., corr. London, Printed by W. and J. Innys, 1722.
405p. x215/R21w/1722

1847 **Ray.**
The wisdom of God manifested in the works of the creation. In two parts . . . 11th ed., corr. London, Printed for W. Innys, 1743.
405p. x215/R21w/1743

1848 **Razumovskiĭ, Gregoriĭ,** d. 1837.
Voyage minéralogique et physique, de Bruxelles à Lausanne, par une partie du pays de Luxembourg, de la Lorraine, de la Champagne, & de la Franche-Comté. Fait en 1782. Lausanne, Mourer, 1783.
118p., tables. x554/R21v

1849 **Razumovskiĭ.**
Essai d'un système des transitions de la nature dans le règne minéral. Lausanne, F. La-Combe, 1785.
184p. 552.4/R21e

1850 **Razumovskiĭ.**
Histoire naturelle du Jorat et de ses environs et celle des trois lacs de Neuchâtel, Morat et Bienne. Lausanne, J. Mourer, 1789.
2v., illus. 570.9494/R21h

1851 **Réaumur, René-Antoine Ferchault de,** 1683-1757.
Remarques sur les coquilles fossiles de quelques cantons de la Touraine, & sur les utilités qu'on en tire, *in* Mémoires de l'Académie Royale des Sciences, Amsterdam, 1720.
p.519-541. 506 PAH/1720

This famous article on the faluns of Touraine was used by Buffon and others to prove that mountains were formed by the sea.

1852 **Reboul, Henri Paul Irénée,** 1763-1839.
Géologie de la période quaternaire et introduction à l'histoire ancienne. Paris, F. G. Levrault, 1833.
222p. 551.79/R24g

Bibliographical footnotes.

1853 **Recupito, Giulio Cesare,** 1581-1647.
Avviso dell' incendio del Vesuvio . . . Tradotto dalla lingua Latina all' Italiana . . . Napoli, Per Egidio Longo, 1635.
264p. uncat.

1854 **Recupito.**
De Vesuviano incendio nuntius . . . Editio tertia. Lovanii, Typias Everardi de Witte, 1639.
180p. uncat.

1855 **Redeker, Conrad.**
. . . Brevis descriptio . . . Bilfeldiani fontis, & usûs ejusdem. Quam, Cum subjunctis selectioribus sanatorum exemplis, Latinè reddidit & nonnullis animadversionibus auxit Nicolaus Hobokenus . . . Arnoldino . . . atque Aulae Steinfurtensis Archiater. Amstelaedami, Sumptibus Joannis Janssonii, 1668.
174p. uncat.

1856 **Reichestzer, Franz.**
Anleitung zur Geognosie insbesondere zur Gebirgskunde nach Werner . . . Wien, Gedruckt in der Degenschen Buchdr., 1812.
292p. 551/R272a

1857 **Remus, Georgius.**
See BREYN, JOHANN PHILIPP, Epistola de melonibus petrefactis montis Carmel . . . 1722.

1858 **Renevier, Eugène,** 1831-1906, jt. author.
See PICTET, FRANÇOIS JULES, Description des fossiles . . . Rhône . . . 1858.

1859 **Renou, Émilien Jean,** 1815-1902.
Géologie de l'Algérie, accompagnée d'une notice minéralogique sur le massif d'Alger par M. Ravergie et d'une description des coquilles fossiles par M. Deshayes. [Paris, Impr. nationale, 1848].
182p., 5 plates (incl. 2 fold. col. maps). q556.5/R28g

(Exploration scientifique de l'Algérie pendant les années 1840, 1841, 1842. Sciences physiques).

The contribution by M. Deshayes is missing in the UI copy.

1860 **Ranovanz, Hans Michael,** 1744-1798.
Mineralogisch-geographische und andere vermischte Nachrichten von den Altaischen Gebürgen Russisch kayserlichen Antheils, . . . mit Kupfern. Reval, Gedruckt mit Lindforsischen Schriften auf Kosten des Verfassers, 1788.
272p., tables, 4 fold. plates (part. col.; incl. maps, diagrs.).
549/R32m

1861 **Repetti, Emanuele,** 1776-1852.
Sopra l'Alpe apuana ed i marmi di Carrara; cenni. [Firenze], Badia fiesolana, 1820.
234p., plates, map, table. 551.43/R29s

Bibliographical footnotes.

1862 Reports of a Committee for investigating the coal and mineral resources of India. Calcutta, G. H. Huttman, Bengal Military Orphan Press, 1838.
94p., plate, geol. map. 555.4/In2

Reports written by secretary J. McClelland.

1863 **Reuss, August Emanuel von,** 1811-1873.
Geognostische Skizzen aus Böhmen. [Prag, etc., etc., C. W. Medan], 1840-44.
2v., 9 fold. plates. 554.33/R319g

Vol. 2 has imprint: Prag, Druck und Verlag des Artistisch-typographischen Instituts von C. W. Medan & comp., 1844.
Each volume has special t.-p.
"Literatur": v.1 p.xi-xii.

1864 **Reuss.**
Kurze Uebersicht der geognostischen Verhältnisse Böhmens. Fünf Vorträge, gehalten im Naturwissenschaftlichen Vereine Lotos im Jahre 1853. Mit drei geologischen Uebersichtskarten. Prag, J. G. Calve'sche Buchhandlung, 1854.
103p., 3 fold. maps. 554.37/R31k

1865 **Reuss, Franz Ambrosius,** 1761-1830.
Naturgeschichte der Biliner Sauerbrunnen in Böhmen. Mit Kupfern. Prag, Schönfeld, 1788.
316p., illus. 613.12/R31n

1866 **Reuss.**
Die Mineralquellen zu Liebwerda in Böhmen. Prag, Gedruckt bei G. Haase, 1811.
207p., 2 fold. plates. 613.12/R31m

1867 **Reynaud de Montlosier, François Dominique de,** 1755-1838.
Essai sur la théorie des volcans d'Auvergne. Nouv. éd. corr. et copiée littéralement sur celle de 1789. Riom, Landriot et Rousset, 1802.
184p. 551.21/R33e/1802

1868 **Ricci, Angelo Maria,** 1776-1850.
De gemmis et de adamante poëmatium mineralogico-chemicum. Neapoli, Ex regio typographio V. Orsini, 1797.
40p., port. 875.1/R35d

1869 **Ricci, Vito Procaccini,** 1765-1845.
Memoria su la grotta di Frasassi nei contorni di Fabriano, dipartimento del Musone dell'italico regno. Senigallia, Lazzarini, 1809.
121p. 551.44/R35m

1870 **Richardson, John,** 1787-1865.
On the frozen soil of North America, &c. (From the Edinburgh New Philosophical Journal for January 1841).
16p. x551.508/M565

Read before the Wernerian Society, Dec. 12, 1840.
Bound with Meteorology; [pamphlets].

1871 **Richardson, William,** 1740-1820.
Inquiry into the origin of the opinion that basalt is a volcanic production, with the motives that induced the author to publish it, . . . and also the arguments and facts from the Giant's Causeway and its vicinity, by which this opinion is proved to be precipitately adopted, and totally unfounded. Dublin, Printed by Graisberry and Campbell, 1805.
73p. 552.2/R39i

An examination and criticism of N. Desmarest's "Mémoire" published in the Histoire de l'Académie royale des sciences, 1771.
Author's presentation copy.

1872 **Richardson.**
Essay on the peat bogs of Ireland; with observations on the modes of reclaiming them directed by a late Act . . . London, V. Griffiths, printer, [1812?].
13p. 553.21/R397e

1873 **Richter, Christian.**
Saxoniae electoralis miraculosa terra; oder, Des weltberühmten Chur-Sachsenlandes bewundernswürdige Erde, wie dieselbe durch des höchsten Gottes sonderbahre Gnade und verliehenen Bergwercks-Verstand, auch unermüdeten Fleiss entdecket worden. Schneeberg, Auf Kosten des autoris, gedruckt bey Joh. Ernst Schulzen, 1732.
[148]p., illus. 553.8/R42s

1874 **Rink, Friedrich Theodor.**
See KANT, IMMANUEL, Immanuel Kant's Physische Geographie . . . 1802.

1875 **Rinman, Sven.**
Åminnelse-tal öfver framledne bergmästaren och Kongl. Vetenskaps academ. ledamot, välborne herr Axel Fred. Cronstedt, på Kongl. Academiens vägnar, hållet i stora Riddarhus-Salen den 6.martii, 1766. Stockholm, Tryckt hos L. Salvius, 1766.
40p. B/C9472r

1876 **Río, Andrés Manuel del,** 1765-1849.
Elementos de orictognosia, ó del conocimiento de los fosiles, dispuestos, segun los principios de A. G. Wérner, para el uso del Real Seminario de Minería de Mexico. México, Impressos por M. J. de Zúñiga y Ontiveros, 1795-1805.
2v., 3 fold. plates. x549/R47e

1877 **Río.**
Elementos de orictognosia, ó del conocimiento de los fosiles, segun el sistema de Bercelio; y según los principios de Abraham Góttlob Wérner, con la sinonimia inglesa, alemana y francesa, para uso del Seminario nacional de minería de México. Parte práctica — 2. ed. Filadelfia, Imprenta de J. F. Hurtel, 1832.
683p., 24 plates. 549/R47e/1832

1878 **Rio, Niccolò da,** 1755-1845.
Catalogo ragionato delle rocce e delle altre produzioni minerali dei Monti Euganei. Tratto dall' orittologia Euganea. Padova, Tip. Cartallier, 1836.
40p. q552/R48c

1879 **Ritchie, Archibald Tucker.**
The dynamical theory of the formation of the earth. London, Longman, Brown, Green and Longmans, 1850.
2v. 550/R51d

1880 **Ritter, Albrecht,** 1683?-1748?
Lucubratiuncula de alabastris Hohnsteinensibus, nonnullisque aliis eiusdem loci rebus naturalibus. [Helmestadi], 1731.
16p., plate. 553.635/R51ℓ

Bound with the author's Lucubratiuncula II. de alabastris Schwartzburgicis, cui subnexa est rerum quarumdam naturalium ejusdem Terrae brevis delineatio. Helmestadi, 1732. 31p., plate.

1881 **Ritter.**
Specimen I [-II] oryctographiae Calenbergicae, sive rerum fossilium quae sub adpellatione rerum naturalium plerumqve veniunt et in ducatu electorali Brunsvico-Luneburgico Calenberg ervuntur historico-physicae delineationis. Sondershusae, 1741-43.
2 pts. in 1v., 2 fold. plates. 560.9435/R51s

1882 **Rivière, Alphonse Ennemond Auguste,** 1805-1877.
Études géologiques faites aux environs de Quimper, et sur

VALLESIAE

DESCRIPTIO,

LIBRI DVO.

DE ALPIBVS

COMMENTARIVS,

IOSIA SIMLERO AVCTORE.

DESCRIBVNTVR vero in his libris, primo quidem regionis situs, populi mores, Reipublicæ forma, vrbes, arces, pagi, montes, flumina, aliaq; memoratu digna: & præterea Vallesianorũ res gestę. Deinde Alpium natura & magnitudo, vetera nomina & itinera: & multa quæ propria & peculiaria sunt Alpibus commemorantur: Auctorum etiam loci plurimi citantur & explicantur.

ACCESSIT HIS APPENDIX *descriptionis Vallesiæ.*

TIGVRI EXCVDEBAT CH. Froschouerus. 1574.

Cat. No. 2060

quelques autres points de la France occidentale . . . Paris, Carilian-Goeury, 1838.
64p., fold. col. plates, fold. map. 554.411/R52e

1883 **Rivière.**
Mémoires divers. Paris, [1838-39].
9 pieces in 1v., illus. (part. col.), 2 fold. maps (1 col.). 554.4/R52m

1884 **Rivière, Guillaume,** 1658-1754.
Mémoire sur quelques singularités du terroir de Gabian, et principalement sur la fontaine de l'huile de pétrole qui y coule. Montpellier, H. Pech, 1717.
28p. x554.484/R52m

Bound with Société royale des sciences de Montpellier, Mémoire . . . au sujet de l'histoire naturelle de la Province de Languedoc . . . 1726.

1885 **Robert, Eugène,** 1806-1879.
Voyages de la Commission Scientifique du Nord, en Scandinavie, en Laponie, au Spitzberg et aux Feröe, pendant les années 1838, 1839 et 1840 . . . Géologie, minéralogie, métallurgie et chimie, par M. Eugène Robert. Paris, A. Bertrand, [1845?].
308p. 554.8/R54g/v.2

1886 **Robert.**
Voyages de la Commission Scientifique du Nord, en Scandinavie, en Laponie, au Spitzberg et aux Feröe . . . Atlas géologique, gravé à l'aqua-tinta d'après les dessins de M. E. Robert. Paris, A. Bertrand, [1845].
Atlas of 21 plates. 554.8/D93g/plates

1887 **Roberts, William F.**
Map of the south or first coal region of Pennsylvania. Philadelphia, ca. 184-.
Col. map on 2 sheets. G3820/.H9/184-/.R6

Scale, 400 perches in one inch.

1888 **Robinet, Jean Baptiste René,** 1735-1820.
De la nature. Amsterdam, E. van Harrevelt, 1766.
4v., front. x100/R55d

Only volume 1 mentions "Troisième édition," but all four volumes have the same publisher and the same year. Each volume has an analytical table of contents.

Volume 1 contains a discussion on the organization of minerals, their growth from seeds, fecundation, and nutrition. Volume 3 gives a history of cosmogonies and Robinet's own cosmogony. Volume 4 contains a theory of the earth according to which the Earth is compared to an animal body.

1889 **Robinson, Samuel.**
A catalogue of American minerals, with their localities; including all which are known to exist in the United States and British provinces, and having the towns, counties, and districts in each state and province arranged alphabetically. With an appendix, containing additional localities and a tabular view. Boston, Cummings, Hilliard, & co., 1825.
316p. 549/R55c

Bibliography: p.v-vi.
Author's autograph presentation copy.

1890 **Robinson, Thomas,** d. 1719.
The anatomy of the earth . . . London, Printed for J. Newton, 1694.
24p. uncat.

With this is a pamphlet reprint, "The anatomy of the earth" . . . A seventeenth century cosmogony, by F. G. North, from the Geological Magazine, LXXI, no. 846, Dec. 1934.

1891 **Robinson.**
New observations on the natural history of this world of matter, and this world of life: in two parts. Being a philosophical discourse, grounded upon the Mosaick system of the creation, and the Flood. To which are added some thoughts concerning Paradise, the conflagration of the world, and a treatise of meteorology: with occasional remarks upon some late theories, conferences, and essays. London, Printed for John Newton, 1696.
218p. uncat.

1892 **Robinson.**
Essay towards a natural history of Westmoreland and Cumberland. Wherein an account is given of their several mineral and surface productions, with some directions how to discover minerals by the external and adjacent strata, and upper covers, etc. To which is annexed, a vindication of the philosophical and theological paraphrase of the Mosaick system of the creation, etc. London, 1709.
95, 118p. x570.9428/R567e

1893 **Rochas, Henry de,** fl. 1619-1648.
La vraye anatomie spagyrique des eaux mineralles et de toutes des choses qui les composent, avec leurs qualitez & vertus, curieusement observées . . . Paris, chez Pierre Billaine, 1636-37.

2v. in 1. uncat.

Vol. 1 is a reissue of his Traicté des observations nouvelles, published in Paris in 1634.

1894 **Roemer, Friedrich Adolph,** 1809-1869.
Die Versteinerungen des Norddeutschen Oolithen-Gebirges. Mit 16 lithographirten Tafeln. Hannover, Hahn, 1836.
218p., 16 plates. q560/R62v

1895 **Roemer.**
Die Versteinerungen des Norddeutschen Oolithen-Gebirges. Ein Nachtrag. Mit 5 lithographirten Tafeln. Hannover, Hahn, 1839.
59p., 5 plates. q560/R62v/sup.

1896 **Rogers, Henry Darwin,** 1808-1866.
See NEW JERSEY. GEOLOGICAL SURVEY, Report . . . 1836.

1897 **Rogers.**
See PENNSYLVANIA, GEOLOGICAL SURVEY, First-sixth annual report(s) . . . [1836-1841].

1898 **Rogers.**
Second annual report of the geological exploration of the State of Pennsylvania by Henry D. Rogers, State Geologist . . . Harrisburg, Printed for Packer, Barrett and Parke, 1838.
91p., front. (section), table. 557/P38re/v.2

1899 **Rogers,** jt. author.
See ROGERS, WILLIAM BARTON, On the physical structure of the Appalachian chain . . . 1842.

1900 **Rogers.**
Address delivered at the meeting of the Association of American Geologists and Naturalists, held in Washington, May, 1844. With an abstract of the proceedings at their meeting. New York and London, Wiley & Putnam, 1844.
58, 43p. 550.8/R63a

1901 **Rogers, Robert,** 1731-1795.
A concise account of North America; containing a description of the several British colonies on that continent, including the islands of Newfoundland, Cape Breton, &c. . . . Also of the interior, or westerly parts of the country, upon the rivers St. Laurence, the Mississippi, Christino, and the Great Lakes. To which is subjoined, an account of the several nations and tribes of Indians residing in those parts, as to their customs,

manners, government, numbers, &c. . . . By Major Robert Rogers. London, Printed for the author, 1765.
264p. x917/R63c

1902 **Rogers, William Barton,** 1804-1882.
Report of the geological reconnoissance of the State of Virginia. Made under the appointment of the Board of Public Works. Philadelphia, Desilver, Thomas & Co., 1836.
143p., fold. front. x557/R63r

Frontispiece wanting.

1903 **Rogers.**
On the physical structure of the Appalachian chain, as exemplifying the laws which have regulated the elevation of the great mountain chains generally, by William Barton Rogers and Henry Darwin Rogers, *in* American Journal of Science and Arts, 1842.
v. 43, p.177-178 (abstract). 505/AJ

Full text appears *in* Reports of the meetings of the Association of American Geologists and Naturalists, 1843. p.474-551. Reprint Arno Press, 1978.

1904 **Rolland du Roquan, Oscar,** 1812?-1863.
Description des coquilles fossiles de la famille des rudistes qui se trouvent dans le terrain crétacé des Corbières. Carcassonne, Impr. de L. Pomiés-Gardel, 1841.
69p., 8 plates. q564.1/R64d

1905 **Romani, Giovanni,** 1757-1822.
Dell'antico corso de'fiumi Po, Oglio ed Adda negli agri cremonese, parmigiano, casalasco e basso mantovano; memoria storico-critica. 2.ed., con aggiunta di osservazioni inedite sul fiume Adda. Milano, G. Silvestri, 1828.
85p. 551.48/R66d/1828

Bibliographical footnotes.
Library owns two copies.

1906 **Romé de l'Isle, Jean Baptiste Louis de,** 1736-1790.
Essai de cristallographie, ou Description des figures géometriques, propres à différens corps du regne minéral, connus vulgairement sous le nom de cristaux, avec figures et développemens. Paris, Chez Didot jeune [etc.], 1772.
427p., 10 fold. plates, 2 fold. tables. 548/R66e

The second edition has title: Cristallographie, ou description des formes propres à tous les corps du règne minéral.

"Table alphabétique des principaux auteurs qui ont écrit sur les cristaux, ou qui sont cités dans cet ouvrage à l'occasion des cristaux": p. xvii-xxviii.

1907 **Romé de l'Isle.**
Cristallographie, ou Description des formes propres à tous les corps du règne minéral, dans l'état de combinaison saline, pierreuse ou métallique . . . 2. ed. . . . Paris, Imprimerie de Monsieur, 1783.
4 v., VIII (i.e. 12) double plates, tables. 552/R66c

1908 **Romé de l'Isle.**
Description méthodique d'une collection de minéraux du cabinet de M. D. R. D. L. Ouvrage où l'on donne de nouvelles idées sur la formation & la décompostion des mines, avec un court exposé des sentimens des minéralogistes les plus connus, sur la nature de chaque espèce, le minéralisateur qui s'y rencontre, & la quantité de métal qu'elle produit. A Paris, Chez Didot jeune, libraire [etc.], 1773.
299p., front. 549/R66d

Bibliography: p.xx-xxv.

1909 **Romé de l'Isle.**
L'action du feu central bannie de la surface du globe, et le soleil rétabli dans ses droits; contre les assertions de MM. le comte de Buffon, Bailly, de Mairan, etc. Stockholm, Paris, P. Fr. Didot le jeune, 1779.
84p. 523.1/R66a

1910 **Romé de l'Isle.**
Observations sur les rapports qui paroissent exister entre la mine dite cristaux d'étain et les cristaux de fer octaedres. Aerford, Georg Adam Keyser, 1786.
11p. 549/F59

Bound with Flade, Christian Gottlob, De re Metallica . . . [1791].

1911 **Romei, Annibale,** 16th cent.
Dialogo . . . Diviso in due giornate. Nella prima delle quali si tratta delle cause universali del terremoto, e di tutte le impressioni, & apparenze, che, con stupor del volgo, nell' aria si generano. Nella seconda, del terremoto, della salsedine del mare, della via lattea, e del flusso, e reflusso del mare s'assegnano cause particolari, diverse d'Aristotele, e da qualunque filosofo sin' ad hora a no habbi scritto. Ferrara, Per Vittorio Baldini, 1587.
108p. uncat.

1912 **Rose, Gustav,** 1798-1873.
Elemente der Krystallographie, nebst einer Tabellarischen Uebersicht der Mineralien nach de Krystallformen. Zweite

ISTORIA
DEL
MONTE VESUVIO

Divisata in due Libri

DA D. IGNAZIO SORRENTINO

Sacerdote Secolare della Torre del Greco.

DEDICATA

All'Ill.mo, e Reverend.mo Signor

D. CELESTINO GALIANO

Arcivescovo di Tessalonica, e Cappellano Magiore del Regno di Napoli &c.

NAPOLI MDCCXXXIV.
Per Giuseppe Severini.

Con licenza de' Superiori.

Cat. No. 2086

Auflage. Mit Zehn Kupfertafeln. Berlin, Ernst Siegfried Mittler, 1838.
175p., 10 plates. 548/R72e/1838

1913 **Rose.**
Éléments de cristallographie. Traduit de l'allemand par Victor Regnault. Paris, L. Hachette, etc., 1834.
2v. 548/R72eFr

Translation of: Elemente der Krystallographie. Contents. vol. 1, text; vol. 2, plates.

1914 **Rose.**
Das krystallo-chemische Mineralsystem. Leipzig, W. Engelmann, 1852.
156p., tables. 549/R72k

Bibliographical footnotes.

1915 **Rosenmüller, Johann Christian,** 1771-1820.
Beschreibung merkwürdiger Höhlen. Ein Beitrag zur physikalischen Geschichte der Erde, hrsg. von Dr. Rosenmüller and Dr. Tillesius. Mit Kupfertafeln. Leipzig, Breitkopf und Härtel, 1799-1805.
2v., 18 plates (10 fold. col.). 551.44/R72b

Plate 10 wanting.

1916 **Rosina, Gaetano.**
Osservazioni e ricerche mineralogico-chimiche sopra alcune valli dell'Ossola . . . coll'aggiunta di un metodo economico per estrarre l'oro da una miniera di quei dintorni riputata finora incoltivabile. Milano, G. Pirotta, 1819.
106p., table. 549/R731o

Bibliographical footnotes.

1917 **Rosinus, Michael Reinhold,** 1687-1725.
Tentaminis de lithozois ac lithophytis olim marinis jam vero subterraneis, prodromus; sive, de stellis marinis quondam nunc fossilibus disquisitio. Hamburgi, Typis N. Sauer, 1719.
88p., 10 plates. 563.6/R73t

Bound with the author's De belemnitis et hisce plerumque insidentibus alveolis animadversiones . . . Francohusae, 1728. 11p.

1918 **Rosinus.**
De belemnitis et hisce plerumque insidentibus alveolis animadversiones . . . Francohusae, 1728.
11p.

Bound with the author's Tentaminis de lithozois . . . 1719.

1919 **Rosnel, Pierre de.**
Le mercure indien; ou, Le trésor des Indes. Première partie dans laquelle est traitté de l'or, de l'argent & du vifargent, de leur formation, de leur origine, de leur usage & de leur valeur. Avec une explication sommaire des titres de l'or & de l'argent, & de leur affinage. Paris, [Impr. de R. Chevillion], 1667.
35, 71, 23p. x549/R732m

1920 **Ross, John,** 1777-1856.
A voyage of discovery, made under the orders of the admiralty, in his majesty's ships Isabella and Alexander, for the purpose of exploring Baffin's Bay, and inquiring into the probability of a north-west passage. London, John Murray, 1819.
252p., illus., fold. plates and maps. x919.8/R73v/1819

1921 **Ross.**
A voyage of discovery. Made under the orders of the admiralty, in his Majesty's ships Isabella and Alexander, for the purpose of exploring Baffin's Bay, and inquiring into the probability of a north-west passage. 2d ed. London, Straham and Spottiswood, 1819.
2v. in 1. 919.8/R73v

1922 **Ross.**
Entdeckungsreise unter den Befehlen der britischen Admiralität mit den königlichen Schiffen Isabella und Alexander um Baffins-Bay auszuforschen und die Möglichkeit einer nord-westlichen Durchfahrt zu untersuchen. Von mehrern Sprach und Sachkundigen aus dem Englischen übersetzt, hrsg. von P. A. Nemnich. Mit der Entdeckungskarte und 27 zum Theil illuminirten Kupfern. Leipzig, F. Fleischer, 1820.
2v., illus. x919.8/R73vGn

1923 **Ross.**
Narrative of a second voyage in search of a north-west passage, and of a residence in the Arctic regions during the years 1829, 1830, 1831, 1832, 1833. Including the reports of . . . James Clark Ross . . . and the discovery of the northern magnetic pole. London, A. W. Webster, 1835.
2v., front. (port., v.2), plates (partly col.), maps (partly fold.), col. plan. x919.8/R73n

[Vol. 1] bound with Braithwaite, John. Supplement to Captain Sir John Ross's narrative of a second voyage, 1835; and Ross, Sir John, Explanation and answer to Mr. John Braithwaite's supplement . . . 1835. 8p.

[Vol. 2] "Appendix . . ." (containing vocabularies in English, Danish and Eskimo, sketch of the Eskimos, report on the natural history, scientific observations, etc.)

1924 **Ross.**
Narrative of a second voyage in search of a north-west passage, and of a residence in the Arctic regions during the years 1829, 1830, 1831, 1832, 1833 . . . Including the reports of Capt. J. C. Ross . . . and the discovery of the northern magnetic pole. Paris, Baudry's European Library [etc.], 1835.
475p., folded map. 919.8/R73n/1835

1925 **Ross.**
Narrative of a second voyage in search of a north-west passage, and of a residence in the Arctic regions, during the years 1829, 1830, 1831, 1832, 1833. Including the reports of James Clark Ross and the discovery of the northern magnetic pole. Paris, A. & W. Galignani, 1835.
541p., front., folded map, 4 maps. 919.8/R73n/1835b

1926 **Ross.**
Narrative of a second voyage in search of a north-west passage, and of a residence in the Arctic regions, during the years, 1829, 1830, 1831, 1832, 1833 . . . Including the reports of . . . J. C. Ross . . . and the discovery of the northern magnetic pole. Philadelphia, E. L. Carey & A. Hart. Baltimore, Carey & Hart, 1835.
456p., front., folded map. 919.8/R73n/1835a

1927 **Ross.**
Die zweite Entdeckungs-Reise des Capitain John Ross nach den Gegenden des Nordpols; und sein Aufenthalt daselbst während der Jahre 1829-1833. Aus dem Englischen von G. W. Becker und J. Sporschil. Mit einer Charte, 2 Ansichten in Stahl gestochen v. Skelton und dem Portrait des Verfassers v. Louis Zoellner. Leipzig, J. J. Weber, 1835.
2v. 919.8/R73nGb

1928 **Roth, Justus Ludwig Adolph,** 1818-1892.
Die Kugelformen im Mineralreich und deren Einfluss auf die Absonderungsgestalten der Gesteine. Dresden und Leipzig, Arnold, 1844.
40p., plates. q549/R74k

Another copy is bound with Bennigsen-Förder, Rudolf von, Erläuterungen zur geognostischen Karte der Umgegend von Berlin, 1843. 554.3155/B43e

1929 **Rotter, Carolus Eduardus.**
De Plumbi vi et Efficacia. Lipsiae, Staritzii, 1828.
28p. 549/F59

Bound with Flade, Christian Gottlob, De re Metallica . . . [1791].

1930 **Rousselot de Surgy, Jacques Philibert,** d. 1737.
Histoire naturelle et politique de la Pensylvanie, et de l'établissement des Quakers dans cette contrée. Traduite de l'allemand. P. M. d. S. censeur royal. Précédée d'une carte géographique. Paris, Ganeau, 1768.
372p. x917.48/R76h

1931 **Roux, William,** jt. author.
See PICTET, FRANÇOIS JULES, Description des mollusques fossiles . . . Genève . . . 1847-53 and 1853.

1932 **Rozet, Claude Antoine,** 1798-1858.
Description géognostique du bassin du Bas-Boulonnais. Paris, Selligue, imprimeur-libraire, 1828.
[19]-123p., fold. map, fold. table. 554.427/R817d

1933 **Rozet.**
Traité élémentaire de géologie. Paris, A. Bertrand, 1835-1837.
2v. & atlas (13 leaves of plates, col. illus., maps). 550.7/R817t/v.1-3

1934 **Ruhnkenius, David,** 1723-1798, praeses.
De terrae motu. Lugduni Batavorum, Apud J. et H. Verbeek, 1767.
23p. 551.22/R85d

1935 **Rumpf, Georg Eberhard,** 1627-1702.
Thesaurus imaginum piscium testaceorum; quales sunt Cancri, Echini, Echinometra, Stellae Marinae, &c. ut & Cochlearum; inter quas numerantur Lunares; Laciniatae; Trochi; Valvatae, sive Semilunares; Valvatae striatae; Cassides tuberosae, verrucosae, laeves & Murices; Globosae; Buccina; Strombi; Volutae; Alatae; Porcellanae majores & minores; Cylindri, &c. Quibus accedunt Conchylia, ut Nautilus, Cornu Ammonis, &c. Conchae univalviae & bivalviae; quarum species sunt Solenes univalvii, Chamae asperae, Chamae laeves, Pectines, Pectunculi, Tellinae, Solenes bivalvii, Musculi, Pinnae, Ostrea, &c. Denique Mineralia; uti Metalla, Lapides & Argillae, variis in locis reperta. Lugduni Batavorum, Apud Petrum vander Aa, Bibliopolam, 1711.
15p., 60 plates. (new acq.)

1936 **Rumpf.**
Thesaurus imaginum piscium testaceorum . . . Hagae — Comitum, Petrum de Houdt, 1739.
14p., 60 plates. q560.9493/B95o

Bound with manuscript notes.
Bound with Burtin, F. X. de, Oryctographie de Bruxelles; ou description des fossiles tant naturels . . . 1784.

1937 **Rumpf.**
D'Amboinsche rariteitkamer, behelzende eene beschryvinge van allerhande zoo weeke als harde schaalvisschen, te weeten raare krabben, kreeften, en diergelyke zeedieren, als mede allerhande hoorntjes en schulpen, die men in d'Amboinsche zee vindt: daar benevens zommige mineraalen, gesteenten, en soorten van aarde, die in d'Amboinsche, en zommige omleggende eilanden gevonden worden. Verdeelt in drie boeken . . . T'Amsterdam, Jan Roman de Jonge, 1741.
340, [43]p., 60 plates, port. q595.3/R86d

"Beschryving en verdeeling der Amboinsche hoornen en schulpen, door den Heer Sipman": p.167-193.

1938 **Rütimeyer, Ludwig,** 1825-1895.
Über das Schweizerische Nummulitenterrain, mit besonderer Berücksichtigung des Gebirges zwischen dem Thunersee und der Emme. Bern, Stämpflische Buchdruckerei, 1850.
120p., diagrs. col. map. 563.1/R93u

1939 **Rutty, John,** 1698-1777.
An essay towards a natural, experimental and medicinal history of the mineral waters of Ireland . . . The whole illustrated with tables exhibiting a clear view of the experiments in concert, and a comparison of the Irish to the English, and other foreign waters. Dublin, Printed for the author, 1757.
478p., tables. x613.38/R939e

Library owns 2 copies; c.2 bound with: The argument of sulphur or no sulphur in waters discussed. Dublin, 1762.

1940 **Rutty.**
A methodical synopsis of mineral waters, comprehending the most celebrated medicinal waters, both cold and hot, of Great-Britain, Ireland, France, Germany, and Italy, and several other parts of the world. Wherein their several impregnating minerals being previously described, and their characteristics investigated, each water is reduced to its proper genus; and beside the particular analysis, the virtues, uses, and abuses of the water are described, in a method entirely new. Inter-

ERASMI STELLAE
LIBONOTHANI
INTERPRAE=
TAMENTI
GEMMA=
RVM LI=
BELLVS
VNI=
CVS.

Sola ſalus ſeruire deo,
Sunt cœtera fraudes,

Cat. No. 2105

spersed with tables, tending to throw a light upon this intricate subject; and abstracts of the principal authors who have treated of mineral waters; and the accounts, dispersed in the acts of the most learned societies in Europe, are collected and properly digested. London, Printed for W. Johnston, 1757.
660p. x613.38/R939m

1941 **Sack, Friederich Wilhelm.**
Geologie, oder Betrachtung der Erde. Breslau, W. G. Korn, 1785.
351p., illus. 550/Sa14g

1942 **Sage, Balthazar Georges,** 1740-1824.
Examen chymique de différentes substances minérales; essais sur le vin, les pierres; les bézoards, & d'autres parties d'histoire naturelle & de chymie; traduction d'une lettre de Monsieur Lehmann, sur la mine de plomb rouge. Paris, Impr. de P. de Lormel, 1769.
151 (i.e. 251)p. 540/Sa18e

1943 **Sage.**
Élémens de minéralogie docimastique. Paris, P. de Lormel, 1772.
276p., table. 549.1/Sa1e

1944 **Sage.**
Élémens de minéralogie docimastique. 2.éd. Paris, Impr. royale, 1777.
2v., tables (1 fold.). 549.1/Sa1e/1777

Bibliographical footnotes.

1945 **Sage.**
Précis des mémoires [et d'autres oeuvres]. Paris, n.p., 1809.
7v. in 1. 504/Sa1p

Each vol. has special t.p.
Lus dans les séances de la première classe de l'Institut, pendant l'année, 1809.

1946 **St. John, Samuel,** 1813-1876.
Elements of geology, intended for the use of students. Hudson, Ohio, [Pentagon Press], 1851.
334p., illus., fold. plate. uncat.

1947 **Saline Coal & Manufacturing Co.**
Extracts from the report of Dr. David Dale Owen on the properties of the Saline Coal Company located in Gallatin County, Illinois, about 6 miles south of Shawneetown, and

the Martha Iron property located in Hardin County, about 18 miles south of Shawneetown. [n.p., 1856?]
16p. 622.33/Sa33rl

1948 **Sampson, George Vaughan.**
A memoir, explanatory of the chart and survey of the county of Londonderry, Ireland. By the Rev. George Vaughan Sampson . . . honorary member of the Geological Society of London . . . London, Printed by W. Bulmer and co. and published by G. and W. Nicol, 1814.
359p., front., 6 plates, map. 630.9415/Sa4s/1814

An enlarged and revised ed. of the author's "Statistical survey of the county of Londonderry," 1802.

1949 **Santorelli, Antonio,** 1583-1653.
Discorsi della natura, accidenti, e pronostici dell' incendio del monte de Somma dell' anno 1631 . . . Napoli, Appresso Egidio Longo, 1632.
58p. uncat.

1950 **Sardinia (Kingdom). Azienda Generale Economica dell'Interno.**
Cenni de statistica mineralogica degli stati di S. M. il Re de Sardegna; ovvero, Catalogo ragionato della raccolta formatasi presso l'Azienda generale dell'interno, per cura di Vincenzo Barelli, Torino. Tipografia de G. Fodratti, 1835.
686p., tables (1 fold.). 549/Sa7c

1951 **Sartorius von Waltershausen, Georg Friedrich Christoph,** 1765-1828.
Geognostiche Beobachtungen und Erfahrungen vorzüglich in Hinsicht des Basaltes; nebst Angabe mehrerer Höhenbestimmungen der vorzüglichsten Orte im Eisenacher Kreis. Eisenach, J. F. Barecke, 1821.
115p. 552.28/Sa77g

1952 **Saucerotte, Antoine Constant.**
Élémens d'histoire naturelle, présentant dans une suite de tableaux synoptiques, accompagnés de figures, un précis complet de cette science . . . Minéralogie. Paris, A. Delalain, 1834.
46p., 3 plates. 549.02/Sa8e

Signed by the author.

1953 **Saunders, William,** 1743-1817.
A treatise on the chemical history and medical powers of some of the most celebrated mineral waters; with practical remarks on the aqueous regimen. To which are added,

Observations on the use of cold and warm bathing. 2d ed., enl. London, Phillips and Fardon, 1805.
570p., fold. table. 613.12/Sa8t

1954 **Saussure, Horace Bénédict de,** 1740-1799.
Voyages dans les Alpes, précédés d'un essai sur l'histoire naturelle des environs de Genève. Neuchâtel, S. Fauche, 1779-96.
4v., illus., plates, fold. maps, tables. 914.94/Sa87v/1779

Vol. 2 has imprint: Genève, Barde, Manget & compagnie, 1786.
Vols. 3-4 have imprint: Neuchâtel, L. Fauche-Borel, 1796.

1955 **Saussure.**
Voyages dans les Alpes, précédés d'un essai sur l'histoire naturelle des environs de Genève. Neuchâtel, S. Fauche, 1780-86.
4v., illus., plates (part. fold.), fold. maps, tables. 914.94/Sa87v/1780

Vols. 3-4 have imprint: Genève, Barde, Manget, 1786.

1956 **Saussure.**
Voyages dans les Alpes, précédés d'un essai sur l'histoire naturelle des environs de Genève. Neuchâtel, L. Fauche-Borel, imprimeur, 1787-96.
8v., fold. plates, maps, tables, diagrs. 914.96/Sa87v

v.1-4 published at Genève.
Vols. 5-8 published in 1796.

1957 **Saussure.**
Voyages dans les Alpes; partie pittoresque des ouvrages de H. B. Saussure. Genève, Chez les principaux libraires; Paris, A. Cherbuliez, 1834.
396p. 914.94/Sa87v/1834

1958 **Saussure.**
Relation abrégée d'un voyage à la cime du Mont-Blanc en août, 1787. Genève, Barde, Manget, [1787?].
31p. 914.449/Sa8r

Facsimile reprints of the French ed., 1787, and the German ed., 1788, by C. Wolf in 1928 are fastened together in the same binder under the call number 914.449/Sa8r/1928.

1959 **Saussure.**
See SENEBIER, JEAN, Mémoire historique sur... Horace Bénédict de Saussure... 1801.

1960 **Savérien, Alexandre,** 1720-1805.
Histoire des progrès de l'esprit humain dans les sciences

exactes et dans les arts qui en dépendent. Paris, Lacombe, 1766.
536p. 509/Sa9h

1961 **Scafe, John.**
King Coal's levee; or, Geological etiquette, with explanatory notes; and The council of the metals. 3d ed. To which is added, Baron Basalt's tour. London, Longman, Hurst, Rees, Orme, and Brown [etc., etc.], 1819.
119p. 550/Sca2k/1819

In verse.

1962 **Scafe.**
A geological primer in verse; with, A poetical geognosy; or feasting and fighting, and sundry right pleasant poems, with notes. To which is added, A critical dissertation on "King Coal's levee," addressed to the professors and students of the University of Oxford. London, 1820.
68p. 550/Sca2g

1963 **Schäffer, Jacob Christian,** 1718-1790.
Kalchartiges Bergmeel in einer Steinkluft ohnweit Regensburg entdecket und beschrieben. Gebruckt bey den Gebrüdern Zunkel; in Commission zu haben bey J. C. Gollner, Leipzig, [1757].
38p. 549.7/Sch1k

Bibliographical footnotes.

1964 **Schenkenberg, F. C. A.**
Die lebenden Mineralogen; Adressen-Sammlung aller in Europa und den übrigen Weltthcilen bekannten Oryktognosten, Geognosten, Geologen und mineralogischen Chemiker... Stuttgart, E. Schweizerbart, 1842 [cover 1843].
167p. 549/Sch27ℓ

1965 **Scheuchzer, Johann Jacob,** 1672-1733.
Specimen lithographiae Helveticae curiosae, quo lapides ex figuratis Helveticis selectissimi aeri incisi sistuntur & describuntur à Johanne Jacobo Scheuchzero. Tiguri, Typis Davidis Gessneri, 1702.
67p., 7 fold. plates. 560.9494/Sch2s

1966 **Scheuchzer.**
ΟΥΡΕΣΙΦΟΙΤΗΣ HELVETICUS, sive, Itinera alpina tria: in quibus incolae, animalia, plantae, montium altitudines barometricae, coeli & soli temperies, aquae medicatae, mineralia, metalla, lapides figurati, aliaque fossilia; & quicquid insuper

in natura, artibus, & antiquitate, per Alpes helveticas, & rhaeticas, rarum sit, & notatu dignum, exponitur, & iconibus illustratur. Londini, impensis Henrici Clements, 1708.
3v. in 1, front. (port.), 43 plates (partly fold.)
x508.494/Sch29o/1708

There is one general t.-p.; each part has special t.-p.: ΟΥΡΕΣΙΦΟΙΤΗΣ HELVETICUS, sive Itineris alpini descriptio prima [-tertia] anno MDCCII-MDCCIV].
Imperfect: port. wanting.

1967 **Scheuchzer.**
Herbarium diluvianum collectum. Tiguri, Literis D. Gessneri, 1709.
44p. 10 plates. q561/Sch2h
One plate included in the paging.

1968 **Scheuchzer.**
Bibliotheca scriptorum historiae naturali omnium terrae regionum inservientium. Historiae naturalis Helvetiae prodromus. Accessit celeberrimi viri Jacobi Le Long, bibliothecarii oratoriani De scriptoribus historiae naturalis Galliae. Tiguri, typis H. Bodmeri, 1716.
241p. 016.57/Sch2b

1969 **Scheuchzer.**
Museum diluvianum quod possidet Joh. Jacobus Scheuchzer. Tiguri, Typis H. Bodmeri, 1716.
107p., front. 560/Sch2m

1970 **Scheuchzer.**
Helvetiae Stoicheiographia. Orographia et Oreographia. Oder Beschreibung der Elementen, Grenzen und Bergen des Schweitzerlands, Der Natur-Historie des Schweitzerlands erster [-dritter] Theil. Zürich, Bodmerische Truckerey, 1716-18.
3v. in 1., fold. plates. 550/Sch2n

Half title: Helvetiae historia naturalis, oder Natur-Historie des Schweitzerlandes.
Each volume has special t.p.; v.3 reads: "Ist der dritte oder eigentlich der sechste Theil der Natur-Geschichten des Schweitzerlands."

1971 **Scheuchzer.**
Herbarium diluvianum collectum, a Johanne Jacobo Scheuchzero . . . Editio novissima duplo auctior . . . Lugduni Batavorum, sumptibus Petri van der Aa, 1723.
5-119p., front. (port.), 14 plates. q561/Sch2h/1723

TRATTATO
DEL LEGNO FOSSILE MINERALE
NVOVAMENTE SCOPERTO

NEL QVALE BREVEMENTE SI ACCENNA LA VARIA, & mutabil natura di detto Legno, rappresentatoui con alcune figure, che mostrano il luogo doue nasce, la diuersità dell'onde, che in esso si vedono, e le sue così varie, e marauigliose forme.

DI FRANCESCO STELLVTI ACCAD. LINCEO
DA FABRIANO.

All'Emin.mo & Reuer.mo Sig. Card.

FRANCESCO
BARBERINO.

IN ROMA, Appresso Vitale Mascardi, MDCXXXVII.

CON LICENZA DE'SVPERIORI.

Cat. No. 2107

1972 **Scheuchzer.**
ΟΥΡΕΣΙΦΟΙΤΗΣ HELVETICUS, sive itinera per Helvetiae alpinas regiones facta annis MDCCII. MDCCIII. MDCCIV. MDCCV. MDCCVI. MDCCVII. MDCCIX. MDCCX. MDCCXI. Plurimis tabulis aeneis illustrata a Johanne Jacobo Scheuchzero. In quatuor tomos distincta. Lugduni Batavorum, typis ac suptibus P. van der Aa, 1723.
4v. in 2. x508.494/Sch29o

1973 **Scheuchzer,** praeses.
Homo diluvii testis et ΘΕΟΣΚΟΠΟΣ publicae *συζητήσει* expositus. Respp. pro examine philosophico . . . Henrico Wolphio [et al.] Tiguri, Typis J. H. Byrgklini, 1726.
24p., fold. plate. 560/Sch2h

At head of title: ΣΥΝΘΕΩ.

1974 **Scheuchzer.**
Physica sacra Johannis Jacobi Scheuchzeri . . . iconibus aeneis illustrata procurante & sumptus suppeditante. Johanne Andrea Pfeffel . . . Augustae Vindelicorum & Ulmae, 1731-35.
4 vols., ports., plates. xq220.85/Sch29p

Plates XLVI-LIX contain geological and paleontological subjects.

1975 **Scheuchzer.**
Kupfer-Bibel, in welcher die Physica Sacra, oder Beheiligte Natur-Wissenschafft derer in Heil. Schrifft vorkommenden Natürlichen Sachen, deutlich erklärt und bewährt . . . Anbey zur Erläuterung und Zierde des Wercks in Künstlichen Kupfer-Tafeln ausgegeben und verlegt durch Johann Andreas Pfeffel . . . Augspurg und Ulm, Gedruckt bey C. V. Wagner, 1731-35.
4 vols., ports., plates. xq220.85/Sch29pG

1976 **Schimper, Wilhelm Philipp,** 1808-1880.
See BUCKLAND, WILLIAM, Die Urwelt und ihre Wunder . . . 1838.

1977 **Schlüter, Christoph Andreas.**
De la fonte, des mines, des fonderies, &c. Tr. de l'allemand de Christophe-André Schlutter. Le tout augmenté de plusieurs procédés & observations; & pub. par M. Hellot . . . Paris, La veuve Pissot [etc.], 1750-53.
2v., 58 plates (53 fold.). 622/Sch38dFh

Copy imperfect: t.p. of v.2 missing.

1978 **Schmeisser, Johann Gottfried, 1767-1837.**
A system of mineralogy, formed chiefly on the plan of Cronstedt. London, C. Dilly, 1795.
2v., plates. x549/Sch4s

1979 **Schmerling, Philippe Charles.**
Recherches sur les ossements fossiles découverts dans les cavernes de la province de Liège. Liège, Collaradin, 1833-34.
2v., atlas (74 leaves of plates). uncat.

Title page lacking.

1980 **Schmidel, Casimir Cristoph,** 1718-1792.
Erz Stuffen und Berg Arten mit Farben genau abgebildet. Nürnberg, J. M. Seligmann, 1753.
36p., 26 col. plates. x549/Sch521e

Bound with his Fossilium metalla et res metallicas concernentium glebae suis coloribus expressae . . . 1753.

1981 **Schmidel.**
Fossilium metalla et res metallicas concernentium glebae suis coloribus expressae quas descripsit et digessit. Norimbergae, J. M. Seligman, 1753.
36p. x549/Sch521e

Bound with his Erz Stuffen und Berg Arten . . . 1753.

1982 **Schmidel.**
Vorstellung einiger merkwürdigen Versteinerungen mit kurzen Anmerkungen versehen. Erlangen, J. J. Palm, 1793.
25, 45p., 24 col. plates. x561/Sch5v/1793

Edited by Valentin Bischoff; the "Fortgesetzte Vorstellung" (45p., 2d group) edited by Joh. Chr. Schreber, has special t.p. Cf. Cat. générale de la Bibl. nat., v. 166 (1944) column 303.
Plates signed: Val. Bischoff ad nat. pinx. sculp. et exc. Norimb.
Perhaps a re-issue of the 1780 ed. with cancel. t.p.'s dated 1793.

1983 **Schmidt, A. R.**
Geognostische Karte von Vorarlberg. Aufgenommen und gezeichnet vom K. K. Markscheider A. R. Schmidt in den Jahren 1839-1841. [n.p.], 1841?
G6493/.V953C5/1841/.S3

Scale not given.
Two sheets; sheet 2 consists of geological profiles.

1984 **Schmieder, Karl Christoph,** 1778-1850.
Topographische Mineralogie der Gegend um Halle in Sachsen; oder, Beschreibung der sich um Halle findenden Mineralien

und Fossilien, nebst genauer Anzeige der Orte. Halle, J. C. Hendel, 1797.
151p. 549.94321/Sch5t

1985 **Schmieder.**
Die Geognosie nach chemischen Grundsätzen. Leipzig, S. L. Crusius, 1802.
368p. 551.1/Sch5g

1986 **Schoolcraft, Henry Rowe,** 1793-1864.
A view of the lead mines of Missouri; including some observations on the mineralogy, geology, geography, antiquities, soil, climate, population, and productions of Missouri and Arkansas, and other sections of the western country. New York, C. Wiley & co., 1819.
299p., front., 2 plates. 557/M66s

1987 **Schoolcraft.**
Narrative journal of travels from Detroit Northwest through the Great Chain of American Lakes to the sources of the Mississippi River in the Year 1820. Albany, E. & E. Hosford, 1821.
419p., illus., front. (map). x917.7/Sch6n

1988 **Schoolcraft.**
A memoir on the geological position of a fossil tree, discovered in the secondary rocks of the river Des Plaines. Read before the American geological society. Albany, Printed by E. and E. Hosford, 1822.
18p. x557/Iℓ6sc

1989 **Schoolcraft.**
Travels in the central portions of the Mississippi Valley; comprising observations on its mineral geography, internal resources, and aboriginal population. Performed under the sanction of government, in the year 1821. New York, Collins and Hannay, 1825.
459p., front., plates, fold. maps. 917.7/Sch6t

1990 **Schoolcraft.**
Narrative of an expedition through the Upper Mississippi to Itasca Lake, the Actual source of this river: embracing an exploratory trip through the St. Croix and Burntwood (or Broule) rivers: in 1832. New York, Harper, 1834.
307p. 917.76/Sch6n

1991 **Schöpf, Johann David,** 1752-1800.
Beyträge zur mineralogischen Kenntniss des östlichen Theils

von Nord-Amerika und seiner Gebürge. Erlangen, J. J. Palm, 1787.
194p. x557/Xsc

1992 **Schöpf.**
Reise durch einige der mittlern und südlichen Vereinigten nord-amerikanischen Staaten nach Ost-Florida und den Bahama-Inseln unternommen in den Jahren 1783 and 1784. Erlangen, J. J. Palm, 1788.
2v., fold. map. x917.3/Sch6r

Translated into English as Travels in the Confederation, 1783-1784, Philadelphia, 1911.

1993 **Schott, Gasper,** 1608-1666.
P. Gasparis Schotti . . . Anatomia physico-hydrostatica fontium ac fluminum Libris VI. explicata, et figuris aeri incisis exornata; quibus praemissâ fontium ac fluminum historiâ, primò eroundem principium & matrix accurato scrutinio inquiritur; deinde modus quo è matrice praecipua (quae mare est) oriuntur, ex Sacrae Scripturae, Philosophiae naturalis, & Hydrostaticae principiis detegitur; postremò variae proprietates ac veluti miracula in variis fontibus, elucescentia discutiuntur. Accedit in fine, appendix de vera origine Nili. Herbipoli, Viduae Jo. Godofr. Schönwetteri Excudit Jobus Hertz, 1663.
433p., illus. uncat.

1994 **Schott.**
See KIRCHER, ATHANASIUS, Iter extaticum coeleste . . . 1671.

1995 **Schröter, Johann Samuel,** 1735-1808.
Lithologisches Reallexikon, in welchem so wohl die Lithographie, als auch die nöthigsten Wahrheiten der Lithogeognosie enthalten sind. Berlin, J. G. Bosse, 1772-88.
8v. in 4., port. 552/Sch7ℓ

Vol. 2-8: Lithologisches Real- und Verballexikon, in welchem nicht nur die Synonymien der deutschen, lateinischen, französischen und holländischen Sprachen angeführt und erläutert, sondern auch alle Steine und Versteinerungen ausführlich beschrieben werden. Frankfurt am Mayn, Varrentrapp Sohn und Wenner, 1779-88.

1996 **Schröter.**
Vollständige Einleitung in die Kenntniss und Geschichte der Steine und Versteinerungen. Altenburg, In der Richterischen Buchhandlung, 1774-84.
4v., plates (part. fold.). 552/Sch7v

1997 **Schultz, Christian.**
Travels on an inland voyage through the states of New-York,

Pennsylvania, Virginia, Ohio, Kentucky and Tennessee, and through the territories of Indiana, Louisiana, Mississippi, and New-Orleans; performed in the years 1807 and 1808; including a tour of nearly six thousand miles. With maps and plates. New-York, Printed by Isaac Riley, 1810.
2v. in 1, fronts. (v.1, port.), plates, fold. maps, fold. plan.
x917.3/Sch8t

Front. of v.1 wanting.

1998 **Schultz, Ernst Christoph,** d. 1810.
Entdeckung einer dem Kreuz-Steine wesentlichen Enstehungs-Art der Kreuz-Figur. An . . . Friedrich Anton Freyherrn von Heinitz. Hamburg, J. P. C. Reuss, [1780].
38p., plate. 549/Sch82e

1999 **Schultze, Christian Friedrich,** 1730-1775.
Kurtze Betrachtung derer Kräuterabdrücke im Steinreiche, worinnen dieselben so wohl in Ansehung ihres Ursprungs, als auch ihres eigenthümlichen Unterscheides und übrigen Eigenschafften, in Erwegung gezogen werden. Nebst beygefügten Kupfertafeln. Dressden, F. Hekel, 1755.
76p., 6 fold. tables. 561/Sch8k

2000 **Schumacher, Christian Friedrich,** 1757-1830.
Versuch eines Verzeichnisses der in den Dänisch-Nordischen Staaten sich findenden einfachen Mineralien mit Tabellen der einfachen Fossilien nach ihren vorwaltenden Bestandtheilen. Kopenhagen, F. Brummer, 1801.
172p., tables. 549/Sch86v

2001 **Schweling, Johann Eberhard,** 1645-1714.
Memorabilium ab orbe condito in nostra tempora usq: terraemotuum historias physicè explicatas, auspice domino ac sospitatore nostro Jesu Christo, . . . publicae commilitum fuorum ventilationi submittit Luderus Abraham, . . . auto ac resp. ad diem III currentis Maji horis solitis in acroaterio theologorum. Bremae, Typis H. Braueri, 1684.
36p. x551.22/Sch9m

2002 **Schwoy, Franz Joseph.**
Topographie vom Markgrafthum Mähren. Wien, gedruckt bey Joseph Hraschanzky, 1793-94.
3v., port. 551.4/Sch9t

2003 Scientific tracts, for the diffusion of useful knowledge . . . complete in one volume. Boston, Light & Stearns, 1836.
308p. uncat.

NICOLAI STENONIS
ELEMENTORVM
MYOLOGIÆ SPECIMEN,

SEV

Muſculi deſcriptio Geometrica.

CVI ACCEDVNT

CANIS CARCHARIÆ DISSECTVM CAPVT,

ET

DISSECTVS PISCIS EX CANVM GENERE.

A D

SERENISSIMVM

FERDINANDVM II.

MAGNVM ETRVRIÆ DVCEM.

F L O R E N T I Æ,

Ex Typographia ſub ſigno STELLÆ. MDCLXVII.

Superiorum Permiſſu.

Cat. No. 2108

Includes, among other papers, the Natural history of water by C. T. Jackson, Granite rock by Samuel Fish, and Theory of the earth by Samuel Fish.

2004 **Scilla, Agostino,** 1639-1700.
La vana speculazione disingannata dal senso. Lettera risponsiva circa i corpi marini, che petrificati si trovano in varij luoghi terrestri. Napoli, Andrea Colicchia, 1670.
168p., 28 plates, front. uncat.

2005 **Scilla.**
De corporibus marinis lapidescentibus quae defossa reperiuntur. Addita dissertatione Fabii Columnae De glossopetris. Ed. altera emendatior. Romae, Sumptibus V. Monaldini, 1752.
84p., illus., 28 [i.e. 30] plates.
560/Sci4c/1752, and xg560.Sci4c/1752

Half-title: Vanae Speculationis sensus moderator.

2006 **Scilla.**
De corporibus marinis lapidescentibus quae defossa reperiuntur. Addita dissertatione Fabii Columnae De glossopetris. Ed. altera emendatior. Romae, Sumptibus V. Monaldini, 1759.
82p., illus., 28 [i.e. 30] plates. 560/Sci4c/1759

Half-title: Vanae Speculationis sensus moderator.

2007 **Scinà, Domenico, 1765-1837.**
Rapporto sulle ossa fossili di Mardolce e degli altri contorni di Palermo. Palermo, Dalla Reale Tipografia di Guerra, 1831.
64p., 2 plates (incl. map). 554.5817/Sc61r

2008 **Scopoli, Giovanni Antonio,** 1723-1788.
Dissertationes ad scientiam naturalem pertinentes. Pars I. Tentamen Mineralogicum de schematibus Metallorum. De Minera Argenti rubra. De Sinopi Hungarica Sinopi dicta. Plantae subterraneae descriptae & delineatae. Pragae, Sumptibus W. Gerle, 1772.
120p., 46 plates. 570/Sco6d

2009 **Scopoli.**
. . . Principia mineralogiae systematicae et practicae, succincte exhibentia structuram telluris, systemata mineralogica, lapidum classes, genera, species, cum praecipuis varietatibus, eorumque characteribus, synonymis, analysi et usu, nec non regulis nonnullis generalibus, ad docimasiam et pyrotechniam metallurgicam pertinentibus. Pragae, Apud W. Gerle, 1772.
228p., port. 549/Sco6p

2010 **Scopoli.**
Anfangsgründe der systematischen und praktischen Mineralogie, kürzlich vorstellend den Bau der Erdkugel, die mineralogischen Lehrgebäude, die Klassen, Geschlechter, Arten und vornehmsten Abänderungen der Steine, dann ihre Kennzeichen, Synonimen, Zerlegung und Gebrauch, nicht weniger einige allgemeine zur Probier- und Schmelzkunst gehörigen Regeln. Aus dem Lateinischen übersetzt von Karl von Meidinger. Prag. W. Gerle, 1775.
192p. 549/Sco6pGf

Translation of Principia mineralogiae systematicae et practicae.

2011 **Scriba, Franciscus Petrus.**
Kurtze doch mit hinlänglicher Deutlichkeit entworffene Beschreibung des vor einigen Jahren neu-entdeckten mineralischen Gesund-Brunnens, in dem Hochfürstlichen Stifft Corvey, ohnweit einem Dorffe Godelheim genannt . . . Wobey zugleich aus Veranlassung noch zweyer andern mineralischen Wassern in eben diesem Stifft, etwas weniges von Erzeugung der mineralischen Wassern in dem Schoos der Erden mit eingeschaltet wird. Huxar, G. E. Schultzen, [1770?].
96, 15p. 613.12/Scr3k

2012 **Scribonius, Wilhelm Adolph,** 16th cent.
See WIDDOWES, DANIEL, Naturall philosophy . . . 1631.

2013 **Scrope, George Julius Duncombe Poulett,** 1797-1876.
Considerations on volcanos; the probable causes of their phenomena, the laws which determine their march, the disposition of their products and their connexion with the present state and past history of the globe; leading to the establishment of a new theory of the earth. London, W. Phillips; Edinburgh, 1825.
270p., illus., fold. col. map. 551.21/Scr5v/1825

Subsequently pub. under title: Volcanos.

2014 **Scrope.**
Volcanos. The character of their phenomena, their share in the structure and composition of the surface of the globe, and their relation to its internal forces. With a descriptive catalogue of all known volcanos and volcanic formations, 2d. ed., rev. and enl. With a map of the volcanic areas of the globe, frontispiece, wootcuts, etc. London, Longman, Green, Longmans, and Robert, 1862.
490p., illus. 551.21/Scr5v

First ed. was published as Considerations on volcanos . . . 1825.

2015 **Scrope.**
Memoir on the geology of central France; including the volcanic formations of Auvergne, the Velay, and the Vivarais. London, Longman, Rees, Orme, Brown, and Green, 1827.
182p., and atlas of 18 (i.e. 19) plates and 2 maps. q554.4/Scr5m

"Published notices on the volcanos of Auvergne": p. [35]-39. Atlas wanting.

2016 **Scrope.**
The geology and extinct volcanos of Central France. Second edition, enlarged and improved. London, John Murray, 1858.
258p., illus., 16 plates, map. 551.2/Scr5

2017 **Seale, Robert F.**
The geognosy of the island St. Helena, illustrated in a series of views, plans and sections; accompanied with explanatory remarks and observations. London, Ackermann and co., 1834.
25p., illus., 11 col. plates. q559.7/Se1g

2018 **Sears, Reuben.**
A poem, on the mineral waters of Ballston and Saratoga, with notes illustrating the history of the springs and adjacent country. Ballston Spa, published by the author, J. Comstock, printer, 1819.
108p. uncat.

2019 **Sedgwick, Adam,** 1785-1873.
On the physical structure of Devonshire, and on the subdivisions and geological relations of its older stratified deposits, &c, by Adam Sedgwick and Roderick Impey Murchison, *in* Transactions of the Geological Society of London, 1840.
2nd ser., v.5, pt. 3, p.633-704, 8 plates. 550.6/GEO

2020 **Sedgwick.**
On the distribution and classification of the older or Palaeozoic deposits of the North of Germany and Belgium, and their comparison with formations of the same age in the British Isles, by Adam Sedgwick and Roderick Impey Murchison, *in* Transactions of the Geological Society of London, 1842.
2nd ser., v.6, pt. 2, p.221-301, 15 plates. 550.6/GEO

2021 **Sedgwick.**
On the classification of the fossiliferous slates of North Wales, Cumberland, Westmoreland and Lancashire, *in* Quarterly Journal of the Geological Society of London, 1847.
v.3, pt. 1, p.133-164. 550.6/GL

2022 **Sendel, Nathanael,** 1686-1757.
See BREYN, PHILIPP, Epistola de melonibus petrefactis montis Carmel . . . 1722.

2023 **Sendel.**
Historia succinorum corpora aliena involventium et naturae opere pictorum et caelatorum, ex Regiis Augustorum cimeliis Dresdae conditis aeri insculptorum conscripta. Lipsiae, apud I. F. Gleditschium, 1742.
328p., 13 plates. xf560/Se55h

2024 **Senebier, Jean,** 1742-1809.
Riflessioni generali sopra i vulcani. Traduzione dal francese. Venezia, Presso P. Q. G. B. Pasquali, 1795.
106p. 551.21/Se5rl

2025 **Senebier.**
See SPALLANZANI, LAZZARO, Voyage dans les Deux Siciles . . . 1795-96.

2026 **Senebier.**
Mémoire historique sur la vie et les écrits de Horace Bénédict de Saussure; pour servir d'introduction à la lecture de ses ouvrages. Genève, Paschoud, [1801].
219p. B/S2591s

"Lu à la Société de physique et d'histoire naturelle de Genève, le 23 prairial an VIII."

2027 **Senger, Wilhelm von.**
Versuch einer Oryctographie der gefürsteten Grafschaft Tirol. Innsbruck, Wagner'schen Schriften, 1821.
94p. 549.94364/Se5v

2028 **Serao, Francesco,** 1702-1783.
Istoria dell'incendio del Vesuvio accaduto nel mese di maggio del'anno MDCCXXXVII, scritta per l'Accademia delle scienze. Napoli, 1738.
122p., fold. plates. x551.21/Se65i

Bibliographical footnotes.

2029 **Serao.**
Istoria dell'incendio del Vesuvio accaduto nel mese di maggio dell'anno MDCCXXXVII. scritta per l'Accademia delle scienze. Napoli, Stamperia de Novello de Bonis, 1738.
163p., 2 fold. plates. x551.21/Se65i/1738a

Italian text and Latin translation in parallel columns.
Bibliographical footnotes.

2030 **Serao.**
Istoria dell'incendio del Vesuvio accaduto nel mese di maggio dell'anno MDCCXXXVII scritta per l'Accademia delle scienze. 2.ed. riv. ed. accresciuta di alquante annotazioni. Napoli, Stamperia di Angelo Vocola, 1740.
226p., fold. plates. x551.21/Se65i2

2031 **Serao.**
Neapolitanae scientiarum academiae De Vesuvii conflagratione quae mense majo anno MDCCXXXVII accidit commentarius. Neapoli, typis Novelli de Bonis, 1738.
118p., fold. plate. x551.21/Se65d

Bibliographical footnotes.
Folded plate wanting.

2032 **Serao.**
Histoire du Mont Vésuve, avec l'explication des phénomènes qui ont coûtume d'accompagner les embrasements de cette montagne. Le tout traduit de l'italien de l'Académie des sciences de Naples. Par M. Duperron de Castera. Paris, Huart, 1741.
361p., 3 fold. plates. 551.21/Se65iFd

2033 **Serao.**
The natural history of Mount Vesuvius, with the explanation of the various phenomena that usually attend the eruptions of this celebrated volcano. Translated from the original Italian, composed by the Royal Academy of sciences at Naples, by order of the king of the two Sicilies. London, Printed for E. Cave, 1743.
231p., 2 fold. plates, fold. tables. x551.21/Se65iE

Translation of Istoria dell'incendio del Vesuvio.

2034 **Serpetro, Nicolò.**
Il mercato delle maraviglie della natura, overo, istoria naturale . . . I Venetia, Per il Tomasini, 1653.
368p. uncat.

2035 **Serres, Marcel de,** 1783-1862.
Géognosie des terrains tertiaires; ou, Tableau des principaux animaux invertébrés des terrains marins tertiaires, du midi de la France. Montpellier, Paris, Chez Pomathion-Durville, 1829.
276p., 6 plates, 3 fold. tables. 560.178/Se6g

2036 **Serres.**
De la simultanéité des terrains de sédiment supérieur. Paris, Impr. de Madame Veuve Agasse, 1830.

NICOLAI STENONIS
DE SOLIDO
INTRA SOLIDVM NATVRALITER CONTENTO
DISSERTATIONIS PRODROMVS.

AD

SERENISSIMVM

FERDINANDVM II.
MAGNVM ETRVRIÆ DVCEM.

FLORENTIÆ

Ex Typographia ſub ſigno STELLÆ MDCLXIX.

SVPERIORVM PERMISSV.

Cat. No. 2110

124p., tables. 551.781/Se6d

"Extrait du tome v de la Géographie-physique de l'encyclopédie méthodique."

2037 **Serres.**
Recherches sur les cavernes à ossemens de Lunel-vieil, Hérault. Dubeuil . . . , Jeanjean . . . [between 1832 and 1839].
[358]ℓ., illus., maps. MS/xq551.44/Se68r

Holograph.
Title page mutilated, with partial loss of map on verso.

2038 **Serres.**
Notice géologique sur le département de l'Aveyron. [Bruxelles, 1845].
95p., col. plate, table. q554.474/Se6n

([Académie royale des sciences, des lettres et des beaux-arts de Belgique. Mémoires couronnés et mémoires des savants étrangers] t.XVIII [7]).

2039 **Severin, L. J.**
Dictionnaire des nomenclatures chimique et minéralogique anciennes, comparées aux nomenclatures chimique et minéralogique modernes, d'après les ouvrages des chimistes et le traité de minéralogie de Mr. Haüy; auquel on a joint trois tableaux synoptiques destinés à offrir les principaux caractères des corps simples, et un quatrième tableau qui présente les caractères des acides; avec trois planches pour les signes chimiques. Paris, Samson, 1807.
232p., fold. tables. 541/Se89d

2040 **Shepard, Charles Upham,** 1804-1886.
Treatise on mineralogy. New Haven, H. Howe, 1832.
256p., diagrs. 549/Sh4t

2041 **Shepard.**
Treatise on mineralogy; second part, consisting of descriptions of the species, and tables illustrative of their natural and chemical affinities . . . New Haven, H. Howe & co., and Herrick & Noyes, 1835.
2v. in 1, diagrs. 549/Sh4tr

2042 **Shepard.**
Geology of upper Illinois. [New Haven], 1838.
p.[3]-30, illus., map. uncat.

Disbound from the American Journal of Science and Arts, v.34, no. 1.

2043 **Shephard, Charles.**
An historical account of the island of Saint Vincent. London, Printed by W. Nicol, 1831.
216, [76]p., front. 972.984/Sh4h

2044 **Short, Thomas,** 1690?-1772.
The natural, experimental, and medicinal history of the mineral waters of Derbyshire, Lincolnshire, and Yorkshire, particularly those of Scarborough. Wherein they are carefully examined and compared, their contents discovered and divided, their uses shewn and explained, and an account given of their discovery and alterations. Together with the natural history of the earths, minerals and fossils through which the chief of them pass. The groundless theories, and false opinions of former writers are exposed, and their reasonings demonstrated to be injudicious and inconclusive. To which are added, large marginal notes, containing a methodical abstract of all the treatises hitherto published on these waters, with many observations and experiments. As also four copperplates representing the crystals of the salts of thirty four of those waters. London, Printed for the author, and sold by F. Gyles, 1734.
359p., 5 plates (fold.). x615.79/Sh8n

Bound with his An essay towards a natural . . . history of the principle mineral waters of Cumberland . . . 1740.

2045 **Short.**
An essay towards a natural, experimental, and medicinal history of the principle mineral waters of Cumberland, Northumberland, Westmoreland, Bishop-prick of Durham, Lancashire, Cheshire, Staffordshire, Shropshire, Worcestershire, Glocestershire, Warwickshire, Northamptonshire, Leicestershire, and Nottinghamshire, particularly those of Neville Holt, Cheltenham, Weatherflack, Hartlepool, Astrope, Cartmall, etc. Wherein they are carefully examined and compared, their mineral contents are discovered and separated, their uses shewn and explained, etc. To which is added a short discourse on cold and tepid bathing, and a table of the temperature of all the warm waters in England, and most of the cold baths, from Carlisle to Glocester and Oxford. Being the second volume of the mineral waters of England. Sheffield, Printed for the author by J. Garnet, 1740.
330p. x615.79/Sh8n

Bound with his The natural, experimental, and medicinal history of the mineral waters of Derbyshire . . . London, 1734.

2046 **Short.**
A general treatise on various cold mineral waters in England, but more particularly on those at Harrogate, Thorp-Arch, Dorst-Hill, Wigglesworth, Nevill-Holt, and others of the like nature. With their principles, virtues and uses. Also a short discourse on solvents of the stone in the kidneys and bladder. London, Printed for the author, 1765.
[1]-144, 129-248p. x613.38/Sh8g

2047 **Shower, John,** 1657-1715.
Practical reflections on the late earthquakes in Jamaica, England, Sicily, Malta, &c. anno 1692. With a particular, historical account of those, and divers other earthquakes. London, Printed for J. Salusbury [etc.], 1693.
206p., port. x551.2/Sh8p

2048 **Shower.**
Practical reflections on the late earthquakes that have happened in Europe and America, but chiefly in the islands of Jamaica, England, Sicily, Malta, etc. With a particular and historical account of them, and divers other earthquakes. The 2d ed. London, Cook, James, and Kingman [etc.], 1750.
98p. x551.2/Sh8p/1750

2049 **Shumard, Benjamin Franklin,** 1820-1869, jt. author.
See YANDELL, LUNSFORD PITTS, Contributions to the geology of Kentucky... 1847.

2050 **Sibbald, Robert,** 1611-1722.
Scotia illustrata sive Prodromus historiae naturalis in quo regionis natura, incolarum ingenia & mores, morbi iisque medendi methodus, & medicina indigena accuratè explicantur... Edinburgi, ex officinâ typographicâ Jacobi Kniblo, J. Solingensis & J. Colmarii, sumptibus auctoris, 1683-84.
3v. in 1, 2 plates, port. xq570.9411/Si1s

Each part has special t.-p.

Contents. — [p.1] Nuncius scoto-britannus, sive Admonitio de atlante scotico seu descriptione Scotiae antiquae et modernae. 1683. — p.2. t.1. De plantis Scotiae. 1684. — p.2. t.2. De animalibus Scotiae... et de mineralibus metallis et marinis Scotiae. 1684.

2051 **Sigaud-Lafond, Joseph Aignan,** 1730-1810.
Dictionnaire des merveilles de la nature, par A. J. S. D. Nouvelle édition, revue, corrigée et considérablement... augmentée par l'auteur. Paris, Chez Delaplace, 1802.
3v. 503/Si2d/1802

2052 **Silbermann,** fl. 1773.
Manuel metallotechnique, ou Recueil de secrets & de curiosités sur les métaux & les minéraux appliqués aux arts & aux métiers. Leipsick, Arkstée & Merkus, 1773.
473p. 553/Si32mF

2053 **Silberschlag, Johann Esaias,** 1721-1791.
Théorie des fleuves, avec l'art de bâtir dans leurs eaux et de prévenir leurs ravages. Ouvrage traduit de l'allemand par M. d'Au. Paris, C. A. Jombert, 1769.
130p., 13 plates. q627.4/Si3Fa

2054 **Silliman, Benjamin,** 1779-1864.
... Sketch of the mineralogy of the town of New Haven ...
506/CONN/v.1/no.1

Connecticut academy of arts and sciences. Memoirs. New Haven, 1810. v.1, p. [83]-96.

2055 **Silliman.**
See BAKEWELL, ROBERT, An introduction to geology... 1829, 1833 and 1839.

2056 **Silliman.**
Report on the coal formation of the valleys of Wyoming and Lackawanna; to which are added miscellaneous remarks and communications. New Haven, printed by Hezekiah Howe, 1830.
29p., illus., fold. map. uncat.

2057 **Silliman.**
See AGASSIZ, LOUIS, Nouvelles études et expériences sur les glaciers actuels ... 1847.

2058 **Silliman.**
See MANTELL, GIDEON ALGERNON, The wonders of geology... 8th ed.... 1864-66.

2059 **Silvertop, Charles.**
A geological sketch of the Tertiary formation in the provinces of Granada and Murcia, Spain. With notices respecting primary, secondary, and volcanic rocks in the same district and sections. London, Longman, Rees, Orme, Brown, Green, and Longman, 1836.
236p., fold. front. (map), 7 profiles. 554.6/Si3g

2060 **Simler, Josias,** 1530-1576.
Vallesiae descriptio, libri duo. De Alpibus commentarius. Describuntur vero in his libris, primo quidem regionis situs,

populi mores, Reipublicae forma, urbes, arces, pagi, montes, flumina, aliaq; memoratu digna: & praeterea Vallesianorũ res geste. Deinde Alpium natura & magnitudo, ve tera nomina & itinera: & multa quae propria & peculiaria sunt Alpibus commemorantur: Auĉtorum etian loci plurimi citantur & explicantur. Tiguri excudebat Ch. Froschoverus, 1574.
209p. uncat.

2061 **Simler.**
Vallesiae et Alpium descriptio. Lugduni Batavorum, ex officina Elzeviriana, 1633.
377p., front. uncat.

2062 **Simond, Louis,** 1767-1831.
Switzerland; or, A journal of a tour and residence in that country in the years 1817, 1818, and 1819; followed by an historical sketch of the manners and customs of ancient and modern Helvetia . . . Boston, Wells and Lilly, 1822.
2v. 914.94/Si5s2

2063 **Simond.**
Voyage en Suisse, fait dans les années 1817, 1818 et 1819. Paris, Treuttel et Wurtz, 1822.
2v. uncat.

2064 **Simpson, William,** fl. 1665-1677.
Hydrologia chymica: or, the chymical anatomy of the Scarbrough, and other spaws in York-shire wherein are interspersed some animadversions upon Dr. Wittie's lately published treatise of the Scarbrough-Spaw. Also a short description of the spaws at Malton and Knarsbrough. And a discourse concerning the original of hotsprings and other fountains: with the causes and cures of most of the stubbornest diseases (either chronical or acute) incident to the body of Man. Also, a vindication of chymical physick . . . Lastly, is subjoyned an appendix of the original of springs; with the author's *ternary* of medicines: and the epilogue to the whole, of the essence of the Scarbrough-Spaw. London, Printed by W. G. for Richard Chiswel, 1669.
374p. uncat.

2065 **Sinclair, George,** d. 1696.
The hydrostatiks; or, The weight, force, and pressure of fluid bodies, made evident by physical, and sensible experiments. Together with some miscellany observations, the last whereof is a short history of coal, and of all the common, and proper

THE

PRODROMUS

TO A

DISSERTATION

Concerning

SOLIDS Naturally Contained within SOLIDS.

Laying a Foundation for the Rendering a Rational Accompt both of the *Frame* and the ſeveral *Changes* of the Maſſe of the EARTH, as alſo of the various *Productions* in the ſame.

By *NICOLAUS STENO.*

Engliſh'd by *H. O.*

LONDON,
Printed by *J. Winter*, and are to be Sold by *Moſes Pitt* at the *White-Hart* in *Little Brittain*, 1671.

Cat. No. 2112

accidents thereof; a subject never treated before. Edinburgh, George Swinton, James Glen, and Thomas Brown, 1672.
319p., illus. uncat.

Title page mutilated.

2066 **Sinclair.**
Natural philosophy improven by new experiments. Touching the mercurial weather-glass, the hygroscope, eclipsis, conjunctions of Saturn and Jupiter. — Lastly, there is a large discourse anent coal, coal-sinks, dipps, risings, and streeks of coal, levels, running of mines, gaes, dykes, damps, and wildfire. Edinburgh, to be sold by Gideon Schaw, 1683.
302p., illus. uncat.

2067 **Sivry, Esprit Pierre de.**
Journal des observations minéralogiques faites dans une partie des Vosges et de l'Alsace . . . Nancy, H. Haener, 1782.
117p. 554.438/Si9j

Ouvrage qui a remporté le prix au jugement des Messieurs de la Société royale des sciences, belles-lettres & arts de Nancy, en 1782.

2068 **Smellie, William,** 1740-1795.
The philosophy of natural history. Edinburgh, Printed for the heirs of Charles Elliot, 1790-99.
2v. x591.5/Sm3p

Vol. 2 has imprint: Edinburgh, Printed for Bell & Bradfute, 1799.

2069 **Smith, Hamilton Lanphere,** 1819-1903.
The world; or First lessons in astronomy and geology, in connection with the present and past condition of our globe. Cleveland, M. C. Younglove, 1848.
324p., front., illus., diagrs., tables. 551/Sm5w

2070 **Smith, John Pye,** 1774-1851.
On the relation between the Holy Scriptures and some parts of geological science. New-York, D. Appleton, 1840.
364p. 213/Sm6o

Half-title: The Congregational lecture. Sixth ser.

2071 **Smith, Thomas,** 1775 (or 1776)-1830.
The wonders of nature and art, or, A concise account of whatever is most curious and remarkable in the world; compiled from historical and geographical works of established celebrity, and illustrated with the discoveries of modern

travellers. Rev., corr., and improved, by James Mease. Philadelphia, Printed by R. Carr for Birch & Small, 1806-07.
14v. x910/Sm67w/1806

Library owns v.8, 11-14 only.

2072 **Smith, William,** 1769-1839.
A delineation of the strata of England and Wales, with part of Scotland; exhibiting the collieries and mines, the marshes and fen lands originally overflowed by the sea . . . [London], 1815.
15 col. double maps. xf554.2/Sm6d

Map in sections, bound.
Imperfect: general map wanting.

2073 **Smith.**
A memoir to the map and delineation of the strata of England and Wales, with part of Scotland . . . London, J. Cary, 1815.
51p., fold. tables. x554.2/Sm6d/sup.

2074 **Smith.**
Strata identified by organized fossils, containing prints on colored paper of the most characteristic specimens in each stratum. London, Printed by W. Arding, 1816.
32p., front., 18 plates. xq551.7/Sm6s

2075 **Smith.**
Stratigraphical system of organized fossils, with reference to the specimens of the original geological collection in the British Museum: explaining their state of preservation and their use in identifying the British strata. [Pt. 1], London, Printed for E. Williams, 1817.
118p., 2 fold. tables. q551.7/Sm6st

No more published.

2076 **Smith.**
Smith's Geological atlas. London, 1819-43.
39 col. maps on 42ℓ. 58x69cm. xf554.2/Sm6g

Binder's title.
Contains all but one (Kent) of the county maps in Smith's A new geological atlas of England and Wales (1819-24). In addition there are 19 maps of other English counties, by John Cary, engraver, dated 1842-43.

2077 **Smith.**
Syllabus of lectures on geology in the Town-Hall, Scarborough, including three lectures on organic fossils by . . . J. Phillips. Scarborough, John Cole, 1824.
4p. x550.715/Sm6s/1824

Caption title.

Facsimile of the first page included in J. M. Edmonds, The geological lecture-courses given in Yorkshire by William Smith and John Phillips, 1824-1825, p.382.

Accompanied by a letter from L. R. Cox of British Museum (Natural history) addressed to George W. White at Dept. of Geology, Univ. of Illinois, Urbana, stating that "It is most probable that your copy and the one at Oxford are the only two which survive." Letter dated March 30, 1955.

2078 **Smith.**
Geological map of Yorkshire. London, J. Cary, 1824.
Col. map. G5753/.Y65/1824/.S6

Scale, 1 inch equals ca. 2.75 miles. [1:169,000]

"A new map of Yorkshire divided into its Ridings and Subdivided into Wapontakes. By John Cary. 1824."

2079 **Smith.**
A new geological map of England and Wales, with the inland navigations exhibiting the districts of coal and other sites of mineral tonnage. London, J. Cary, 1824.
Col. map. G5750/.C5/1824/.S6

Scale, 1 inch equals 15 British Statute miles.

2080 **Smith.**
Synopsis of geological phenomena. Oxford, Eng., 1832.
1 folded sheet, 54 x 36 cm., folded to 27 x 22 cm.
x550.18/Sm68s

Mounted on cloth.

2081 **Smith.**
See PHILLIPS, JOHN, Memoirs of William Smith . . . 1844.

2082 **Smyth, John Ferdinand Dalziel.**
Tour in the United States of America; containing an account of the present situation of that country; the population, agriculture, commerce, customs, and manners of the inhabitants . . . with a description of the Indian nations . . . likewise improvements in husbandry . . . London, For G. Robinson [etc.], 1784.
2v. x917.3/Sm9t

2083 **Soldani, Ambrogio,** ca. 1736-1808.
Saggio orittografico; ovvero, Osservazioni sopra le terre nautilitiche ed ammonitiche della Toscana. Con appendice o indice latino ragionato de'piccoli testacei, e d'altri fossili d'origine marina per schiarimento dell'opera. Siena, Stamp. di V. Pazzini, 1780.
146p., 25 plates. 564.5/So4s

2084 **Solenander, Reiner,** 1524-1601.
. . . De caloris fontium medicatorum causa, eorúmque temperatione, libri duo, et philosophis et medicu perutiles. Lugduni, apud Joannem Franciscum de Gabiano, 1558
238p. uncat.

2085 **Sopwith, Thomas,** 1803-1899.
An account of the mining districts of Alston Moor, Weardale and Teesdale, in Cumberland and Durham; comprising descriptive sketches of the scenery, antiquities, geology, and mining operations, in the upper dales of the rivers Tyne, Wear, and Tees. Alnwick, Printed by and for W. Davison, 1833.
183p., front. (map), illus., diagr. 914.28/So6a

2086 **Sorrentino, Ignazio.**
Istoria del Monte Vesuvio, divisata in due libri. Napoli, Per G. Severini, 1734.
224p. 551.21/So61

2087 **Soulavie, Jean Louis Giraud,** 1752-1813.
Histoire naturelle de la France méridionale; ou, Recherches sur la minéralogie du Vivarais, du Viennois, du Valentinois, du Forez, du Velai, de l'Uségeois, du Comtat-Vénaissin, des Diocèses de Nismes, Montpellier, Agde, &c. Sur la physique de la mer Méditerranée; sur les méteores, les arbres, les animaux, l'homme & la femme de ces contrées. Nismes, C. Belle, Imprimeur du Roi, 1780-84.
7v., plates (part. fold.), maps (part. fold.). 549.944/S08h

Subtitle varies.
Vols. 3-7 have imprint: Paris, J. F. Quillau.
Library owns v.1-6.

2088 **Soulavie.**
See HAMILTON, WILLIAM, Oeuvres complettes . . . and Neuere Beobachtungen . . . Vulkane Italiens . . . 1781 and 1784.

2089 **Soulavie.**
Les classes naturelles des minéraux et les époques de la nature correspondantes à chaque classe; ouvrage qui a remporté le second accessit sur la question proposée par l'Académie impériale des sciences de St. Pétersbourg, pour le prix de 1785. St. Pétersbourg, Impr. de l'Académie impériale des sciences, 1786.
561p., plates. 549/So8c

2090 **Sowerby, James,** 1757-1822.
British mineralogy; or, Coloured figures intended to elucidate

the mineralogy of Great Britain. London, Printed by R. Taylor and Co., 1804-17.
5v., 550 col. plates. 549.942/So9b

Vol. 5 printed by Arding and Merrett.

2091 **Sowerby.**
A short catalogue of British minerals, according to a new arrangement. Pt. I. Combustibles and earths. London, Printed by B. Meredith; sold by the author, 1811.
72p. 549.942/So9s

No more published?

2092 **Sowerby.**
Exotic mineralogy, or, Coloured figures of foreign minerals, as a supplement to British mineralogy, I. London, Printed by Benjamin Meredith, 1811-20.
2v., col. illus. x549/So93e

Library owns v.1 only.
Vol.1: ill. 93 numbered 73; lacks ill. 73.

2093 **Sowerby.**
The mineral conchology of Great Britain; or coloured figures and descriptions of those remains of testaceous animals or shells which have been preserved at various times and depths in the earth. London, various printers, 1812-1829.
7v., col. plates. 564/So9m

v.7 was not completed; it consists of 8 parts, and has no title page or index.
v.5-7, "continued by J. D. C. Sowerby."
For papers on the dates of publication of the 113 parts, *See* E. Renevier, "Bull. Soc. Vaudoise Sci. Nat.," tom. IV, 1855, p. 318; and E. R. Sykes, "Proc. Malac. Soc. Lond.," vol. VII, 1906, p. 19.

2094 **Spada, Giovanni Giacomo,** 1680-1774.
Corporum lapidefactorum agri Veronensis catalogus quae apud Joan. Jacobum Spadam . . . asservantur. Editio altera multo auctior, cui accedunt annotationes, & marmorum quae in eodem agro reperiuntur elenchus. Veronae, typis D. Ramanzinii bibliopolae apud S. Thomam, 1744.
80p., 10 plates (incl. map). 560.94534/Sp1c/1744

2095 **Spallanzani, Lazzaro,** 1729-1799.
Opere. Milano, Società de' classici italiani, 1825-26.
6v. 570.4/Sp190

These are the complete works, the first three volumes contain his travels in the Two Sicilies.

ΘΕΟΦΡΑΣΤΟΥ ΤΟΥ ΕΡΕΣΙΟΥ ἍΠΑΝΤΑ.

THEOPHRASTI ERESII

Græce & Latine opera omnia.

DANIEL HEINSIVS

Textum Græcum locis infinitis partim ex ingenio partim e libris emendauit: hiulca ſuppleuit, male concepta recenſuit: interpretationem paſsim interpolauit.

Cum Indice locupletiſſimo.

LVGDVNI BATAVORVM,

Ex Typographio HENRICI ab HAESTENS.

IMPENSIS

Iohannis Orlers, And. Cloucq, & Ioh. Maire.

ANNO CIↃ. IↃCXIII.

Cat. No. 2164

2096 **Spallanzani.**
Viaggi alle Due Sicilie e in alcune parti dell'Appennino . . . Pavia, Stamperia di B. Comini, 1792-97.
6v., 11 fold. plates. 508.45/Sp19v

Vol.6 has title: Opuscoli sopra diversi animali, che servono di appendice ai Viaggi alle Due Sicilie . . .

2097 **Spallanzani.**
Voyage dans les Deux Siciles et dans quelques parties des Appenins. Berne, E. Haller, 1795-96.
4v. in 2, fold. plates. 508.45/Sp19vFs

Title page of v.3, pt.2 reads: Voyages dans les Deux Siciles . . . Avec une rélation de l'éruption de Vésuve, arrivée le 15 juin 1794.

Translation and introduction by Jean Senebier.

2098 **Spallanzani.**
Travels in the Two Sicilies, and some parts of the Apennines. Tr. from the original Italian of the Abbé Lazzaro Spallanzani . . . London, G. G. and J. Robinson, 1798.
4v., 11 fold. plates. x508.45/Sp19vE

2099 **Spallanzani.**
See FORTIS, ALBERTO, Mémoires . . . 1802.

2100 **Spallanzani.**
See MARTINENGHI, GIOVANNI, Lettera di Giovanni Martinenghi . . . dell'Ombra vendicata di Lazzaro Spallanzani . . . 1803.

2101 **Staring, Winand Carel Hugo,** 1808-1877.
Proef eener Nederlandsche kunstspraak (terminologie), voor de aardkunde of geologie. Deventer, A. Ter Gunne, 1844.
95p. 550.14/St2p

2102 **Steel, John Honeywood,** 1780-1838.
A report on the geological structure of the county of Saratoga, in the State of New-York; together with remarks on the nature and properties of the various soils and modes of culture; intended to illustrate the natural history and domestick resources of the county. Published by order of the executive committee of the Society for the Promotion of Agriculture and Domestick Manufactures. Saratoga Springs, G. M. Davison, printer, 1822.
56p. x557/N48st

2103 **Steele, Andrew.**
The natural and agricultural history of peat moss or turf-bog; to which are annexed corroborative writings, correspondence and observations, on the qualities of peat or fen earth, as a soil and manure, and on the methods used in Scotland for converting moss soils into arable and pasture grounds, plantations of trees, &c. Edinburgh, W. & D. Laing; [etc., etc.], 1826.
401p. 631.13/St32n

2104 **Steffens, Henrich,** 1773-1845.
Beyträge zur innern Naturgeschichte der Erde. 1. Th. Freyberg, Im Verlag der Crazischen Buchhandlung, 1801.
317p. 550/St3b

No more published.

2105 **Stella, Erasmus,** 1450-1521.
Erasmi Stellae libonothani interpraetamenti gemmarum libellus unicus. [Nurēberge, Federicus Pypus, 1517].
48p. uncat.

The author is Erasmus Stueler.

2106 **Steller, Georg Wilhelm,** 1709-1746.
Ausführliche Beschreibung von sondernbaren Meerthieren, mit Erläuterungen und nöthigen Kupfern versehen. Halle, C. C. Kümmel, 1753.
[20], 218p., fold. plate. 591.92/St3g

"Zur Einleitung. Anatomie eines Meerkalbes, von Johann Adam Kulmus": p. [1]-35.

2107 **Stelluti, Francesco,** 1577-1646.
Trattato del legno fossile minerale nuovamente scoperto. Nel quale brevemente si accenna la varia, & mutabil natura di detto legno, rappresentatovi con alcune figure, che mostrano il luogo dove nasce, la diversità dell'onde, che in esso si vedono, e le sue così varie, e maravigliose forme. Roma, Vitale Mascardi, 1637.
12p., 13 plates. q uncat.

2108 **Steno, Nicolaus,** 1638-1686.
Elementorum myologiae specimen, seu musculi descriptio geometrica. Cui accedunt canis carchariae dissectum caput, et dissectus piscis ex canum genere. Ad serenissimum Ferdinandum II, Magnum Etruriae Ducem. Florentiae, Ex Typographia sub signo Stellae, 1667.
123p. illus. 7 plates. x612.74/St4e

2109 **Steno.**
Elementorum myologiae specimen, seu musculi descriptio geometrica. Cui accedunt canis carchariae dissectum caput, et dissectus piscis ex canum genere. Ad serenissimum Ferdinandum II, Magnum Etruriae Ducem. Amstelodami, apud Johan. Janssonium à Waesberge, & Viduam Elizei Weyerstraet, 1669.
148p. illus. 7 plates. x612.74/St4e/1669

2110 **Steno.**
De solido intra solidum naturaliter contento dissertationis prodromus. Ad serenissimum Ferdinandum II . . . Florentiae, Ex Typographia sub signo Stellae, 1669.
78p., fold. diagr. x550/St42s

Plate is preceded by folded leaf with descriptive letterpress. Bound in vellum.

2111 **Steno.**
De solido intra solidum naturaliter contento dissertationis prodromus . . . Lugd. Batav., Apud Jacobum Moukee, 1679.
115p., 2 fold. plates. x550/St42s/1679

2112 **Steno.**
The prodromus to a dissertation concerning solids naturally contained within solids. Laying a foundation for the rendering a rational accompt both of the frame and the several changes of the masse of the earth, as also of the various productions in the same. English'd by H[enry] O[ldenburg]. London, Printed by J. Winter, and are to be sold by Moses Pitt, 1671.
112p., 1 fold. plate. uncat.

2113 **Steno.**
See MANNI, DOMENICO MARIA, Vita del letteratissimo Monsig. Niccolò Stenone de Dánimarca . . . 1775.

2114 **Stephens, Walter,** d. 1808.
Notes on the mineralogy of part of the vicinity of Dublin, taken principally from the papers of the late Rev. Walter Stephens, A.M. London, Printed by William Phillips, George Yard, 1812.
57p., col. map. 549.941/J23mi

Bound with Jameson, Robert, A mineralogical description of the country of Dumfries, 1805.

2115 **Stevens, Henry Newton,** 1855-1930.
Lewis Evans, his map of the middle British colonies in America; a comparative account of ten different editions published

between 1755 and 1807. London, H. Stevens, son, and Stiles, 1905.
41p. xf912.73/Ev1

In case with Evans, Lewis, 1700?-1756. Geographical, historical, political, philosophical and mechanical essays . . . Philadelphia, . . . 1755. See Evans for contents of case.

2116 **Stoer, Johann Gottlieb,** 1705-
De privilegio metallicorum commentatio, quam auctoritate magnifici Jctorum Ordinis, priusquam summos in utroque jure capesseret honores more majorum die x. Julii M DCC XLI. publice ventilandam, exhibuit Joannes Gottlieb Stoer. Erfordiae, Prelo Heringiano, [1741].
50p. 622/St63d

2117 **Stork, William.**
A description of East-Florida, with a Journal, kept by John Bartram of Philadelphia, botanist to His Majesty for the Floridas; upon a journey from St. Augustine up the river St. John's as far as the lakes. With explanatory botanical notes . . . 3d ed., much enlarged and improved . . . London, Sold by W. Nicoll [etc.], 1769.
40p., 35p. 2 fold maps (incl. front.), fold. plan. x917.59/St74a/1769

Dedication signed: William Stork.
1st edition (London, 1766) published under title: An account of East-Florida . . .

2118 **Stotter, Michael.**
Die Gletscher des Vernagtthales in Tirol und ihre Geschichte. Innsbruck, In der Wagner'schen Buchhandlung, 1846.
75p., folded col. map. 551.312/St76g

2119 **Strahlenberg, Philip Johann Tabbert,** 1676-1747.
Historie der Reisen in Russland, Siberien, und der Grossen Tartarey. Mit einer Landcharte und Kupferstichen welche die Geographie und Antiquität erläutern . . . Leipzig, G. Kiesewetter, 1730.
438p., illus., fold. maps. 915/St8h

Chapter XIII contains in alphabetical order a description of minerals, mining districts, curiosities, and antiques. This description precedes travels by Gmelin and Pallas.

2120 **Strange, John,** 1732-1799.
De'monti colonnari e d'altri fenomeni vulcanici dello Stato Veneto; memoria. Milano, G. Marelli, 1778.
[70p.], 11 double plates. q551.21/St8d

2121 **Strauss, Johann Christoph,** 1645-1718.
Beschreibung des Carls-Bades . . . In lateinischer Sprache herausgegeben und nunmehro aus verselben zum gemeinen Besten ins Deutsche übersetzt. Leipzig, Verlegts Joh. Friedrich Gleditsch, zu finden by . . . Joh. George Bachman . . . 1695.
166p. uncat.

2122 **Strickland, Hugh Edwin,** 1811-1853.
See MURCHISON, RODERICK IMPEY, Outline of the geology . . . of Cheltenham . . . 1845.

2123 **Stringer, Moses,** fl. 1698-1713.
Opera mineralia explicata: or, the mineral kingdom, within the dominions of Great Britain, display'd. Being a compleat history of the ancient corporations of the city of London, of and for the mines, the mineral and the battery works. With all the original grants, leases, instruments, writs of privilege and protection, by sea and land, from arrest (except in the mineral courts); or being prest, or serving juries and parish-offices: as also the records of the said mineral courts, from the conquest, down to this present year, 1713. Likewise proposals for new settlements and plentiful provision for all the industrious poor, be their number ever so great. London, printed for Jonas Brown, 1713.
368p. x338.6/St860

2124 **Strong, Henry K.**
Report to the legislature of Pennsylvania, containing a description of the Swatara mining district, illustrated by diagrams. Presented by Mr. M'Elwee, and printed by order of the House of representatives. Harrisburg, Printed by Boas & Coplan, 1839.
61p., 8 plates, fold. map. x622.33/St88r

2125 **Struve, Heinrich Christian Gottfried von,** 1772-1851.
Mineralogische Beiträge vorzüglich in Hinsicht auf Würtemberg und den Schwarzwald. Gotha, Ettingereche Buchhandlung, 1807.
202p. 549.94347/St8m

2126 **Struve.**
Beiträge zur Mineralogie und Geologie des Nördlichen Amerika's. Nach Amerikanischen Zeitschriften. Hamburg, Bei Perthes & Besser, 1822.
124p. 557.9/St8b

DERNIERES DECOUVERTES DANS L'AMERIQUE SEPTENTRIONALE

de M. DE LA SALE;

Mises au jour par M. le Chevalier TONTI, Gouverneur du Fort Saint Loüis, aux Islinois.

A PARIS AU PALAIS,

Chez JEAN GUIGNARD, à l'entrée de la Grand' Salle, à l'Image saint Jean.

M. DC. LXXXXVII.

Avec Privilege du Roy.

Cat. No. 2181

2127 **Struve, Henri,** 1751-1826.
Versuch einer neuen Theorie der Salzquellen und des Salzfelsen, vorzüglich in Bezug auf die Bernischen Salzwerke, nebst einer Reise nach der Salzgegend im Gouvernement Aehlen. Aus dem Französischen übersetzt [von J. S. Wyttenbach] mit sehr vielen Verbesserungen und Zusätzen des Verfassers. Bern, Haller, 1789.
170p., fold. diagrs. 553.63/St8nGw

2128 **Struve,** jt. author.
See VANBERCHEM-BERTHOUT, J. P., Principes de minéralogie . . . d'après les leçons du professeur Werner . . . [1795?].

2129 **Struve.**
Méthode analytique des fossiles, fondée sur leurs caractères extérieurs. Lausanne, Aux dépens de l'auteur, & se vend chez Lacombe, 1797.
166p., 2 col. plates. 569/ST89m

2130 **Struve.**
Recueil de mémoires sur les salines et leur exploitation. Avec figures. Genève, J. J. Paschoud, 1803.
187p., fold. plates. 622.363/St8r

2131 **Strzelecki, Paul Edmund de,** 1796-1873.
Physical description of New South Wales and Van Diemen's land. Accompanied by a geological map, sections, and diagrams, and figures of the organic remains. London, Longman, Brown, Green, and Longmans, 1845.
462p., illus., tables, front., plates, maps (part. fold.), fold. geol. map. 919.44/St8p

2132 **Studer, Bernhard,** 1794-1887.
See DÉSOR, ÉDOUARD, Nouvelles excursions . . . Alpes, de M. Agassiz et de ses compagnons . . . 1845.

2133 **Studer.**
Geologie von der Schweiz. Bern, Stampflische Verlagshandlung; [etc., etc.], 1851-53.
2v., illus., fold. map, diagrs. 554.94/St9g

2134 **Stukeley, William,** 1687-1765.
The philosophy of earthquakes, natural and religious. Or, An inquiry into their cause, and their purpose . . . The 2d ed. To which is added, part II. on the same subject. London, C. Corbet, 1750.
2 parts in 1v. x551.22/St92p/1750a

2135 **Stukeley.**
The philosophy of earthquakes, natural and religious. Or, An inquiry into their cause, and their purpose . . . 3d ed. To which is added, part III. on the same subject. London, A. and C. Corbett, 1756.
3 pts. in 1v. x551.22/St92p/1756

2136 **Stulli, Lucca,** 1722-1828.
Sulle detonazioni della isola di Meleda; lettere del Dr. L. Stulli. Ragusa, A. Martecchini, 1823.
54p. 551.22/St9s

2137 **Stütz, Andreas.**
Physikalisch-Mineralogische Beschreibung des Gold- und Silber-Bergwerkes zu Szekerembe bey Nagyag in Siebenbürgen, nebst einer Zugabe über einige problematische Mineralien Siebenbürgens. Wien, C. F. Wappler, 1803.
164p., illus., 2 fold. maps. 622.342/St9p

2138 **Svedenstierna, Erik Tomas,** 1765-1825.
Resa igenom en del af England och Skottland ären 1802 och 1803. Stockholm, Tryckt hos C. Delén, 1804.
329p. 914.2/Sv2r

2139 **Swan, John,** fl. 1635.
Speculum mundi or, a glasse representing the face of the world: shewing both that it did begin, and must also end: the manner how, and time when, being largely examined. Whereunto is joyned an Hexameron, or a serious discourse of the causes, continuance, and qualities of things in nature, occasioned as matter pertinent to the work done in the six dayes of the worlds creation . . . Cambridge, printed by the printers to the University of Cambridge, 1635.
504p. [26]p. x570/Sw2s

The library owns two copies of the second edition (x570/Sw2s2) and one copy of the third (uncat.) and fourth edition (uncat.). The titles have minor changes.

The author mentions the origin of springs, fountains, earthquakes and their effects. He discusses metals, minerals, fossils, precious stones, asphalt, amber, naphta, and salts.

2140 **Swedenborg, Emanuel,** 1688-1772.
Opera philosophica et mineralia. Tres tomi. [Dresde et Lipsiae, sumptibus Friderici Hekelii, 1734.
3v., plates (part. fold.), port., fold. maps, diagrs. q500/Sw3o

General half-title in v.1 only: each volume has special t.-p.

Each volume is in three parts, pts. II and III each with special t.-p.

Contents — [v. 1] Principia rerum naturalium; sive, Novorum tentaminum phaenomena mundi elementaris philosophice explicanti. — [v. 2] Regnum subterraneum sive minerale: De ferro deque modis liquiationum ferri per Europam passim in usum receptis; deque conversione ferri crudi in chalybem . . . &c. — [v. 3] Regnum subterraneum sive minerale: De cupro et orichalco: deque modis liquationum cupri . . . de secretione ejus ab argento: de conversione in orichalcum . . . pariter de chymicis praeparatis, et cum cupro factis experimentis, &c.

2141 **Swedenborg.**
Miscellaneous observations connected with the physical sciences. Translated from the Latin by Charles Edward Strutt. London, W. Newberry, 1847.
168p., 9 plates. 500/Sw3mEs

Translation of Miscellanea observata.

2142 **Switzer, Stephen.**
An introduction to a general system of hydrostaticks and hydraulicks, philosophical and practical. Wherein the most reasonable and advantageous methods of raising and conducting water, for the watering of noblemens and gentlemens seats, buildings, gardens, &c. are carefully (and in a manner not yet publish'd in any language) laid down. A physico-mechanical enquiry into the original and rise of springs, and of all the hypotheses relating thereto; as also the principles of waterworks, and the draughts and descriptions of some of the best engines for raising and distributing water, for the supply of county seats, cities, towns corporate, &c. Deduc'd from the theory of Archimedes, Gallileo, Torricelli, Boyle, Wallis, Plot, Hook, Marriotte, Desaguliers, Derham, Hawsbee, and others. Reduc'd to practice by Vitruvius, Bockler, de Caus, and other architects among the ancient Romans, Italians, French, Flemmings, and Dutch, and much improv'd by later practice and experience. Illustrated and explain'd by sixty copper cuts done by the best hands . . . 1729, London, T. Astley, S. Austen and L. Gilliver.
2 v. x628.1/Sw61

2143 **Tänzel, Johannes Guilielmus Friderico.**
Hungaria seu disputationes miscellaneae de natura Hungariae in alma, Catholica, et episcopali universitate dilingana . . . Dilingae, typis Joannis Caspari Bencard, [1700].
114p., illus., front. uncat.

2144 **Tardin, Jean.**
Histoire naturelle. De la fontaine qui brusle pres de Grenoble. Avec la recherche de ses causes, & principes, & ample traicté

des feux sous-terrains . . . Tournon, pour Guillaume Linocier, 1618.
380p. uncat.

2145 **Tata, Domenico.**
Descrizione del grande incendio del Vesuvio successo nel giorno otto del mese di agosto del corrente anno 1779. Napoli, V. Mazzola-Vocola, 1779.
38p. 551.21/T18d

Dedication by Domenico Tata.
Bibliographical footnotes.

2146 **Tauber, Andreas.**
Mineralogische Beschreibung des Plauischen Grundes bis Tharand. Nürnberg, In der Frauenholzischen Kunsthandlung, 1799.
44p., plates. q914.321/B378p

Bound with Becker, Wilhelm Gottlieb, Der Plauische Grund bei Dresden . . . 1799.

2147 **Tavernier, John Baptista.**
The six voyages of John Baptista Tavernier, a Noble Man of France now living, through Turky into Persia, and the East-Indies, finished in the year 1670. Giving an account of the state of those countries. Illustrated with divers sculptures. Together with a new relation of the present Grand Seignor's Seraglio, by the same author. Made English by J. Phillips. To which is added a description of all the kingdoms which encompass the Euxine and Caspian Seas. By an English traveller, never before printed. London: printed for R.L. and M.P. . . . 1678.
597p., illus. xq915/T19s

Contains some observations and illustrations of diamonds, rubies, topaz, and pearls and descriptions of corals, musk, and bezoar stones.

2148 **Tavernier.**
Collections of travels through Turky into Persia, and the East-Indies. Giving an account of the present state of those countries. As also a full relation of the five years Wars between Aureng-Zebe and his brothers . . . together with a relation of the kingdom of Japan and Tunkin . . . being The Travels of Monsieur Tavernier Bernier, and other great men: adorned

with many copper plates. London, printed for M.P. and sold by George Yonke . . . and William Ewzey, 1688.
893p., illus. xq915/T19c/1688

2149 **Tavernier.**
De zes Reizen . . . die hy, gedurende de tijt van veertigjaren, in Turkÿen, Persiën, en in d'Indiën langs alle de wegen, die derwaarts streeken, gedaan heeft . . . Amsterdam, Weduwe van Johannes van Someren, 1682.
2v., fold. plates, illus. uncat.

2150 **Taylor, Richard Cowling,** 1789-1851.
On the geology of east Norfolk; with remarks upon the hypothesis of J. W. Robbards respecting the former level of the German ocean. London, Printed for J. Cochran, 1827.
48p., col. profiles. 554.261/T21o

2151 **Taylor.**
Report on the surveys, undertaken with a view to the establishment of a Rail Road, from the coal and iron mines near Blossburg or Peters's camp to the state line at Lawrenceville, in the county of Tioga and the state of Pennsylvania, and Mineralogical Report On the Coal Region in the Environs of Blossburg. Philadelphia, printed by Mifflin & Parry, 1833.
56p., fold. sections, map. uncat.

2152 **Taylor.**
Two reports: on the coal lands, mines and improvements of the Dauphin and Susquehanna coal company, and of the geological examinations, present condition and prospects of the Stony Creek coal estate, in the townships of Jackson, Rush, and Middle Paxtang, in the county of Dauphin, and of East Hanover township, in the county of Lebanon, Pennsylvania. With an appendix, containing numerous tables and statistical information, and various maps, sections, and diagrams, chiefly in illustration of coal and iron. Addressed to the board of directors of the Dauphin and Susquehanna coal company, and to the trustees of the Stony Creek coal estate. Philadelphia, E. G. Dorsey, printer, 1840.
74p., front. (fold. diagr.), fold. plates, 3 fold. maps. 553.2/T21t

2153 **Taylor.**
Statistics of coal. The geographical and geological distribution of mineral combustibles or fossil fuel, including, also, notices and localities of the various mineral bituminous substances, employed in arts and manufactures, illustrated by maps and diagrams, embracing, from official reports of the great coal-

LA PHYSIQUE OCCULTE,

Ou TRAITE' de la BAGUETTE DIVINATOIRE,

Et de ſon utilite pour la découverte des ſources d'eau, des minières, des treſors cachez, des voleurs & des meurtriers fugitifs.

Avec des Principes qui expliquent les phenomé-nes les plus obſcurs de la NATURE.

Par M. L. L. DE VALLEMONT, *Pr. D. en Th.*

Augmenté en cette Edition, d'un Traité de la Connoiſſance des Cauſes Magnetiques des Cures Sympathiques, des Tranplantations & comment agiſſent les Philtres.

Par un Curieux de la Nature.

Suivant la Copie de Paris.

A AMSTERDAM,
Chez ADRIAN BRAAKMAN, dans le Beurſtraat, prés du Dam, à l'enſeigne de la Ville d'Amſterdam, 1696.

Cat. No. 2214

producing countries, the respective amounts of their production, consumption and commercial distribution, in all parts of the world; together with their prices, tariffs, duties and international regulations. Accompanied by nearly four hundred statistical tables, and eleven hundred analyses of mineral combustibles, with incidental statements of the statistics of iron manufactures, derived from authentic authorities. Philadelphia, J. W. Moore, 1848.
754p., fold. maps, diagrs. 553.2/T21s

2154 **Taylor.**
Statistics of coal: including mineral bituminous substances employed in arts and manufactures; with their geographical, geological, and commercial distribution, and amount of production and consumption on the American continent. With incidental statistics on the iron manufacture. 2d. ed. revised and brought down to 1854 by S. S. Haldeman. Philadelphia, J. W. Moore, 1855.
640p., fold. maps. 553.2/T21

2155 **Taylor, Silas,** 1624-1678.
The History and Antiquities of Harwich and Dovercourt, topographical, dynastical and political. First collected by Silas Taylor alias Domville, Gent. keeper of the King's stores there, and now much enlarged in all its parts with notes and observations relating to Natural History. Illustrated with many copper plates representing the Cliff itself, the fossils contained therein, and other principal things, by Samuel Dale. London, for C. Davis, and T. Green, 1730.
464p., illus. x942.67/T218h

2156 **Telesio, Bernardino,** 1509-1568.
. . . De his, quae in Aere fiunt; & de terraemotibus, liber unicus. Neapoli, Apud Iosephum Cacchium, 1570.
14ℓ. uncat.

2157 **Tennant, James,** 1808-1881.
A stratigraphical list of British fossils; arranged under the principal divisions of the British strata, with a few elementary remarks on their character and localities. London, Society for Promoting Christian Knowledge, 1847.
132p. 560.942/T25s

"List of publications" with each subject.

2158 **Tennessee. Geologist and Mineralogist.**
Geological report to the . . . General Assembly of the state

of Tennessee. 3d (1835)—9th (1847). Nashville, S. Nye & Co., printers to the House [etc.], 1835-48.
7v., plates (part. fold.), fold. maps (part. col.), fold. table. x557.68/T256g

Biennial.
Library owns 3rd (1835), 4th (1837), and 6th (1841).
Author is Gerald Troost, 1776-1850.

2159 **Terrae Filius,** pseud.
See WESTON, STEPHEN, Bp. of Exeter, 1747-1830.

2160 **Tessin, Carl Gustaf,** 1695-1770.
Museum Tessinianum, opera illustrissimi comitis, dom. Car. Gust. Tessin . . . Hans Excellence . . . Carl Gust. Tessins Naturalie Samling. Holmiae, apud L. Salvium, 1753.
123p., 12 plates. xq549/T28m

2161 **Testa, Domenico,** 1746-1832.
Memoria intorno alle produzioni fossili de'monti Euganei. Padova, 1791.
68p., Risposta, 19p. 567/T28ℓ

With this are bound his Lettera su i pesci fossili del monte Bolca. [Milano, 1793] 103p.; Terza lettera su i pesci fossili del monte Bolca. [Milano, 1793] 50p.

2162 **Testa.**
Lettera su i pesci fossili del monte Bolca. Milano, Imperial monistero di S. Amb. Mag., 1793.
103p. 567/T28ℓ

Bibliographical footnotes.
With this are bound his Terza lettera su i pesci fossili del monte Bolca. [Milano, 1793] 50p.; Memoria intorno alle produzioni fossili de'monti Euganei. Padova, 1791. 68p., Risposta, 19p.

2163 **Testa.**
Lettere recentemente pubblicate sui pesci fossili veronesi con annotazioni inedite agli estratti delle medesime. Verona, Stamperia Ramanzini, 1794.
187p., 6 fold. plates. 567/T28ℓe

Bibliographical footnotes.

2164 **Theophrastus,** 372?-284? B.C.
Peripateticorum post Aristotelem principis, pleraque antehac latinè nunquam, nun graecè & latinè simul edita. Interpretibus Daniele Furlano Cretensi, Adriano Turnebo . . . Hanoviae, typis Wechelianis, apud Claudium Marnium, & haeredes Ioannis Aubrii, 1605.
397p. q uncat.

This work includes De labidibus liber and books on meteorology.

2165 **Theophrastus.**
Graece & latine opera omnia. Daniel Heinsius, Textum graecum locis infinitis partim ex ingenio partim e libris emendavit . . . Lugduni Batavorum, Iohannis Orlers, And. Cloucq. & Ioh. Maire, 1613.
508p. q uncat.

De lapidibus liber is interpreted by Daniele Furlano.

2166 **Theophrastus.**
See LAET, JOANNES DE, De gemmis et lapidibus libri duo quibus praemittitur Theophrasti liber de lapidibus . . . 1647.

2167 **Theophrastus.**
Traité des pierres de Théophraste, traduit du grec; avec des Notes physiques & critiques, traduites de l'anglois, de M. Hill; auquel on a ajouté deux lettres du même auteur, l'une au docteur Parsons, sur les couleurs du saphir & de la turquoise; & l'autre à M. Folkes . . . sur les effets des différens menstrues sur le cuivre. A Paris, Chez J.-T. Herissant, 1754.
287p. 881/T5ℓF

2168 **Theophrastus.**
Theophrastus's history of stones. With an English version, and notes, including the modern history of the gems described by that author; and of many other of the native fossils. To which are added two letters: I. On the colours of the sapphire and turquoise. II. Upon the effects of different menstruums on copper. Both tending to illustrate the doctrine of the gems being coloured by metalline particles. The Second edition; enlarged by the addition of a Greek index of all the words in Theophrastus. Also observations on the new Swedish acid, and of the stone from which it is obtained; and with an idea of a natural and artificial method of fossils. By Sir John Hill. London, Printed for the author, 1774.
342p. x881/T5ℓh

2169 **Thévenot, Melchisédech,** 1620?-1692.
Recueil de voyages. Paris, Chez Estienne Mishallet, 1681.
169p., illus. (incl. 3 fold. maps). x910.8/T84r

Includes early map of Mississippi River and includes description of its discovery by Marquette.

2170 **Thomas, David,** 1776-1859.
Travels through the Western country in the summer of 1816, including notices of the natural history, topography, commerce, antiquities, agriculture, and manufactures: with a map

of the Wabash country, now settling. Auburn, N.Y., Printed by David Rumsey, 1819.
320p. x917.4/T36t

2171 **Thomson, Thomas,** 1773-1852.
Travels in Sweden, during the autumn of 1812. London, R. Baldwin, 1813.
457p., front., 6 plates, 2 port., 5 fold. maps. q914.85/T38t

2172 **Thomson, William,** 1746-1817.
Prospects and observations; on a tour in England and Scotland: natural, economical, and literary. By Thomas Newte, esq. [pseud.]. London, Printed for G. G. J. and J. Robinson, 1791.
440p., front., plates, fold. map. x914.2/N48p

First published, 1789, in shorter form, with title: Tour in England and Scotland, by an English gentleman.

2173 **Tilas, Daniel,** 1712-1772.
Stenrikets historia, utförd i det wid Praesidii aflaggande häldne Talet in för Kongl. Swenska Wetenskaps-Academien den 14. April 1742. Stockholm, L. L. Grefing, [1742].
32p. 549/T45s

2174 **Tilesius, A. von.**
Naturhistorische Abhandlungen und Erläuterungen besonders die Petrefactenkunde betreffend von Dr. A. von Tilesius . . . Cassel, Bei Johann von Christan Krieger und Compagnie, 1826.
154p., 8 col. plates. q560/T46n

2175 **Tillesius,** jt. author.
See ROSENMÜLLER, JOHANN CHRISTIAN, Beschreibung merkwürdiger Höhlen . . . 1799-1805.

2176 **Tilly, de.**
Mémoire sur l'utilité, la nature et l'exploitation du charbon minéral. Paris, A. M. Lottin, 1758.
131p., illus. 553.2/T46m

2177 **Tobalducci, Claudio,** 16th cent.
Delli dialogi della quantita et del numero delle sfere terrestri et celesti . . . Il primo della Terra. Roma, per il Santi, 1588.
108p., illus. uncat.

2178 **Toderini, Giovanni Battista,** 1728-1799.
Dissertazione sopra un legno fossile, che tutto sciogliesi in cenere rossa. Pubblicata con due proprie lettere: 1. Su l'in-

duramento di molti Bachi da seta. 2. Su l'aurora boreale. Modena, Nella stamperia di G. Montanari, 1770.
188p. 561.21/T56d

2179 **Tondi, Matteo,** 1762-1835.
Elementi di orittognosia. Napoli, Presso Angelo Trani, 1817.
2v. 550/T61e

2180 **Tondi.**
Elementi di oreognosia. Napoli, Tip. C. Cataneo, 1824.
521p., 3 plates. 552/T61e

2181 **Tonti, Henri de.**
Dernières découvertes dans l'Amérique septentrionale de M. de la Salle; mises au jour par M. le Chevalier Tonti, Gouverneur du Fort Saint Loüis, aux Illinois. Paris, Jean Guignard, 1687.
335p. uncat.

2182 **Tonti.**
An account of Monsieur de la Salle's last expedition and discoveries in North America. Presented to the French King, and published by the Chevalier Tonti, Governour of Fort St. Louis, in the Province of the Illinois. Made English from the Paris original. Also the adventures of the Sieur de Montauban, Captain of the French Buccaneers on the Coast of Guinea, in the year 1695. London, printed for J. Jonson and S. Buckley, and R. Knaplock, 1698.
255p. uncat.

2183 **Torre, Giovanni Maria della,** 1713-1782.
Scienza della natura. Napoli, S. Porsile, règio stampatore, 1748-49.
2v., plates, port. 500/T63s

2184 **Torre.**
Storia e fenomeni del Vesuvio. Napoli, Presso G. Raimondi, 1755.
120p., 8 fold. plates. 551.21/T63s

Includes Supplemento. [n.p., 1761/]. 15p., fold. plate. Bound with main work.

2185 **Torre.**
Histoire et phénomènes du Vésuve. Traduction de l'italien par M. l'abbé Péton. Paris, J.-T. Hérissant, 1760.
399p., 5 fold. plates, fold. map. 551.21/T63sFp

Translation of Storia e fenomeni del Vesuvio.

DE' CORPI MARINI,

Che su' Monti si trovano;

Della loro Origine, e dello stato del Mondo avanti il Diluvio, nel Diluvio, e dopo il Diluvio:

LETTERE CRITICHE

DI ANTONIO VALLISNERI,

Pubblico Primario Professore di Medicina Teorica nell' Università di Padova;

CON LE ANNOTAZIONI:

Alle quali s' aggiungono tre altre Lettere Critiche contra le Opere del Sig. Andry, Francese, e suoi Giornali.

SECONDA EDIZIONE

Con nuove Giunte, Annotazioni, e Raccolta d'Osservazioni spettanti all'Istoria Medica, e Naturale.

A SUA ECCELLENZA

LA SIGNORA CONTESSA

D. CLELIA GRILLA-BORROMEA.

IN VENEZIA, MDCCXXVIII.

Per Domenico Lovisa.

CON LICENZA DE' SUPERIORI.

Cat. No. 2216

2186 **Torre.**
Histoire et phénomènes du Vésuve. Traduit de l'italien par M. l'abbé Péton. Paris, E. Onfroy, 1776.
399p., 5 fold. plates, fold. map. 551.21/T63sFp/1776

Translation of Storia e fenomeni del Vesuvio.

2187 **Torre.**
Relazione prima dell'eruzione del Vesuvio dagli 11. agosto fino ai 18. settembre 1804. del Duca della Torre. [Napoli?, 1804?].
61p. 551.21/T63r

2188 **Torrubia, José,** 1698-1761.
Vorbereitung zur Naturgeschichte von Spanien. Mit vierzehn Kupfertafeln versehen welche viele Fossilien vorstellen, die in den spanischen Ländern verschiedener Welttheile gefunden werden. Aus dem Spanischen übersetzt, und mit Anmerkungen begleitet, nebst Zusätzen, und Nachrichten, die neueste portugiesische Litteratur betreffend; von Christoph Gottlieb von Murr. Halle, bey J. J. Gebauers Wittwe und J. J. Gebauer, 1773.
147p., 14 plates. uncat.

2189 **Tosi, Domenico.**
Dialoghi. Parte prima nella quale si discorre della figura della terra. Del modo di difendere le campagne, e le città dalle inondazioni de' fiumi. Del modo di conservare il grano, e tutte le altre biade; e del modo d'insegnare à mutoli l'uso della favella. Firenze, Stamperia Paperiniana. 1762.
95p., fold. plate. 551.1/T63d

2190 **Toulmin, George Hoggart.**
The antiquity and duration of the world. London, Printed for T. Cadell, 1780.
200p. x550/T64a

Published in 1785 under title: The eternity of the world, and in 1789 under title: The eternity of the universe. Cf. Brit. Mus. Cat.

2191 **Toulmin.**
The eternity of the world. London, Printed for T. Cadell, 1785.
133p. x550/T64a/1785

Published in 1780 under title: The antiquity and duration of the world, and in 1789 under title: The eternity of the universe. Cf. Brit. Mus. Cat.

2192 **Toussaint von Charpentier,** 1728-1805.
See CHARPENTIER, JOHANN FRIEDRICH WILHELM TOUSSAINT VON.

2193 **Townsend, Joseph,** 1739-1816.
A journey through Spain in the years 1786 and 1787; with particular attention to the agriculture, manufactures, commerce, population, taxes, and revenue of that country; and remarks in passing through a part of France. 2d. ed. London, C. Dilly, 1792.
3v., plates. x914.6/T66j

Rock types along the route are described.

2194 **Townsend.**
The character of Moses established for veracity as an historian, recording events from the creation to the deluge. Bath, Printed by M. Gye and published by Longman, Hurst, Rees, Orme and Brown, London, 1813-15.
2v. x550/T664c

Vols. 1-2 were published separately in 1824. v.1: Geological and mineralogical researches, during a period of more than fifty years, in England, Scotland, Ireland, Switzerland, Holland, France, Flanders, and Spain; wherein the effects of the Deluge are traced, and the veracity of the Mosaic account is established; v.2: Etymological researches . . .

2195 **Townson, Robert,** 1763-1797.
Travels in Hungary; with a short account of Vienna in the year 1793. Illustrated with a map and sixteen other copperplates. London, Printed for G. G. and J. Robinson, 1797.
506p., 16 [i.e.17] leaves of plates (3 folded), illus., col. map.
x914.39/T666t

Appendix: Entomologia, p.[453]-476; Regnum vegetabile, p.[477]-494; Elements of mineralogy, by Dr. Townson, 2p.

Index, p.[495]-506.

2196 **Townson.**
Philosophy of mineralogy. London, Printed for the Author, 1798.
219p., 2 plates (1 fold., col.). x549/T666p

2197 **Trebra, Friedrich Wilhelm Heinrich von,** 1740-1819.
Erfahrungen vom Innern der Gebirge, nach Beobachtungen gesammlet und herausgegeben . . . Dessau und Leipzig, Auf Kosten der Verlagskasse für Gelehrte und Künstler, 1785.
244p., col. illus., 8 fold. plates (part. col.). q549/T71e

2198 **Trebra.**
Observations de M. de Trébra, sur l'intérieur des montagnes, précédées d'un plan d'une histoire générale de la minéralogie, par M. de Veltheim; avec un discours préliminaire et des notes de feu Dietrich. Paris, Impr. de Didot jeune, [1799/1800].
308p., col. illus., col. plates (part. fold.), fold. map, fold. tables. f549/T71eF

Originally published in German under title: Erfahrungen vom Innern der Gebirge.
Bibliography: p. [lvii]- lxiij.

2199 **Trego, Charles B.,** 1794-1874.
A geography of Pennsylvania: containing an account of the history, geographical features, soil, climate, education, government, finances, productions, trade, rail roads, canals, &c. of the state; with a separate description of each county, and questions for the convenience of teachers. To which is appended, a travellers' guide, or table of distances on the principal rail road, canal and stage routes in the state. Illustrated by a map of the state and numerous engravings. Philadelphia, Edward C. Biddle. Stereotyped by C. W. Murray & Co., 1843.
384p., illus., front., fold. map. 917.48/T716g

2200 **Trimmer, Joshua,** 1795-1857.
Practical geology and mineralogy; with instructions for the qualitative analysis of minerals. Philadelphia, Lea and Blanchard, 1842.
[25]-527p., front., illus., plates, diagrs. 550/T73p

2201 **Troil, Uno von,** 1746-1803.
Letters on Iceland: containing observations of the civil, literary, ecclesiastical, and natural history; antiquities, volcanos, basaltes, hot springs; customs, dress, manners of the inhabitants, &c. &c. made, during a voyage undertaken in the year 1772, by Joseph Banks, . . . , assisted by Dr. Solander, . . . Dr. B. Lind, . . . Dr. Uno von Troil, . . . Written by Uno von Troil, . . . also Professor Bergman's curious observations and chemical examination of the lava and other substances produced on the island. With a new map of the island, and a representation of the remarkable boiling fountain called by the inhabitants, geyser. Second edition, corrected and improved. London, Printed for J. Robson, etc., 1780.
400p., front., 1 fold. map. x914.91/T74ℓ

"Catalogue of writers on Iceland," p.xvi-xxiv.

2202 **Troost, Gerald,** 1776-1850.
See TENNESSEE GEOLOGIST AND MINERALOGIST, Geological Report . . . 1835-48.

2203 **Troschel, Heinrich Gottlob Nepomuck.**
Erforderliche Nachrichten von dem Biliner Sauer-Brunnen nach der neuesten Aufsuchung des wahren reinen Quell-Wassers. [Pirna], 1762.
242p. 553.7/T75e

Bound with: Unterricht von dem Augustusbrunnen bey Radeberg. Dresden, 1766, 15p., and Wipacher, David. De thermis Ribariensibus in Hungaria. Lipsiae, 1768. 95p.

2204 **Tuomey, Michael,** 1805-1857.
Report on the geology of South Carolina. Columbia, S. C., Printed and published, for the state, by A. S. Johnston, 1848.
293p., front., illus., 2 fold. maps (incl. geol.). 557/So8tu

2205 **Tyson, Philip Thomas,** 1799-1877.
Geology and industrial resources of California. To which is added, the official reports of Genls. Persifer[,] F. Smith and B. Riley — including the reports of Lieuts. Talbot, Ord, Derby and Williamson, of their explorations in California and Oregon; and also of their examinations of routes for rail road communication eastward from those countries. Baltimore, W. Minifie & co., 1851.
127, 37p., 9 fold. plates, 3 fold. maps (incl. front.). 557/C128t

Reprinted, with introduction, from U. S. 31st Cong., 1st sess., Senate Ex. doc. no. 47.

2206 **Ulstadius, Philippus.**
Coelum philosophorum de secretis naturae. Liber. Argentorati, Ioannis, Grienynger, 1528.
61p. uncat.

Discussion of extraction of minerals and metals and distillation.

2207 **Unger, Franz Joseph Andreas Nicolas,** 1800-1870.
Synopsis plantarum fossilium. Lipsiae, apud Leopoldum Voss, 1845.
330p. 561/Un3s

"Literatura nostri sevi": p. vii-xviii.

2208 **Ungern-Sternberg, Wilhelm Heilwig Karl Robert August von,** 1777-1847.
Werden und seyn des vulkanischen Gebirges. Carlsruhe, G. Braun, 1825.
320p., 8 fold. plates, fold. table. 551.21/Un3w

2209 **U.S. Geological Corps.**
Geological map of Wisconsin, Iowa and Minnesota, exhibiting, also, the extension of the Iowa Coal-field into Missouri, and its relation to the Illinois Coal-field. By J. G. Norwood and Col. Chas. Whittlesey, 1851.
Col. map, 114 x 64 cm. G4130/.C5/1851/.U5

Scale, 1 inch equals ca. 15 miles.
Geological map of the north shore of Lake Superior.
"Constructed from observations by the United States Geological Corps, under instructions of the Secretary of the Treasury. By David Dale Owen, M. D. Principal Geologist. Middle, Northern and North-western Minnesota by J. G. Norwood, M.D. Assistant Geologist. The greater part of the southern shore of Lake Superior, in Wisc. by Col. Chas. Whittlesly."

2210 **U.S. War Department.**
Message from the President of the United States, transmitting a report of the Secretary of War: on the number, value, & position of the copper mines on the southern shore of Lake Superior. — [s.l.: s.m.], 1822 (Washington: Gales & Seaton).
33p. x622.09/Un36m

James Monroe was President.

2211 **Ure, Andrew,** 1778-1857.
New system of geology, in which the great revolutions of the earth & animated nature are reconciled at once to modern science & sacred history. London, 1829.
619p., illus., plates. 550/Ur2

2212 **Ussher, James,** 1580-1656.
Annales veteris testamenti a prima mundi origine deducti; una cum rerum asiaticarum et aegyptiacarum chronico, a temporis historici principio usque ad Maccabaicorum initia producto. London, 1650.
554p. xq930/Us7a/v.1

The library owns the following other editions: London, 1650 (uncat.); London, 1654 (xq930/Us7an/1654); London, three copies, 1658 (xq930/Us7aE/1658); Paris, 1673 (q uncat.).

2213 **Valetim, Gabriel Gustav,** 1810-1883.
Anatomie des Echinodermes. Anatomie du genus *Echinus,* See AGASSIZ, LOUIS, Monographie d'échinodermes . . . 1838-42.

2214 **Vallemont, Pierre le Lorrain de,** 1649-1721.
La physique occulte ou traité de la baguette divinatoire, et de son utilité pour la découverte des sources d'eau, des miniéres, des tresors cachez, des voleurs & des meurtriers fugitifs. Avec

LEZIONE ACCADEMICA

INTORNO L'ORIGINE DELLE FONTANE,

Con le Annotazioni per chiarezza maggiore della medesima,

DI ANTONIO VALLISNERI,

Pubblico Primario Professore di Medicina Teorica nella Università di Padova, Medico di Camera di S. M. C. C. ec.

SECONDA EDIZIONE.

Con la Giunta di varie Lettere Differtatorie, un'altra Lezione Accademica, Osservazioni, Ragioni, ed Esperienze nuove, dimostranti la verità del proposto Sistema, con la Risposta alle Obbiezioni del Sig. Dottore N N. compilata da GASTON-GIUSEPPE GIORGI, Medico, e Fisico Fiorentino,

E CONSACRATA AI MARCHESI

ALESSANDRO, E SCIPIONE FRATELLI MAFFEI,

L'uno General Comandante delle Truppe di Baviera, e Tenente Marescìallo nelle Armate Imperiali, l'altro Gentiluomo di Camera del Re di Sardegna, e Cavalier della Chiave d'Oro dell'Elettor di Baviera.

IN VENEZIA. MDCCXXVI.

Appresso Pietro Poletti.

All'Eloquenza in Merceria di S. Salvatore.

CON LICENZA DE' SUPERIORI, E PRIVILEGIO.

Cat. No. 2217

des principes qui expliquent les phenoménes les plus obscurs de la nature. Augmentée en cette edition, d'un Traité de la connoisance des causes magnetiques, des cures sympathiques, des transplantations & comment agissent les philtres. Par un curieux de la nature. Amsterdam, A. Bkaakman, 1696.
456p., illus. uncat.

2215 **Vallisnieri (or Vallisneri), Antonio,** 1661-1730.
De' corpi marini, che su' monti si trovano; della loro origine; e dello stato del mondo avanti'l Diluvio, nel Diluvio, e dopo il Diluvio. Lettere critiche . . . con le annotazioni, alle quali s'aggiungono tre altre lettere critiche contra le opere del Sig. Andry, Francese e suoi giornali. A sua Eccellenza la Signora Contessa D. Clelia Grillo-Borromea. Venezia, Domenico Loviso, 1721.
254p. 570/V24d

Four letters out of fourteen concern life sciences. Letters on natural history refer to fossil fish on Monte Bolca and theories of the earth to explain their presence.

2216 **Vallisnieri.**
De' corpi marini, che su' monti si trovano; della loro origine; e dello stato del mondo avanti'l Diluvio, nel Diluvio, e dopo il Diluvio. Lettere critiche . . . con le annotazioni, alle quali s'aggiungono tre altre lettere critiche contra le opere del Sig. Andry, Francese e suoi giornali. Seconda Edizione. Con nuove giunte, annotazioni, e raccolta d'osservazioni spettanti all'Istoria medica e naturale. A sua Eccellenza la Signora Contessa D. Clelia Grillo-Borromea. Venezia, Domenico Loviso, 1728.
272p. 570/V24d/1728

Four letters out of seventeen concern life sciences.

2217 **Vallisnieri.**
Lezione accademica intorno all'origine delle fontane, con le annotazioni per chiarezza maggiore della medesima. 2. ed. Con la giunta di varie lettere dissertatorie, un'altra lezione accademica, osservazioni, ragioni, ed esperienze nuove, dimostranti la verità del proposto sistema, con la risposta alle obbiezioni del Sig. Dottore N. N. compilata da Gaston-Giuseppe Giorgi. . . . Venezia, P. Poletti, 1726.
407p., illus. 551.49/V24ℓ/1726

2218 **Valmont de Bomare, Jacques Christophe,** 1731-1807.
Minéralogie, ou, Nouvelle exposition du règne minéral. Ouvrage dans lequel on a tâché de ranger dans l'ordre le plus

naturel les individus de ce règne, & où l'on expose leurs propriétés & usages méchaniques; avec un dictionnaire nomenclateur et des tables synoptiques. Paris, Vincent, 1762.
2v., fold. tables. x549/V24m/1762

2219 **Valmont de Bomare.**
Minéralogie; ou, Nouvelle exposition du règne minéral . . . avec un lexicon ou vocabulaire, des tables synoptiques, & un dictionnaire minéralogico-géographique. 2.ed. Paris, Vincent, 1774.
2v., tables. 549/V24m
Bibliographical footnotes.

2220 **Valmont de Bomare.**
Dictionnaire raisonné universel d'histoire naturelle; contenant l'histoire des animaux, des végétaux et des minéraux, et celle des corps célestes, des météores, & des autres principaux phénomenes de la nature; avec l'histoire et la description des Drogues simples tirées de trois regnes; et le détail & leurs usages dans la médecine, dans l'économie domestique & champêtre, & dans les arts & métiers: . . . plus une table concordante des noms latins, & le renvoi aux objets mentionnés dans cet ouvrage. Nouv. éd., rev. & considérablement augm. par l'auteur. Paris, Brunet, 1775.
9v., front. 503/V24d/1775

2221 **Valmont de Bomare.**
Dictionnaire raisonné, universel d'histoire naturelle, contenant l'histoire des animaux, des végétaux et des minéraux, et celle des corps célestes, des méteores, et des autres principaux phénomenes de la nature; avec l'histoire des trois regnes, et le détail des usages de leurs productions dans la médecine, dans l'économie domestique et champêtre, et dans les arts et métiers; Une table concordante des noms latins, et le renvoi aux objets mentionnés dans cet ouvrage. Lyon, Bruyset Ainé et Cie, 1800.
15 v., front. 503/V24

2222 **Vanberchem-Berthout, J. P.**
Principes de minéralogie; ou, Exposition succinte des caractères extérieurs des fossiles, d'après les leçons du professeur Werner; augmentées d'additions manuscrites fournies par cet auteur. Par J. P. Vanberchem-Berthout . . . et Henri Struve . . . Paris, Chez Reynier, [1795?].
176p., plates, fold. table. 549/W49p

2223 **Van Rensselaer, Jeremiah.**
An essay on salt, containing notices of its origin, formation,

geological position and principal localities, embracing a particular description of the American salines; with a view of its uses in the arts, manufactures and agriculture. Delivered as a lecture before the New York lyceum of natural history. New York, O. Wilder and J. M. Campbell, 1823.
80p. 553.63/V35e

Author's autographed presentation copy.

2224 **Van Rensselaer.**
Lectures on geology; being outlines of the science, delivered in the New-York Athenaeum, in the 1825. New-York, E. Bliss & E. White, 1825.
358p. 550/V35ℓ

2225 **Vanuxem, Lardner,** 1792-1848.
Geology of New York, Part 3, comprising the geology of the Third Geological District. Albany, White and Visscher, 1842.
306p., illus. q508.742/N48/v.4, pt.3

Natural History of New York, v.4, pt.3.

2226 **Van Zandt, Nicholas Biddle.**
A full description of the soil, water, timber & prairies of each lot or quarter section of the military lands between the Mississippi & Illinois rivers. Washington City, P. Force, 1818.
127p. x917.73/V39

2227 **Varen, Bernhard,** 1622-1650.
Geographia generalis, in qua affectiones generales telluris explicantur. Amstelodami, Apud L. Elzevirium, 1650.
3v. in 1, 786p., diagrs. x910/V42g

2228 **Varen.**
Geographia generalis. In qua affectiones generales telluris explicantur. Amstelodami, Elzeviriana, 1664.
748p., diagr., fold. tables. x910/V42g/1664

2229 **Varen.**
Geographia generalis, in qua affectiones generales telluris explicantur. Amstelodami, ex officina Elzeviriana, 1671.
784 (i.e. 748)p., diagrs., 4 fold. plates. x910/V42g/1671

Error in paging: p. 748 numbered 784.
Library owns 2 copies; c.2 imperfect: leaves [9]-[16] from table of contents bound before leaf [5] in dedication.

2230 **Varen.**
Geographia generalis, in qua affectiones generales telluris explicantur, summâ curâ quam plurimis in locis emendata, & XXXIII

schematibus novis, aere incisis, unà cum tabb. aliquot quae desiderabantur aucta & illustrata. Ab Isaaco Newton . . . Cantabrigiae, Joann Hayes, 1672.
511p., diagrs. x910/V429/1692

2231 **Varen.**
Geographia generalis, in qua affectiones generales telluris explicantur, summâ curâ quam plurimis in locis emendata, & XXXIII schematibus novis, aere incisis, unà cum tabb. aliquot quae desiderabantur aucta & illustrata. Ab Isaaco Newton . . . Editio secunda auctior & emendatior. Cantabrigiae, ex officina J. Hayes, sumptibus Henrici Dickinson, 1681.
511p., 5 fold. diagr. x910/V42g/1681

2232 **Varen.**
A compleat system of general geography: explaining the nature and properties of the Earth; viz. It's figure, magnitude, motions, situation, contents, and division into land and water, mountains, woods, deserts, lakes, rivers, with particular accounts of the different appearances of the Heavens in different countries; the seasons of the year over all the globe; the tides of the sea; bays, capes, islands, rocks, sand-bars and shelves. The state of the atmosphere; the nature of exhalations; winds, storms, tornados, etc. The origin of springs, mineral-waters, burning mountains, mines, etc. The uses and making of maps, globes, and sea-charts. The foundations of dialling; the art of measuring heights and distances; the art of ship-building, navigation, and the ways of finding the longitude at sea. Originally written in Latin . . . since improved and illustrated by Sir Isaac Newton and Dr. Jurin; and now translated into English; with additional notes, copper plates, an alphabetical index, and other improvements . . . by Mr. Dugdale. The whole revised and corrected by Peter Shaw. London, Printed for Stephen Austen, 1733.
2v. x910/V42gEd

2233 **Varen.**
Géographie générale, composée en latin. Revue par Isaac Newton, augmentée par Jacques Jurin, traduite en anglois d'après les éditions latines données par ces auteurs avec des additions sur les nouvelles découvertes, & présentement traduite de l'anglois en françois. Paris, Vincent, 1755.
4v., illus., fold. map. 910/V42gF/1755

Title of original: Geographia generalis.

2234 **Varley, Delvalle (Lowry).**
Conversations on mineralogy. With plates, engraved by Mr. and Miss Lowry, from original drawings. First American from the 1st London edition. Philadelphia, U. Hunt, 1822.
2v., fold. col. front., 11 fold. plates. 549/V42c/1822

Library lacks v.2.

2235 **Vaughan, Thomas.**
A brief natural history intermixed with variety of philosophical discourses; and observations of the burnings of Mount Aetna. With refutations of such vulgar errours as our modern authors have omitted . . . London, printed for Matthew Smelt, 1669.
120p. uncat.

2236 **Velo, Giovanni Battista di,** 1752-1819.
Dei passaggi alpini con un opuscolo sulle fisiche rivoluzioni delle Alpi. Milano, G. G. e Destefanis, 1804.
218p. 551.43/V546d

Bound with: Della interruzione di continuità nella gran catena delle Alpi. Milano, G. G. e Destefanis, 1803. 64p.

2237 **Veltheim, August Ferdinand von,** 1741-1801.
Grundriss einer Mineralogie. Braunschweig, Im Verlage des Fürstl. Waisenhaus, 1781.
[28]p. xq549/V54g

Contains inserted pages of handwritten notes.

2238 **Veltheim.**
Ueber der Herren Werner und Karsten Reformen in der Mineralogie; nebst Anmerkungen über die ältere und neuere Benennung einiger Stein-Arten. Helmstedt, C. G. Fleckeisen, 1793.
84p. 549/V54u

Bound with Nose, Karl Wilhelm, Ueber einige Ereignisse in der mineralogischen Literature unserer Tage. Frankfurt am Mayn, 1793. 56p.

2239 **Veltheim.**
See TREBRA, FRIEDRICH WILHELM HEINRICH VON, Observations . . . précédées d'un plan d'une histoire générale de la minéralogie, par M. de Veltheim . . . [1799/1800].

2240 **Venegas, Miguel,** 1680-1764?
Natuurlyke en burgerlyke historie van California. Behelzende eene naauwkeurige beschryving van dat gewest . . . De gewoontens van desselfs inwoonders, hunnen godsdienst [en] regeering . . . Mitsgaders de berichten van verscheidene rei-

Natürliche und bürgerliche

Geschichte

von

Californien

nebst einer neuen Charte dieses Landes und der benachbarten Meere.

Aus dem Englischen übersetzt
und herausgegeben
von

Johan Christoph Adelung.
Sächsisch-Gothaischen Rath und der Churfürstl. Mainzischen Academie zu Erfurt Beysitzer.

Erster Theil.

Lemgo,
in der Meyerschen Buchhandlung, 1769.

Cat. No. 2241

zen en tochten . . . Uyt het oorsprongkelyk spaans van Miguel Venegas, Jesuit te Mexico, te Madrid in't jaar 1758 uytgecomen in't engels, en nu in't nederduyts vertaald door J. J. D. . . . Haerlem, Gedrukt by Johannes Enschedé, 1761-62.
2v., plates, fold. map. 979.4/V55nDd

2241 **Venegas.**
Natürliche und bürgerliche Geschichte von Californien, nebst einer neuen Charte dieses Landes und der benachbarten Meere. Aus dem englischen übersetzt und hrsg. von Johan Christoph Adelung . . . Lemgo, Meyerschen Buchhandlung, 1769-70.
3v. in 1, maps. 979.4/V55nGa

2242 **Venegas.**
Histoire naturelle et civile de la Californie, contenant une description exacte de ce pays . . . les moeurs de ses habitans, leur religion, leur gouvernement, & leur façon de vivre avant leur conversion au Christianisme; un détail des différens voyages, & tentatives qu'on a faites pour s'y établir, & reconnoître son golfe & la côte de la mer du Sud. Enrichie de la carte du pays & des mers adjacentes. Tr. de l'anglais, par M. E.** . . . Paris, Durand, 1767.
3v., fold. map. 979.4/V55nFe

Translated by M. A. Eidous.

2243 **Venturi, Giovanni Battista,** 1746-1822.
Memoria intorno ad alcuni fenomeni geologici, del cavaliere Giambattista Venturi. Pavia, Presso Fusi, 1817.
39p. 550.8/V56m

"Letta all'Instituto [di scienze ed arti in Milano] . . . il giorno 4. luglio 1816."

"Inserita nel t. X, bim. IV, 1817, del Giornale di fisica ecc. di Pavia."

Bound with the author's Due lettere, [Pavia, 1817].

2244 **Vermont. State Geologist.**
Report on the Geological Survey of the State of Vermont, by Edward Hitchcock. Montpelier, E. P. Walton, printer, 1857.
12p. 557/V59r

2245 **Vermont. State Geologist.**
Report on the geological survey of the state of Vermont, by Edward Hitchcock, state geologist . . . Burlington, Daily times office print, 1858.
13p. 557/V59re

2d annual report by E. Hitchcock.
Geological pamphlets, v.16, no. 5.

2246 **Verneuil, Edouard de,** 1805-1873, jt. author.
See MURCHISON, RODERICK IMPEY, The geology of Russia . . . 1845.

2247 **Villa, Antonio.**
Sulla costituzione geologica e geognostica della Brianza e segnatamente sul terreno cretaceo; memoria di Antonio e Giovanni Battista Villa . . . Milano, Presso gli editori dello Spettatore industriale, 1844.
46p., fold. map, fold. diagrs. 554.4/V71s

2248 **Vogt, Karl Christoph,** 1817-1895.
See DESOR, ÉDOUARD, Agassiz geologische Alpenreisen . . . 1844.

2249 **Vogt.**
Lehrbuch der Geologie und Petrefactenkunde. Zum Gebrauche bei Vorlesungen und zum Selbstunterrichte. Theilweise nach L. Élie de Beaumont's Vorlesungen an der École des mines. Braunschweig, F. Vieweg, 1846-47.
2v., illus., maps. 550/V86/1846

2250 **Vogt.**
Agassiz' und seiner Freunde geologische Alpenreisen in der Schweiz, Savoyen und Piemont. 2. stark verm. Aufl. Frankfurt am Main, Literarische Anstalt (J. Rütten), 1847.
672p., illus. (some col.), map (fold. col.). 554.947/V86a

2251 **Voight, Johann Carl Wilhelm,** 1752-1821.
Ueber Aquamarine und Topaz, *See* HEYER, JUST CHRISTIAN HEINRICH, Chemische Versuche mit Bernstein . . . 1787.

2252 **Volkmann, Georg Anton,** b. 1663.
Silesia subterranea; oder, Schlesien, mit seinen unterirdischen Schätzen, Seltsamheiten welche dieses Land mit andern gemein oder zuvor aus hat . . . Leipzig, M. G. Weidmann, 1720.
344p., plates (part. fold.). 553/V88a

2253 **Vollmer, Johann Jakob Wilhelm.**
See KANT, IMMANUEL, Physiche Geographie . . . [1803]-1805.

2254 **Volney, Constantin François Chasseboeuf,** 1757-1820.
Travels through Syria and Egypt, in the years 1783, 1784, and 1785 . . . Tr. from the French . . . London, G. G. J. and J. Robinson, 1787.
2v., fold. plates, fold. maps, fold. plan. x916.2/V88vE

2255 **Volney.**
Tableau du climat et du sol des Etats-Unis d'Amérique. Suivi d'éclaircissemens sur la Floride, sur la colonie Française au Scioto, sur quelques colonies Canadiennes et sur les sauvages. Enrichi de quatre planches gravées, dont deux cartes géographiques et une coupe figurée de la chute de Niagara. Paris, Courcier & Dentu, 1803.
2v., 2 fold. plates, 2 fold. maps. x917.3/V88t

Plate of Niagara and map of the United States in color.

2256 **Volney.**
Tableau du climat et du sol des Etats-Unis d'Amérique. Suivi d'éclaircissemens sur la Floride, sur la colonie française au Scioto, sur quelques colonies canadiennes et sur les sauvages. Enrichi de quatre planches gravées, dont deux cartes géographiques et une coupe figurée de la chûte de Niagara. Paris, Courcier & Dentu, 1803.
2v. in 1, 2 fold. plates, 2 fold. maps. 21cm. x917.3/V88t/1803

Plates and maps in black and white.

2257 **Volney.**
A view of the soil and climate of the United States of America: with supplementary remarks upon Florida; on the French colonies on the Mississippi and Ohio, and in Canada; and on the aboriginal tribes of America. Translated, with occasional remarks, by C. B. Brown. Philadelphia, Pub. by J. Conrad & Co., 1804.
446p., 3 fold. plates. x917.3/V88tEb

2258 **Volney.**
Reisen durch die Vereinigten Staaten von Nordamerika. Mit einer Charte und Kupfer. Hamburg und Mainz, G. Vollmer, 1804.
2v. in 1, plates, map. 917.3/V88t:G

Magazin der neuesten und besten ausländischen Reisebeschreibungen. 5.-6. bd.

2259 **Volta, Giovanni Serafino.**
Anfangsgründe der analytischen und systematischen Mineralogie. Nebst zweien Briefen des Herrn Bozza und Volta über die allgemeine Revoluzion der Erde und, über die Versteinerungen des Veronesischen Gebiets . . . Aus dem Italienischen übersetzt und mit neuen Zusätzen des Verfassers und Anmerkungen vermehrt von Karl Freyherrn von Meidinger. Wien, Bey A. Doll. 1793-
288p. 549/V88aGm

2260 **Volta.**
Ittiolitologia veronese del Museo Bozziano ora annesso a quello del Conte Giovambattista Gazola e di altri gabinetti di fossili veronesi. Con la versione latina. Verona, Stamperia Giuliari, 1796-[1809].
323p. and atlas of 76 plates. f567/V88i

Library owns text and the first 11 plates only. Text bound as c.1 and plates as c.2. Copy 2 also contains duplicate text material.

2261 **Vossius, Isaac,** 1618-1689.
. . . De motu marium et ventorum . . . Hagae-Comitis, Adriani Vlacq, 1663.
123p., diagrs. 551.47/V93d

2262 **Vreeswyck, Goosen van,** 1626-after 1689.
Het cabinet der Mineralen, Metalen, en Berg-eerst. Amsterdam, J. Janssonius, 1670.
56p. x669/V95c

2263 **Wad, Gregers,** 1755-1832.
Tabulae synopticae terminorum systematis oryctognostici Werneriani, latine, danice, et germanice. Hafniae, F. Brummer, 1798.
25p. q549/W11t

Based on the system of Abraham Gottlob Werner.

2264 **Wagner, Andreas Johann,** 1797-1861.
Beschreibung einer fossilen Schildkröte und etlicher anderer Reptilien-Ueberreste aus den lithographischen Schiefern und dem Grünsandsteine von Kelheim. München, Verlag der k. Akademie, in Commission bei G. Franz, 1853.
24p., 3 plates. 568.1/W12b

Aus den Abhandlungen der k. bayer. Akademie d. Wiss. II. Cl., VII. Bd., I. Abth.

2265 **Wagner, Johann Jakob.**
Historia naturalis helvetiae curiosa, in VII sectiones compendiose digesta . . . Tiguri, Joh. Henrici Lindinneri, 1680.
390p. uncat.

Chapter I is a general description of Switzerland; Chapter II describes the Alps in respect to their shape, height, glaciers, and caves. Chapters III treats lakes, rivers, and mineral springs in that region while Chapters IV and V discuss animals and plants. Chapter VI, under the title of De fossilibus, gives a systematic description of organic and inorganic fossils (50p). The last Chapter VII is entitled De meteoris.

2266 **Walch, Johann Ernest Emanuel,** 1725-1778.
See KNORR, GEORG WOLFGANG, Sammlung von Merkwürdigkeiten der Natur . . . 1755-1773, and subsequent translations in French and Dutch.

2267 **Walch.**
Das Steinreich, systematisch entworfen. Mit vielen Kupfern. Neue sehr verm. Aufl. Halle, J. J. Gebauer, 1764-69.
2v. in 1, 24 plates. [v.1, 1769]. 552/W14s

2268 **Walcher, Joseph,** 1718-1803.
Nachrichten von den Eisbergen in Tyrol. Wien, J. Kurzböcken, 1773.
99p., illus., fold. plates, table. 551.31/W14n

2269 **Walchner, Friedrich August,** 1799-1865.
Handbuch der gesammten Mineralogie in technischer Beziehung, zum Gebrauche bei seinen Vorlesungen und zum Selbstudium, mit besonderer Berücksichtigung der mineralogischen Verhältnisse des Grossherzogthums Baden. Carlsruhe, Ch. Th. Groos, 1829-32.
2v., diagrs., profiles (part. col.) 549/W14h

2270 **Walchner.**
See OKEN, LORENZ, Allgemeine Naturgeschichte für alle Stände . . . 1833-41.

2271 **Walchner.**
Der practische Naturforscher, ein unentbehrliches Hand- und Hülfsbuch für Freunde der Naturwissenschaften. Dritte Abteilung, der Petrefactologie, Karlsruhe, C. Macklot, 1843.
488-599p. 560/W14p

2272 **Walckenaer, Charles Athanase,** 1771-1852.
Cosmologia, o Descrizione generale della terra, considerata ne' suoi rapporti astronomici, fisici, istorici, politici e civili. Recata la prima volta dal francese in italiano e corredata di note da Guiseppe Ciancio. Napoli, Tramater, 1827.
2v. 551.4/W14clc

Translation of Cosmologie, ou Description générale de la terre . . .

2273 **Walcott, John,** fl. 1778-1822.
Descriptions and figures of petrifications, found in the quarries, gravel-pits, &c. near Bath, collected and drawn, by John Walcott. Bath, Printed for the author, by S. Hazard, London; sold by J. Matthews . . . and all the booksellers in Great Britain and Ireland, 1779.

TABLEAU
DU
CLIMAT ET DU SOL
DES ETATS-UNIS
D'AMÉRIQUE.

SUIVI d'éclaircissemens sur la Floride, sur la colonie Française au Scioto, sur quelques colonies Canadiennes et sur les Sauvages.

Enrichi de quatre Planches gravées, dont deux Cartes Géographiques et une coupe figurée de la chûte de Niagara.

PAR C.-F. VOLNEY,

Membre du Sénat conservateur, de l'Institut national de France, Membre honoraire de la Société philosophique Américaine de Philadelphie; de la Société Anglaise-asiatique de Calcutta: des Athenées d'Avignon, d'Alençon, etc.

TOME PREMIER.

A PARIS,

CHEZ { COURCIER, Imprimeur-Libraire, quai des Augustins, n°. 71.
DENTU, Imprimeur-Libraire, Palais du Tribunat, Galleries-de-Bois, n° 240.

AN XII.—1803.

51p., 15 leaves of plates, illus. (cuts). x560/w142d

Some of the plates are signed by the engraver, J. Collyer.

2274 **Waldauf von Waldenstein, Joseph,** b. 1779.
Die neuesten Beobachtungen und Erfahrungen von Garnier, Hericart de Thury, Baillet, Omalius d'Halloy, Flachat, Beurrier, v. Bruckmann über die Anlage der artesischen Brunnen. Als Anhang und Nachtrag zur Uebersetzung der ersten Ausgabe von Garnier's Preisschrift: Ueber die Anwendung des Bergbohrers zur Aufsuchung von Brunnenquellen. Wien, In F. Beck's Universitäts-Buchhandlung, 1831.
194p., illus. 628.11/W14n

2275 **Walker, John and Charles.**
Geological map of England, Wales and part of Scotland, showing also the Inland Navigation by means of Rivers & Canals, with their elevation in feet above the sea, together with the Rail Roads & Principal Roads. London, J. and C. Walker, 1837.
Col. map, 135 x 97cm. G5740/.C5/1837/.W3

Scale, [1] inch equals 18 miles.

Insets: "Geological and mineralogical section from the lands end to the German Ocean" and "Geological & mineralogical section from the St. George's channel to the German Ocean."

2276 **Walker.**
Geological map of England, Wales and part of Scotland, showing also the inland navigation with the railroads and principal roads. London, J. & C. Walker, 1838.
G5740/.C5/1838/.W3

Col. map.
Scale, ca. 1 inch equals 15 English miles.

2277 **Wallerius, Johan Gottschalk,** 1709-1785.
Mineralogia, eller Mineralriket, indelt och beskrisvit. Stockholm, I. Salvii, 1747.
479p., fold. plate. 549/W15m

2278 **Wallerius.**
. . . Mineralogie, oder Mineralreich, von ihm eingeteilt und beschrieben. Ins Deutsche übersezt von Johann Daniel Denso . . . Berlin, C. G. Nicolai, 1750.
600p., 16ℓ., fold. plates. 549/W15mGd/1750

2279 **Wallerius.**
Mineralogie, oder Mineralreich, von ihm eingeteilt und beschrieben. Ins Deutsche übersetzt von Johann Daniel

Denso . . . 2. verb. und verm. Aufl. Mit allergnädigsten Freiheiten. Berlin, Bey Friedrich Nicolai, 1763.
24, 600p., 16ℓ., fold. plates. 549/W15mGd/1763

2280 **Wallerius.**
Minéralogie; ou, Description générale des substances du règne minéral. Ouvrage tr. de l'allemand . . . Paris, J.-T. Hérissant [etc.], 1759.
2v., 4 fold. plates. 549/W15mF

Vol. 2 contains a treatise on hydrology.

2281 **Wallerius,** praeses.
Dissertatio . . . de origine montium. Upsaliae, 1758.
18p. 551.43/W15d

Diss. — Uppsala (L. Ekstrand, respondent)

2282 **Wallerius,** praeses.
Dissertatio . . . de montibus ignivomis. Upsaliae, 1760.
14p. 551.21/W15d

Diss. — Uppsala (G. J. Gerdin, respondent)

2283 **Wallerius.**
Lucubrationum academicarum specimen p:um de systematibus mineralogicis et systemate mineralogico rite condendo. Holmiae, Impensis direct. L. Salvii, 1768.
158p. 549/W15ℓ

2284 **Wallerius.**
Systema mineralogicum, quo corpora mineralia in classes, ordines, genera et species suis cum varietatibus divisa, describuntur, atque observationibus, experimentis et figuris aeneis illustrantur. Holmiae, Impensis L. Salvii, 1772-75.
2v., port., fold. plates. 549.1/W15s

2285 **Wallerius.**
. . . Systema mineralogicum, quo corpora mineralia in classes, ordines, genera et species suis cum varietatibus divisa, describuntur, atque observationibus, experimentis et figuris aeneis illustrantur. Tom. I-[II]. . . . Editio alter correcta. Vindobonae, In officina Krausiana, 1778.
2v., front. (port.), 2 fold. plates. 549.1/W15s/1778

Includes bibliographies.

2286 **Wallerius.**
Tankar om verldenes i synnerhet jordenes, danande, och ändring. Stockholm, H. Fougt, 1776.

183p., 1 fold. plate. 550/W16mSW

Translation of Meditationes physico-chemicae de origine mundi.

2287 **Wallerius.**
De l'origine du monde, et de la terre en particulier; ouvrage dans lequel l'auteur développe ses principes de chymie & de minéralogie, & donne, en quelque manière, un abrégé de tous ses ouvrages. Traduit par J. B. D.** [i.e. Jean Baptiste Dubois]. Varsovie, J. F. Bastien, 1780.
360p., fold. plates. 550/W16mFd

2288 **Warburton, William,** Bp. of Gloucester, 1698-1779.
Julian. Or, A discourse concerning the earthquake and firey eruption, which defeated that emperor's attempt to rebuild the temple at Jerusalem. In which the reality of a divine interposition is shewn; the objections to it are answered; and the nature of that evidence which demands the assent of every reasonable man to a miraculous fact, is considered and explained. The 2d ed., with additions. London, Printed for J. and P. Knapton, 1751.
320p. x933/W19j/1751

2289 **Warburton.**
Dissertation sur les tremblemens de terre et les éruptions de feu qui firent échouer le projet formé par l'Empéreur Julien de rebâtir le temple de Jérusalem. Paris, N. M. Tilliard, 1764.
2v. 933/W19jFm

Translation of Julian. Or, A discourse concerning the earthquake and firey eruption.

2290 **Warren, Erasmus.**
Geologia: or, A discourse concerning the earth before the deluge, wherein the form and properties ascribed to it, in a book intituled the Theory of the earth, are excepted against; and it is made appear, that the dissolution of that earth was not the cause of the universal flood. Also a new explication of that flood is attempted. London, Printed for R. Chiswell, 1690.
359p., illus. x213/W25g

Written in refutation of Thomas Burnet's Telluris theoria sacra.

2291 **Wasser, Ioh. Tob.**
Salium origine. Ienae Litteris Fickelscherrianis, 1726.
14p. 549/F59

Bound with Flade, Christian Gottlob, De re Metallica . . . [1791].

2292 **Watson, White,** fl. 1773-1831.
A delineation of the strata of Derbyshire, forming the surface from Bolsover in the east to Buxton in the west, by a plate, designed from a tablet, composed of the specimens of each stratum within the above line, with an explanatory account of the same; together with a description of the fossils found in these strata; and also of the nature and quality of the respective soils. Sheffield, Printed by W. Todd., 1811.
76p., fold. front., plate. x554.251/W33d

2293 **Webster, John,** 1610-1682.
Metallographia: or, An history of metals. Wherein is declared the signs of ores and minerals both before and after digging, the causes and manner of their generations, their kinds, sorts and differences; with the description of sundry new metals, or semi metals, and many other things pertaining to mineral knowledge. As also, the handling and shewing of their vegetability, and the discussion of the most difficult questions belonging to mystical chymistry, as of the philosophers gold, their mercury, the liquor alkahest, aurum potabile, and such like. Gathered forth of the most approved authors that have written in Greek, Latine, or High Dutch; with some observations and discoveries of the author himself. London, Printed by A. C. for W. Kettilby, 1671.
388p. x546.3/W39m/1671 and uncat.

Error in paging, nos. 78-79 reversed.

2294 **Webster, John White,** 1793-1850.
A description of the island of St. Michael, comprising an account of its geological structure; with remarks on the other Azores or Western Islands . . . Boston, R. P. & C. Williams, 1821.
[9]-244p., front., plates, fold. maps. 914.699/W39d

2295 **Webster, Thomas,** 1773-1844.
See ENGLEFIELD, HENRY CHARLES, A description of the . . . Isle of Wight . . . 1816.

2296 **Wecker, Johan Jakob,** 1528-1586.
Les secrets et merveilles de nature, recueillis de divers autheurs, & divisez en XVII livres . . . Traduicts en françois, & corrigez outre l'impression precedente selon la traduction latine. Avec une table tres-ample. Lyon, Barthelemi Honorati, 1586.
1008p. uncat.

Contains discussion of metals, precious stones, pearls.

2297 **Wecker.**
De secretis libri XVII. Ex variis authoribus collecti, methodiceque digesti, & rectium iam aucti . . . Basileae ex officina per E.A., 1587.
902p. BALDWIN/921/uncat.

2298 **Wende, Georg,** fl. 1701, praeses.
Examen Suecici lapidis, qui anno superiori miris superbire figuris dicebatur. Thorunii, Typis amplissimi Senatus Thoruniensis, 1701.
[16]p. 549/W48e

Disputatio — Gymnasium, Torun, Poland (J. A. Rachlitz, proponent).

2299 **Werner, Abraham Gottlob,** 1749-1817.
Von den äusserlichen Kennzeichen der Fossilien. Leipzig, S. L. Crusius, 1774.
302p., 8 fold. tables. x549/W49v

Library owns 2 copies; cop.2 has blank pages inserted on which are hand written notes by Werner (refer to Werner, A. G., On the external characters of minerals, Transl. by Albert V. Carozzi, Univ. Illinois Press, 1962, p. xiv).

2300 **Werner.**
Von den äusserlichen Kennzeichen der Fossilien. Wien, Gedruckt bey Johann Thomas, Edlen von Trattnern, 1785.
237p., fold. tables. x549/W49v/1785

2301 **Werner.**
A' köveknek és értzeknek külso megesmértetö jegyeikröl. Irott, szép, és igen hasznos, Könyvetskéje, mellyet, Hazájanak, és a' tanuló Ifjuságnak, lehetö Hasznára Magyarra forditott és a' ket Magyar Hazabéli, 's mas Idegen Kö 's Ertz Nemekkelis, a' Peldákban megbövitett, Benkö Ferentz. R. P. Göttingába, 1782. Esztendöbe. Kolo' svaratt, Nyomt. A' Reform. Koll. Betüivel, 1784.
213p. x549/W49v9Hb

Translation, in Hungarian, of Von den äusserlichen Kennzeichen der Fossilien.

2302 **Werner.**
Traité des caractères extérieurs des fossiles, traduit de l'allemand par le traducteur des Mémoires de chymie de Scheele. Dijon, Impr. L. N. Frantin, 1790.
350p., fold. tables. 549/W49vFg

Translation by Madame Guyton de Morveau of Von den äusserlichen Kennzeichen der Fossilien.

A' Bányáſz Tudomány, és a' Linſiai Gazdaságról értekező, Tudós Tárſaság' Nemes Tagjának

VVERNER ÁBRAHÁM URNAK

A'

KÖVEKNEK ÉS ÉRTZEKNEK

Külső

MEGESMÉRTETŐ JEGYEIKRÖL

Irott, ſzép, és igen haſznos, Könyvetskéje; mellyet, Hazájának, és a' Tanuló Ifjuságnak, lehető Haſznára Magyarra forditott és a' két Magyar Hazabéli, 's más Idegen Kő 's Ertz Nemekkel-is, a' Példákban meg-bővitett,

BENKŐ FERENTZ.

R. P.

Göttingába, 1782. Eſztendőbe.

KOLO'SVÁRATT,

Nyomt. A' Reform. Koll. Betűivel, 1784. Eſzt.

Cat. No. 2301

2303 **Werner.**
A treatise on the external characters of fossils. Translated from the German of Abraham Gottlob Werner. Dublin, 1805.
312p. 549/W49vEw

Translation of: Von den äusserlichen Kennzeichen der Fossilien.

2304 **Werner.**
Kurze Klassifikation und Beschreibung der verschiedenen Gebürgsarten, *in* Česká Společnost Nauk, Prague. Abhandlungen der Böhmischen Gesellschaft der Wissenschaften, 1786.
v.2, p.272-297. x063.7/CESB

Includes col. folded map with title, Petrographische Karte.

2305 **Werner.**
Neue Theorie von der Entstehung der Gänge: mit Anwendung auf den Bergbau, besonders den freibergischen. Freiberg: Gerlachische Buchdrukkerei, 1791.
256p. x551.8/W49n

2306 **Werner.**
Nouvelle théorie de la formation des filons. Application de cette théorie à l'exploitation des mines particulièrement de celles de Freiberg. Traduit de l'allemand et augmenté d'un grand nombre de notes. Freiberg, Craz, 1802.
311p., port. x551.8/W49nFa

Translation by J. F. d'Aubuisson de Voisins, of Neue Theorie zur Entstehung der Gänge: mit Anwendung auf den Bergbau.

2307 **Werner.**
New theory of the formation of veins; with its application to the art of working mines. Translated from the German. To which is added, an appendix, containing notes illustrative of the subject: by Charles Anderson. Edinburgh, A. Constable, 1809.
259p., port. 551.8/W49nEa

2308 **Werner.**
Ausführliches und sistematisches Verzeichnis des Mineralien-Kabinets des weiland kurfürstliche sächsischen Berghauptmans Herrn Karl Eugen Pabst von Chain. Freiberg, In der Crazischen Buchhandlung, 1791-92.
2v. in 1. 549/P11a

2309 **Werner.**
Oryktognosie, oder Handbuch für die Liebhaber der Mineralogie, vermittelst welchem die Mineralien aus ihrer äusser-

lichen Beschaffenheit leicht zu erkennen, von einander zu unterscheiden, und anderen kenntlich zu machen sind. Leipzig, Bey S. L. Crusius, 1792.
274p. uncat.

2310 **Werner.**
See VENBERCHEM-BERTHOUT, J. P., Principes de minéralogie . . . d'après les leçons du professeur Werner . . . [1795?].

2311 **Werner.**
See RIO, ANDRÉS MANUEL DEL, Elementos de orictognosia . . . 1795-1805.

2312 **Werner.**
See WAD, GREGERS, Tabulae synopticae terminorum systematis oryctognostici Werneriani . . . 1798.

2313 **Werner.**
See LUDWIG, CHRISTIAN FRIEDRICH, Handbuch der Mineralogie nach A. G. Werner . . . 1803-04.

2314 **Werner.**
. . . Letztes Mineral-System. Aus dessen Nachlasse auf Oberbergamtliche Anordnung herausgegeben und mit Erläuterungen versehen. Freyberg und Wien, Cras und Gerlach und bey Carl Gerold, 1817.
58p. x549/W49ℓ

2315 **Werner.**
Herrn Bergrath Werner's Leztes Mineral-System; nun nach den neuesten und lezten Entdeckungen hrsg. und mit neuen Beobachtungen und Zusätzen vermehrt von Christoph Mayr. Nebst einem Anhang die Beschreibung der nothwendigsten Theile eines Apparates enthaltend, um die Mineralien auf die zweckmässigste Weise zu sammeln und zu untersuchen. Wien, L. Grund, 1820.
238p. x549/W49ℓ/1820

2316 **Werner.**
Unterhaltungen aus der Naturgeschichte des Mineralreichs, Band 1, Neue Ausg. Augsburg, J. A. Schlosser, 1834.
786p. 552/W49u/1834

2317 **Werner.**
See HASSE, TRAUGOTT LEBRECHT, Denkschrift zur Erinnerung an die Verdienste des in Dresden am 30. Juni 1817 verstorbenen K.S. Bergrath's Werner . . . 1848.

2318 **Weston, Stephen,** Bp. of Exeter, 1747-1830.
Werneria; or, Short characters of earths, with notes according to the improvements of Klaproth, Vauquelin, and Haüy, by Terrae Filius. London, C. and R. Baldwin, 1805.
105p. 549/W522w

In verse.

2319 **Wettersteint, Leopold.**
A full account of the great and terrible earthquake in Germany, Hungary, and Turky. Which was one of the greatest and most dreadful earthquakes that ever was in the world. With a most exact and particular account of the terrible eruptions of fire, conflagration of mountains, devastation of whole towns and villages. As also a learned discourse of the nature, causes and kinds of earthquakes: with a view of all the several opinions of the most eminent philosophers in the world, as well ancient as modern. To which is annexed, some reflections and probable conjectures of the consequents and significations of this calamitous accident. Translated into English by Rich. Alcock . . . London, Printed for R. C., 1673.
[3]-8p. uncat.

2320 **Whiston, William,** 1667-1752.
Memoirs of the life and writings of Mr. William Whiston. Containing memoirs of several of his friends also. Written by himself. London, Printed for the author and sold by Mr. Whiston, 1749-50.
3v. xB/W5761w1

2321 **Whiston.**
A New Theory of the Earth, From its Original, to the Consummation of all Things. Wherein The creation of the world in six days, the universal deluge, and the general conflagration, as laid down in the Holy Scriptures, are shewn to be perfectly agreeable to Reason and Philosophy. With a large introductory discourse concerning the genuine nature, stile, and extent of the Mosaick history of the creation. London, Printed by R. Roberts for Benj. Tooke, 1696.
388p., front., illus. uncat.

2322 **Whiston.**
A new theory of the earth, from its original, to the consummation of all things. Wherein the creation of the world in six days, the universal deluge, and the general conflagration, as laid down in the Holy Scriptures, are shewn to be perfectly agreeable to reason and philosophy. With a large introductory

discourse concerning the genuine nature, stile, and extent of the Mosaick history of the creation. 3d ed., with great additions, improvements and corrections. London, Printed for B. & S. Tooke, 1722.
95, 460p., illus., front. x550.1/W57n/1722

2323 **Whiston.**
A new theory of the earth, from its original, to the consummation of all things. Wherein the creation of the world in six days, the universal deluge, and the general conflagration, as laid down in the Holy Scriptures, are shewn to be perfectly agreeable to reason and philosophy. With a large introduction concerning the genuine nature, style, and extent of the Mosaick history of the creation. 5th ed. To which is added an appendix, containing a new theory of the deluge. London, printed for John Whiston, 1737.
95, 478p. 550.1/W57n5

2324 **Whiston.**
See KEILL, JOHN, An examination of Dr. Burnet's theory of the earth; with some remarks on Mr. Whiston's New theory of the earth . . . 1734.

2325 **White, Gilbert,** 1720-1793.
The Natural History and Antiquities of Selbourne, in the County of Southampton: with engravings, and an appendix. London, Printed by T. Bensley, for B. White and son, 1789.
468p., front., illus. x591.942/W58n

2326 **Whitehurst, John,** 1713-1788.
The works of John Whitehurst, F.R.S. With memoirs of his life and writings. London, Printed for W. Bent, 1792.
[379]p., front. (port.), illus., 10 plates (8 fold.), tables. x500/W587/1792

Various paging.
Ed. by Charles Hutton.

2327 **Whitehurst.**
An inquiry into the original state and formation of the earth; deduced from facts and the laws of nature. To which is added an appendix, containing some general observations on the strata in Derbyshire. With sections of them, representing their arrangement, affinities, and the mutations they have suffered at different periods of time. Intended to illustrate the preceding inquiries, and as a specimen of subterraneous geography. London, Printed for the author and W. Bent, by J. Cooper, 1778.
199p., 7 plates (part. fold.). x550/W58i

2328 **Whitehurst.**
An inquiry into the original state and formation of the earth; deduced from facts and the laws of nature. The 2d ed., considerably enlarged, and illustrated with plates. London, Printed for W. Bent, 1786.
283p., front. (port.), 7 plates (part. fold.). x550/W58i/1786

2329 **Whitney, Josiah Dwight,** 1819-1896.
See FOSTER, JOHN WELLS, Geological map of Isle Royale . . . [1847?].

2330 **Whitney.**
See FOSTER, JOHN WELLS, Report on the geology and topography of a portion of the Lake Superior land district . . . 1850-51.

2331 **Whittlesley, Charles,** 1808-1886.
See U.S. GEOLOGICAL SURVEY. Geological map of Wisconsin, Iowa and Minnesota . . . 1851.

2332 **Widdowes, Daniel.**
Naturall philosophy: or A description of the world, and of the severall creatures therein contained: viz. Of angels, of mankinde, of the heavens, the starres, the planets, the foure elements, with their order, nature, and government: as also of minerals, mettals, plants, and precious stones; with their colours, formes, and vertues. The second edition, corrected and enlarged. Printed at London by Tho. Cotes, 1631.
65p. x500/W633n/1631

Translated and abridged from the "Rerum naturalium doctrina methodica" of Gulielmus Adolphus Scribonius by Daniel Widdowes.
The epistle dedicatory signed, I. Wyddowes, alias Woodhouse.
Imperfect: lacks p. 11-18.
Upper margins closely trimmed.
With this are bound several religious works.

2333 **Widenmann, Johann Friederich Wilhelm,** 1764-1798.
Handbuch des oryktognostischen Theils de Mineralogie. Mit einer Farbentabelle und einer Kupfertafel. Leipzig, S. L. Crusius, 1794.
1040, 20p., 2 fold. plates (1 col.). uncat.

"Ich habe daher die Schriften des Herrn Werners ganz benutzt, und mir nur da erlaubt von ihm abzugehen, wo mich meine individuelle Ueberzeugung eines andern belehrte." — p.vii.

2334 **Widenmann.**
La Orictognosia, escrita en aleman por Juan Federico Guil-

A NEW

THEORY

OF THE

EARTH,

From its ORIGINAL, to the CONSUMMATION of all Things.

WHEREIN

The CREATION of the World in Six Days,

The Universal DELUGE,

And the General CONFLAGRATION,

As laid down in the Holy Scriptures,

Are shewn to be perfectly agreeable to REASON and PHILOSOPHY.

With a large Introductory Discourse concerning the Genuine Nature, Stile, and Extent of the *Mosaick* History of the CREATION.

By *WILLIAM WHISTON*, M. A. Chaplain to the Right Reverend Father in God, *JOHN* Lord Bishop of *NORWICH*, and Fellow of *Clare-Hall* in *Cambridge*.

LONDON:
Printed by *R. Roberts*, for *Benj. Tooke* at the *Middle-Temple-Gate* in *Fleet-street*. MDCXCVI.

Cat. No. 2321

lermo Widenmann. Y traducida por Christiano Herrgen. Madrid, La Imprinta Real por P. J. Pereyra, 1797-98.
2v., 2 fold. plates. 549/W63hSPh

Based on the system of Abraham Gottlob Werner.

2335 **Wilckens, Christian Friedrich,** 1721-1784.
Nachricht von seltenen Versteinerungen, vornemlich des Thier-Reiches, welche bisher noch nicht genau genug beschrieben und erkläret worden, mit Kupfern. In drey Sendschreiben an seine Gönner und Freunde abgefasset. Berlin, G. A. Lange, 1769.
82p., 8 fold. plates. 560/W64n

2336 **Wild, François Samuel,** 1743-1802.
Essai sur la montagne salifère du gouvernement d'Aigle, situé dans le canton de Berne. Genève, L'Auteur, Impr. de Barde, Manget, 1788.
350p., plate, fold. map. 553.63/W64e

2337 **Wilkinson, Charles Henry.**
Analytical researches into the properties of the Bath waters, being the first comparative analysis made of all the hot springs . . . Bath [Eng.], Printed by Wood and Cunningham, 1811.
150p. 615.853/W65a

Bound with the author's Enquiry into the natural history, chemical properties and medical virtues of the rock oil, or green mineral naphtha, of Barbados. London, 1830. 76p.

2338 **Wilkinson.**
Enquiry into the natural history, chemical properties and medical virtues of the rock oil, or green mineral naphtha, of Barbados. London, 1830.
76p. 615.853/W65a

Bound with the author's Analytical researches into the properties of the Bath waters . . . 1811.

2339 **Williams, John,** 1730-1795.
The natural history of the mineral kingdom. In three parts. Part I. Of the natural history of the strata of coal, and of the concomitant strata. Part II. Of the natural history of mineral veins, and other beds and repositories of the precious and useful metals. Part III. Of the natural history of the prevailing strata, and of the principal and most interesting phaenomena upon and within the surface of our globe. Edinburgh, Printed for the author, by T. Ruddiman, 1789.
2v. x553.1/W673n

2340 **Williams.**
The natural history of the mineral kingdom, relative to the strata of coal, mineral veins, and the prevailing strata of the globe. In two volumes. The 2d ed. With an appendix, containing a more extended view of mineralogy and geology. Illustrated with engravings by James Millar... Edinburgh, Printed for Bell & Bradfute; [etc., etc.], 1810.
2v., 2 plates (diagrs.). 553.1/W673n/1810

2341 **Williams.**
Naturgeschichte der Steinkohlen-Gebirge. Aus dem Englischen übersetzt und mit Anmerkungen versehen von A. A. J. W. Reichsfreeiherrn von Danckelman. Dresden, Hilscher, 1798.
406p. 553.2/W672n

Translation of The natural history of the mineral kingdom, pt. 1: Of the natural history of coal.

2342 **Williams, W. Thorne.**
Bonner's map of the State of Georgia, with the addition of its geological features. Savannah, 1849.
G/3921/.C5/1849/.W5

Col. map, 53 x 47cm.
Scale, 1:1,140,000.

2343 **Willis, Richarde.**
See EDEN, RICHARD, History of trauayle in the West and East Indies . . . 1577.

2344 **Willsford, Thomas.**
Natures secrets: or, the admirable and wonderfull history of the generation of meteors. Particularly describing, the temperatures and qualities of the four elements, the heights, magnitudes, and influences of the fixt and wandring stars: the efficient and finall causes of comets, earthquakes, deluges, epidemicall diseases, and prodigies of precedent times . . . London, printed for Nath. Brook, 1658.
199p. uncat.

2345 **Winchilsea, Heneage Finch,** d. 1689.
A true and exact relation of the late prodigious earthquake & eruption of Mount Aetna, or, Monte-Gibello; as it came in a letter written to His Majesty from Naples by the Right Honorable the Earle of Winchilsea . . . who . . . was an eywitness [sic] of that dreadfull spectacle. Together with a more particular narrative of the same, as it is collected out of

severall relations sent from Catania. [London], Printed by T. Newcomb, 1669.

30p., fold. plate. x551.21/W721t

Cop. 1 imperfect: plate wanting. Cop. 2 imperfect: plate mutilated.

2346 **Winchilsea.**

An exact relation of the famous earthquake and eruption of Mount Aetna, or Mont-Gibello, A.D. 1669, being one of the greatest recorded in history; now reprinted from a scarce pamphlet, containing a letter from the Earl of Winchilsea to His Majesty King Charles the Second; together with a more particular narrative, collected from several other authentic materials. To which also is now added a genuine letter [signed W. B. E.] written to a late noble peer purposing to describe the last great eruption, as well as a journey to the very summit of Mount Aetna, in the year 1766. London, 1775.

28, 34p. x551.21/W721t/1775

2347 **Winthrop, John,** 1714-1779.

A lecture on earthquakes, read in the Chapel of Harvard-College in Cambridge, N.E. November 26th 1755. On occasion of the great earthquake which shook New-England the week before. Boston; New-England: Printed and sold by Edes & Gill, 1755.

38p. x551.2/W73ℓ

2348 **Winthrop.**

An account of the earthquake felt in New England, and the neighbouring parts of America. In a letter to Tho. Birch, D.D. Secret. R.S., *in* Philosophical Transactions of the Royal Society, London, 1757.

v.50, Part I, p.1-18. 506 RO/1757

2349 **Wipacher, David.**

De thermis Ribariensibus in Hungaria. Lipsiae, 1768.

95p. 553.7/T75e

Bound with Troschel, Henrich Gottlob Nepomuck, Erforderliche Nachrichten von dem Biliner Sauer-Brunnen nach der neuesten Aufsuchung des wahren reinen Quell-Wassers . . . 1762.

2350 **Witham, Henry Thornton Maire,** 1779-1844.

Observations on fossil vegetables, accompanied by representations of their internal structure, as seen through the microscope. Edinburgh, W. Blackwood; London, T. Cadell [Neill & co., printers], 1831.

48p., 6 plates. 561/W77o

2351 **Witham.**
The internal structure of fossil vegetables found in the carboniferous and oolitic deposits of Great Britain, described and illustrated. Edinburgh, A. & C. Black; [etc., etc.], 1833.
84p., 16 plates (part. col.), front. 561.21/W77i

2352 **Wolski, A. N.**
Mémoire sur le gisement du bassin anthraxifère, dans le département de Maine et Loire et sur les relations géologiques avec divers terrains qui l'avoisinent et qui le couvrent. [Angers, Cosnier et Lachèse, imprimeurs du Congrès scientifique, 1844].
44p., fold. map, profiles. 553.2/W84m

Extrait du 2e volume du compte-rendu de la 11e session du Congrès scientifique de France.
Errata slip mounted on p. 44.

2353 **Wood, William,** d. 1639.
New Englands prospect. A true, lively, and experimentall description of that part of America, commonly called New England . . . London, Tho. Cotes, for Iohn Bellamie, 1635.
83p., fold. map. uncat.

[2d ed.]

2354 **Woodbridge, William Channing,** 1794-1845.
A system of universal geography, on the principles of comparison and classification. 2d ed., illustrated with maps and engravings; and accompanied by an atlas . . . Hartford, O. D. Cooke & co., 1827.
336p., front., illus., plates, maps. x910/W85s/1827

Note: Much on geology, including a geological map of the U.S., p.78 and a diagram of strata, opposite p.20.
Atlas missing.
With this is bound: Willard, Emma. Ancient geography. Hartford, 1827.

2355 **Woodbridge.**
A system of universal geography, on the principles of comparison and classification. 3d ed., illustrated with maps and engravings; and accompanied by an atlas . . . Hartford, O. D. Cooke & co., 1829.
336p., front., illus., plates, maps. x910/W85s/1829

Atlas wanting.
With this is bound: Willard, Emma. Ancient geography. Hartford, 1829.

2356 **Woodbridge.**
A system of universal geography, on the principles of comparison and classification. 4th ed., illustrated with maps and engravings; and accompanied by an atlas . . . Hartford, O. D. Cooke, 1831.
336p., illus., plates, maps. x910/W85s/1831

Atlas wanting.
Bound with Willard, Emma. Ancient geography. Hartford, 1831.

2357 **Woodbridge.**
Modern atlas, physical, political, and statistical; exhibiting on separate maps, the physical and political characteristics of countries, the prevailing religions, forms of government and degrees of civilization, the climates and productions of the earth, and the comparative size of countries, towns, rivers and mountains; accompanied by geographical and statistical tables. Hartford, Belknap and Hamersley, 1843.
6p., 18 maps (part. col., part. double). xq912/W851m/1843

Cover title: Modern atlas, to accompany Woodbridge and Willard's System of universal geography.

2358 **Woods, George.**
An account of the past and present state of the isle of Man; including a topographical description; a sketch of its mineralogy; an outline of its laws, with the privileges enjoyed by strangers; and a history of the island. London [etc.], R. Baldwin [etc.], 1811.
365p., front. (col. geol. map). uncat.

The last page (365) is numbered 36.

2359 **Woodward, John,** 1665-1728.
An essay towards a natural history of the earth; and terrestrial bodies, especially minerals: as also of the sea, rivers, and springs. With an account of the universal deluge; and of the effects that it had upon the earth. London, Printed for R. Wilken, 1695.
277p. uncat.

2360 **Woodward.**
An essay towards a natural history of the earth and terrestrial bodies, especially minerals, as also of the sea, rivers, and springs, with an account of the universal deluge, and of the effects that it had upon the earth. 2d ed. London, Printed by T. W. for R. Wilken, 1702.
277p. 570/W87e/1702

2361 **Woodward.**
An essay towards a natural history of the earth and terrestrial bodies, especially minerals: as also of the sea, rivers, and springs. With an account of the universal deluge: and of the effects that it had upon the earth. The third edition. London, Printed for A. Bettesworth and W. Taylor . . . R. Gosling . . . and J. Clarke . . . 1723.
304p. 570/W87e/1723

2362 **Woodward.**
Géographie physique; ou, Essay sur l'histoire naturelle de la terre, tr. de l'anglois, de Monsieur Wodward, . . . par M. Noguez . . . avec la réponse aux observations de M. le docteur Camerarius; plusieurs lettres écrites sur la même matière; & la distribution méthodique des fossiles, tr. de l'anglois, du même M. Wodward, par le R. P. Niceron, Barnabite. Paris, Briasson, 1735.
389p. 570/W87eFn

Translation of An essay towards a natural history of the earth and terrestrial bodies.

2363 **Woodward.**
Fossils of all kinds, digested into a method suitable to their mutual relation and affinity; with the names by which they were known to the ancients, and those by which they are at this day known; and notes conducing to the setting forth the natural history and the main uses, of some of the most considerable of them, as also several papers tending to the further advancement of the knowledge of minerals, of the ores of metalls, and of subterranean productions. London, 1728.
131p., plates (part. fold.), fold. table. 560/W87p

2364 **Woodward.**
An attempt towards a natural history of the fossils of England; in a catalogue of the English fossils in the collection of J. Woodward, M.D. Containing a description and historical account of each; with observations and experiments, made in order to discover, as well the origin and nature of them, as their medicinal, mechanical, and other uses . . . London, Printed for F. Fayram, 1729, 1728.
2v. in 1. x560.942/W874a

Vol. 1 issued in 2 pt., v.2, in 5 pt., each separately paged.

2365 **Woodward, Samuel,** 1790-1838.
A synoptical table of British organic remains; in which, all the edited British fossils are systematically and stratigraphically arranged, in accordance with the views of the geologists of the present day; and a reference given to their localities, strata, and engraved figures. Accompanied by a lithograph of the fossil turtle in the Norfolk and Norwich Museum. London, Longman, Rees, Orme, Brown, and Green; [etc., etc.], 1830.
50p., front. 560.942/W87s

"Authors and works referred to": 1p. following p. xi.

2366 **Woodward.**
An outline of the geology of Norfolk. London, Longman; printed and sold by J. Stacy, Norwich, 1833.
54p., illus., fold. col. geol. map, tables. 554.261/W87o

2367 **Wrede, Erhard Georg Friedrich.**
Geologische Resultate aus Beobachtungen über einen Theil der Südbaltischen Länder. Halle, Renger, 1794.
204p., diagrs. 554.74/W92g

2368 **Wulfen, Franz Xaver von,** 1728-1805.
Abhandlung vom kärnthenschen pfauenschweifigen Helmintholith oder dem sogenannten opalisirenden Muschelmarmor. Der königlich preussischen Gesellschaft naturforschender Freunde in Berlin gewidmet. Erlangen, J. T. Palm, 1793.
124p., 32 col. plates, fold. diagrs. 553.51/W95a

2369 **Yandell, Lunsford Pitts,** 1805-1878.
Contributions to the geology of Kentucky. By Lunsford P. Yandell, M.D., and Benjamin F. Shumard, M.D. Louisville, Prentice and Weissinger, 1847.
36p., fold. plates, tables. 557/K41y

2370 **Young, George,** 1777-1848.
A geological survey of the Yorkshire coast; describing the strata and fossils occurring between the Humber and the Tees, from the German ocean to the plain of York . . . assisted by John Bird, artist. Whittay, Printed at the office of G. Clark [etc., etc.], 1822.
235 (i.e. 335)p., illus., 17 col. plates, col. map, col. profile.
q554.27/Y85g

2371 **Young.**
A geological survey of the Yorkshire coast; describing the strata and fossils occurring between the Humber and the Tees, from the German ocean to the plain of York . . . assisted

by John Bird, artist . . . 2d. ed., much improved and enlarged, embellished with more than one hundred new figures. Whitby, Printed at the office of R. Kirby [etc., etc.], 1828.
367p., illus., 17 col. plates, fold. col. map, col. profile.
554.27/Y85g/1828

2372 **Zannichelli, Gian Girolami,** 1662-1729.
De ferro ejusque nivis praeparatione dissertatio physico-chymica, in qua varia de ipso metallo explicantur. Venetiis, Apud A. Poleti, 1713.
79p., illus. 546.621/Z16d

2373 **Zanon, Antonio,** 1696-1770?
Della formazione ed uso della torba e d'altri fossili combustibili. Venezia, Appresso M. Fenzo, 1767.
62p. 553.2/Z17d

2374 **Zanon.**
Della marna e di alcuni altri fossili atti a render fertili le terre. Venezia, Appresso M. Fenzo, 1768.
240p. 631.822/Z17d

Bibliographical footnotes.

2375 **Zappe, Joseph Rudolph,** b. 1751.
Mineralogische Handlexikon; oder, Alphabetische Aufstellung und Beschreibung aller bisher bekannten Fossilien. Wien, A. Doll. 1804.
586p. 549.03/Z16m

2376 **Zauschner, Johann Baptist Joseph,** 1737-1799.
. . . Dissertatio de sale a mineralogis haud descripto, opera ejus invento, eruditis communicando ex occasione acidularum ad Pragam recens ab eodem detectarum. a loco na Wénećku Coronariarum, a perillustribus dominis possessoribus Ziegerinarum nomine insignitarum. Pragae, typis Joannis Josephi Clauser, 1768.
183p. 549.4/Z12d

2377 **Zeiller, Charles René,** b. 1847.
Explication de la carte géologique de la France, *See* FRANCE. SERVICE DE LA CARTE GÉOLOGIQUE DE LA FRANCE, Explication . . . v.4, pt. 2, 1878-79.

2378 **Zippe, Franz Xavier Maximilian,** 1791-1863.
Die Steinkohlen, ihr Werth, ihre Wichtigkeit im Allgemeinen, und ihre Verbreitung in Böhmen. Prag. Faltikte, 1842.
55p., fold. col. geol. map. uncat.

2379 **Zuccolo, Vitale,** 1556-1630.
Dialogo delle cose meteorologiche . . . In cui si dichiarano tutte le cose maravigliose che si generano nell'aere, & alcune mirabili proprietà de fonti, fiumi, e mari, secondo la dottrina d'Aristotele con le opinioni d'altri illustri scrittori. Venetia, Paolo Megietti, 1590.
165p. uncat.

2380 **Zúñiga, Juan de,** fl. 1756.
See FEIJÓO Y MONTENEGRO, BENITO JERÓNIMO, El terremoto y su uso, dictamen de Benito Feijóo . . . 1756.

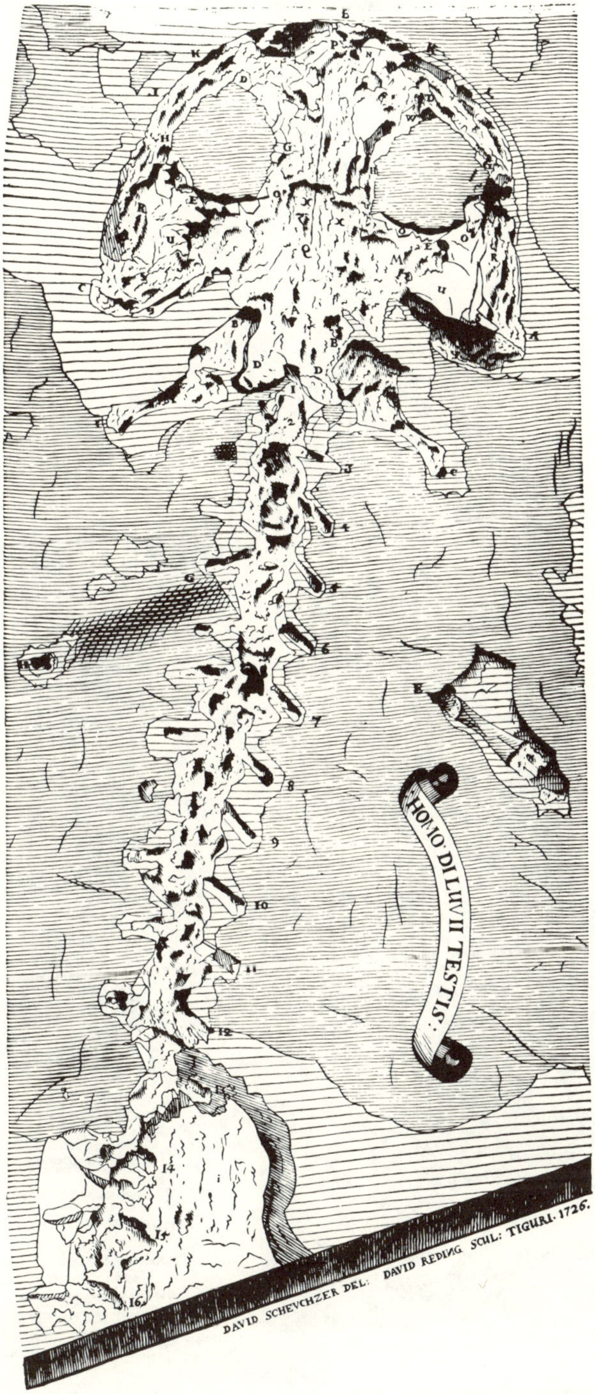

Illustration of Johann SCHEUCHZER, Homo diluvii testis, 1726.

Frontispiece of Johann BERINGER, Lithographiae wirceburgensis, 1726

Index